# AI-Enabled Social Robotics in Human Care Services

Sandeep Kautish
*Lord Buddha Education Foundation, Nepal*

Nirbhay Kumar Chaubey
*Ganpat University, India*

S.B. Goyal
*City University, Malaysia*

Pawan Whig
*Vivekananda Insitute of Professional Studies, India*

A volume in the Advances in Computational Intelligence and Robotics (ACIR) Book Series

IGI Global
PUBLISHER of TIMELY KNOWLEDGE

Published in the United States of America by
IGI Global
Engineering Science Reference (an imprint of IGI Global)
701 E. Chocolate Avenue
Hershey PA, USA 17033
Tel: 717-533-8845
Fax: 717-533-8661
E-mail: cust@igi-global.com
Web site: http://www.igi-global.com



Library of Congress Cataloging-in-Publication Data

Names: Kautish, Sandeep, 1981- editor. | Chaubey, Nirbhay Kumar, 1971- editor. | Goyal, S. B., 1978- editor. | Whig, Pawan, 1980- editor.
Title: AI-enabled social robotics in human care services / edited by: Sandeep Kautish, Nirbhay Kumar Chaubey, S.B. Goyal, and Pawan Whig.
Description: Hershey PA : Engineering Science Reference, [2023] | Includes bibliographical references. | Summary: "AI-Enabled Social Robotics in Human Care Services addresses recent advances in the latest technologies, new research results, and developments in the area of social robotics and AI and the latest developments in the field and future directions which can be beneficial to human society and human care services. Covering topics such as agriculture waste management systems, elder care, and facial emotion recognition, this premier reference source is an essential resource for AI professionals, computer scientists, robotics engineers, human care professionals, students and educators of higher education, librarians, researchers, and academicians"-- Provided by publisher.
Identifiers: LCCN 2023000655 (print) | LCCN 2023000656 (ebook) | ISBN 9781668481714 (hardcover) | ISBN 9781668481721 (paperback) | ISBN 9781668481738 (ebook)
Subjects: LCSH: Robotics--Industrial applications. | Robotics--Social aspects. | Artificial intelligence--Industrial applications. | Human services--Data processing.
Classification: LCC TJ211 .A23 2023 (print) | LCC TJ211 (ebook) | DDC 629.8/92028563--dc23/eng/20230208
LC record available at https://lccn.loc.gov/2023000655
LC ebook record available at https://lccn.loc.gov/2023000656

This book is published in the IGI Global book series Advances in Computational Intelligence and Robotics (ACIR) (ISSN: 2327-0411; eISSN: 2327-042X)

British Cataloguing in Publication Data
A Cataloguing in Publication record for this book is available from the British Library.

# Advances in Computational Intelligence and Robotics (ACIR) Book Series

Ivan Giannoccaro
University of Salento, Italy

ISSN:2327-0411
EISSN:2327-042X

**MISSION**

While intelligence is traditionally a term applied to humans and human cognition, technology has progressed in such a way to allow for the development of intelligent systems able to simulate many human traits. With this new era of simulated and artificial intelligence, much research is needed in order to continue to advance the field and also to evaluate the ethical and societal concerns of the existence of artificial life and machine learning.

The **Advances in Computational Intelligence and Robotics (ACIR) Book Series** encourages scholarly discourse on all topics pertaining to evolutionary computing, artificial life, computational intelligence, machine learning, and robotics. ACIR presents the latest research being conducted on diverse topics in intelligence technologies with the goal of advancing knowledge and applications in this rapidly evolving field.

**COVERAGE**

- Adaptive and Complex Systems
- Evolutionary Computing
- Intelligent Control
- Brain Simulation
- Computational Logic
- Automated Reasoning
- Natural Language Processing
- Machine Learning
- Artificial Life
- Computational Intelligence

IGI Global is currently accepting manuscripts for publication within this series. To submit a proposal for a volume in this series, please contact our Acquisition Editors at Acquisitions@igi-global.com or visit: http://www.igi-global.com/publish/.

The Advances in Computational Intelligence and Robotics (ACIR) Book Series (ISSN 2327-0411) is published by IGI Global, 701 E. Chocolate Avenue, Hershey, PA 17033-1240, USA, www.igi-global.com. This series is composed of titles available for purchase individually; each title is edited to be contextually exclusive from any other title within the series. For pricing and ordering information please visit http://www.igi-global.com/book-series/advances-computational-intelligence-robotics/73674. Postmaster: Send all address changes to above address. 

## Titles in this Series

*For a list of additional titles in this series, please visit: http://www.igi-global.com/book-series/*

***Handbook of Research on AI and Knowledge Engineering for Real-Time Business Intelligence***
Kamal Kant Hiran (Aalborg University, Denmark) K. Hemachandran (Woxsen University, India) Anil Pise (University of the Witwatersrand, South Africa) and B. Justus Rabi (Christian College of Engineering and Technology, India)
Engineering Science Reference • © 2023 • 359pp • H/C (ISBN: 9781668465196) • US $380.00

***Applied AI and Multimedia Technologies for Smart Manufacturing and CPS Applications***
Emmanuel Oyekanlu (Drexel University, USA)
Engineering Science Reference • © 2023 • 377pp • H/C (ISBN: 9781799878520) • US $270.00

***Exploring Future Opportunities of Brain-Inspired Artificial Intelligence***
Madhulika Bhatia (Amity University, India) Tanupriya Choudhury (University of Petroleum and Energy Studies, India) and Bhupesh Kumar Dewangan (School of Engineering, Department of Computer Science and Engineering, O.P. Jindal University, India)
Engineering Science Reference • © 2023 • 215pp • H/C (ISBN: 9781668469804) • US $270.00

***Developments in Artificial Intelligence Creativity and Innovation***
Ziska Fields (University of Johannesburg, South Africa)
Engineering Science Reference • © 2023 • 300pp • H/C (ISBN: 9781668462706) • US $270.00

***Applying AI-Based IoT Systems to Simulation-Based Information Retrieval***
Bhatia Madhulika (Amity University, India) Bhatia Surabhi (King Faisal University, Saudi Arabia) Poonam Tanwar (Manav Rachna International Institute of Research and Studies, India) and Kuljeet Kaur (Université du Québec, Canada)
Engineering Science Reference • © 2023 • 300pp • H/C (ISBN: 9781668452554) • US $270.00

IGI Global
PUBLISHER of TIMELY KNOWLEDGE

701 East Chocolate Avenue, Hershey, PA 17033, USA
Tel: 717-533-8845 x100 • Fax: 717-533-8661
E-Mail: cust@igi-global.com • www.igi-global.com

# Table of Contents

# Detailed Table of Contents

The rate of ageing in the world is extremely fast. Elderly people are becoming more and more prevalent all over the world. The need for services like hospitals, retirement communities, and assisted living facilities has increased as the proportion of older persons in the population rises. For legislators, medical experts, and society as a whole, recovering command over this populace's assumptions will introduce new difficulties. With the aid of smart robot technology, older people can live independent, satisfying lives while remaining safe and secure. As a result, the most notable aspect of this paper is an occy robot. An AI model is used to process facial recognition and speech data, and occy assists elderly people with daily tasks. Occy is more effective for home appliances and healthcare systems. Occy may do tasks like equipment control in the home, safety, and wellbeing sensing, travelling outside, emergency situations and everyday other tasks.



Facial expression recognition (FER) is an important research area in the fields of computer vision and artificial intelligence due to its application in academics as well as in industry. Research shows that using facial images/videos for recognition of facial expression is better because visual expressions carry major information through which emotions can be conveyed. Past research on FER has focused on the study of seven basic emotions; however, many more facial expressions are exhibited by humans that are considered compound emotions. State of art results shows machine learning and deep learning-based approaches are powerful over conventional FER approaches. This chapter focuses on surveying past work done in the field of real-world compound facial emotion recognition and implementing various learning frameworks such as machine learning and deep learning for real-world facial emotion recognition systems for detecting compound emotion using the facial expression image dataset RAF-DB for a real wild scenario.

Robotics is developing at a rate that exceeds the understanding of science at this time. As their integration into society develops, stronger intellectual underpinnings will be needed to reach greater, more dependable, and safer capabilities. Robots are utilized in a variety of fields today, including industry, health, science, space exploration, construction, food packaging, and even surgery. Robotics is the design, creation, and application of devices (robots) to carry out tasks that have historically been completed by humans. Robots are frequently utilised in fields where repetitive, uncomplicated activities must be completed, and in fields where work must be done under hazardous conditions for humans. Robots make it easier to respond to disasters, improve physical capabilities, work in settings where human interaction is required, and permit exploration beyond Earth's orbit. Robots, unlike humans, don't become bored, therefore in many scenarios they can boost production, efficiency, quality, and consistency of products.



The Industrial Revolution 4.0 has produced a wide range of innovative possibilities for the employment of artificial intelligence and computer-guided devices in rehabilitation. In recent times, medical field has shown a keen interest in Human-Robot Interventions (HRI) due to their application in treatment and rehabilitation of various diseases. Social robots are designed to interact with people in a manner that is consistent with human social psychology because of their capacity to elicit social and emotional responses from users. They work in a human-centric environment by speaking, moving, making gestures, or utilizing facial expressions to communicate with their users while adhering to a set of social norms. We outline some intriguing possibilities for social robots in healthcare-related applications in this chapter. Some of these include treatment of social anxiety, dementia, autism spectrum disorder, depression, alzheimer's disease, anorexia nervosa, neurodevelopmental issues, cancer, diabetes and upper limb rehabilitation after stroke.

Legged robots are a class of biologically inspired robots that use articulated leg mechanisms for locomotion. Legged motion is very complex and requires specialized actuation mechanisms and complicated motion control systems to operate. Traditional legged robots were controlled by purely physics-based, however, recent developments of artificial intelligence (AI), and machine learning (ML) techniques have opened new opportunities to train locomotion skills in a legged robot in a much more efficient way than the traditional physics-based controllers. In this chapter, the authors study how machine learning techniques are used to train quadruped robots in basic locomotion skills, evaluate training accuracy, training speed, and also discussed performance, simulation environment, trajectory control, and how the authors achieved accurate tracking of trajectories. Furthermore, this chapter delves into the details of the actual quadruped robot that the authors built to evaluate locomotion policies and some challenges that were faced in building the real robot.



In this chapter, the integration of two rapidly evolving concepts for integration of robotics and cloud computing has been illustrated. Intelligent robots should be preferred over low-cost conventional robots for dynamic and sophisticated applications, the internet, cloud computing, and artificial intelligence principles. "Task offloading" is one of the popular cloud computing techniques for extending the constrained capabilities of robots. The various task features of cloud robotics have also been explained. Literature on various optimization techniques used for task offloading (task offloading, path planning, and access points) in cloud-enhanced robots has been included. The data handling framework and genetic algorithm-based task offloading mechanism of cloud robotic systems have been illustrated. The system architecture and task-flow chart for smart city and manufacturing control applications have been graphically illustrated.

The growth in the sector of autonomous robots in the field of transportation has increased since the 2000s. Improving the dynamics of a robot is a continuous process to tailor its user experience. This field involves the use of planetary and extra-terrestrial robots' (called rovers) autonomous navigation capabilities. It allows analysis of terrain irregularities, climate and weather monitoring, sample collection, etc. As rovers require a significant investment, therefore, it is essential that the rover performs autonomously according to the expectations while ensuring its own safety. It is achieved by the use of complex mathematical models, image analysis techniques, machine learning (ML) and deep learning (DL) models, and allowing execution of the required tasks efficiently. Further, it provides insight on various aspects of rovers' navigation such as intended missions of rover, ML and DL models, comparison in terms of precision and accuracy, merits, and demerits along with future scope.



In this chapter, the precision agricultural and waste management systems using artificial intelligent techniques are described. The fourth agricultural revolution integrates cross-industry technology to increase precision agriculture output and efficiency. Predictive analytics can be used to incorporate massive amounts of data generated by wireless smart networks and the internet of things. "Smart farming" aims to improve both the quantity and quality of agricultural products. The internet of things (IoT) is being used to improve agro-waste management, as well as crop classification and disease detection. Garbage collection and wireless sensor networks on smart bins are employed in the IOT to improve waste management.

# Preface

## INTRODUCTION

Artificial Intelligence (AI) and robotics have been making significant advancements in recent years, and their integration into various sectors of society has opened up new possibilities for innovation and progress. In particular, the application of AI and robotics in human care services has gained attention in research, development, and deployment. AI-enabled social robotics has the potential to revolutionize human care services by providing personalized and responsive care to individuals in need.

The main goal of this book is to provide a comprehensive overview of AI-enabled social robotics in human care service. This book is a collaborative effort of various experts in the fields of AI, robotics, human care, and ethics. We aim to provide a broad understanding of the theoretical, technical, and practical aspects of AI-enabled social robotics in human care service.

This book is divided into three parts. Part 1 provides an overview of the theoretical foundations of AI-enabled social robotics in human care service. In this part, we explore the history of social robotics and AI, the basic concepts of human care service, and the ethical considerations of using AI-enabled social robotics in human care service.

Part 2 focuses on the technical aspects of AI-enabled social robotics in human care service. We discuss the various technologies that are currently being used in social robotics, such as sensors, actuators, and artificial intelligence algorithms. We also delve into the challenges that arise when integrating these technologies into human care service.

Part 3 discusses the practical applications of AI-enabled social robotics in human care service. We explore the current state of the art in AI-enabled social robotics, and we provide case studies of AI-enabled social robotics in various human care service contexts, including elder care, mental health care, and rehabilitation.

This book is intended for a broad audience, including researchers, students, practitioners, and policymakers. We hope that this book will provide a solid foundation for anyone who is interested in understanding the potential of AI-enabled social robotics in human care service.

The integration of AI-enabled social robotics in human care service has the potential to transform the way we provide care to individuals in need. AI-enabled social robotics can provide personalized and responsive care to individuals, while also providing important data and insights to care providers. However, the integration of AI-enabled social robotics in human care service also raises important ethical considerations. We must ensure that AI-enabled social robotics are designed and implemented in a way that is safe, reliable, and respectful of human dignity.

This book provides an important contribution to the growing body of literature on AI-enabled social robotics in human care service. We hope that this book will serve as a valuable resource for anyone who is interested in exploring the potential of AI-enabled social robotics in human care service, and we look forward to the continued development and deployment of these technologies in the years to come.

## Chapter Descriptions

Chapter 1 discusses the complex and continuous nature of the learning process, which encompasses the acquisition of new knowledge, the development of skills, the organization of information, and the discovery of new facts and theories. It highlights the role of AI algorithms in psychometric analysis, which helps in interpreting learner responses and assessing their understanding and performance. The chapter emphasizes the importance of socializing learning materials and resources to maximize the learning process. It also explores the use of evaluation and learning analytics to evaluate learners' behavior and enhance the learning process. Overall, the chapter provides insights into the different aspects of the learning process and how technology can enhance and support it.

Chapter 2 is decentralized interacting with elder-care tasks AI supporting robot has the potential to significantly improve the quality of life for elderly individuals by providing personalized care and support. This robot, whose name is occy. Occy is enables it to learn and adapt to individual needs and preferences over time. Occy refers to its use of secure technology, which allows for secure and transparent sharing of data. This ensures that sensitive information, such as medical records and personal preferences, is kept private and only accessible to authorized individuals.

Chapter 3 based on Facial emotion recognition (FER) is an important research area in computer vision and artificial intelligence due to its application in academics and the industry. Although FER can be conducted primarily using multiple sensors. But, research shows that using facial images/videos to recognize facial expressions is better because visual expressions carry important information through which emotions can be conveyed. Past research on FER has focused on studying seven basic emotions. However, humans exhibit compound emotions as well. This chapter focuses on implementing various learning frameworks detecting compound emotion for FER systems for a real wild scenario on Real-world Affective Faces Database (RAF-DB).

Chapter 4 Discusses Robotics is developing at a rate that exceeds the understanding of the science at this time. As their integration into society develops, stronger intellectual underpinnings will be needed to reach greater, more dependable, and safer capabilities. Robots are utilised in a variety of fields today, including industry, health, science, space exploration, construction, food packaging, and even surgery. Robotics is the design, creation, and application of devices (robots) to carry out tasks that have historically been completed by humans. Robots are frequently utilised in fields where repetitive, uncomplicated activities must be completed, and in fields where work must be done under hazardous conditions for humans. Robots make it easier to respond to disasters, improve physical capabilities, work in settings where human interaction is required, and permit exploration beyond Earth's orbit. Robots, unlike humans, don't become bored, therefore in many scenarios they can boost production, efficiency, quality, and consistency of products.

Chapter 5 explores the innovative possibilities of utilizing social robots in the field of healthcare, specifically in the context of treatment and rehabilitation. With the advent of Industrial Revolution 4.0, artificial intelligence and computer-guided devices have made significant advancements in the medical field. Human-Robot Interventions (HRI) have garnered interest in recent times due to their potential in treating various diseases. Social robots, designed to interact with people in a human-like manner, have the capacity to elicit social and emotional responses from users. This chapter discusses the potential applications of social robots in treating social anxiety, dementia, autism spectrum disorder, depression, Alzheimer's disease, anorexia nervosa, neurodevelopmental issues, cancer, diabetes, and upper limb rehabilitation after stroke.

In this chapter 6 we look at the challenges of legged locomotion and the approaches that can be taken to address these challenges. We explain how machine learning can be a useful tool in achieving legged locomotion in robots. We describe in detail how we built a real quadruped robot and used it for testing locomotion policies. The experimental setup and simulation environments are also covered. Later in the chapter we focus on a particularly interesting problem of accurate trajectory control. We explain the role of accurate tracking of trajectories and how we achieved very high positional accuracy.

In this chapter 7, Intelligent robots should be preferred over conventional robots for dynamic and sophisticated applications, the internet, cloud computing, and artificial intelligence principles. "Task Offloading" is one of the popular cloud computing techniques for extending the constrained capabilities of robots. The data handling framework, genetic algorithm-based task offloading mechanism, system architecture, and task-flow chart for smart city and manufacturing control applications have been graphically illustrated.

Chapter 8 discusses The use cases of ML/DL models for autonomous navigation of planetary rovers have been compiled and analysed for a through perspective. The amount and quality of research spanning in this specific area can be sub-categorized by multiple parameters like safety, efficiency, etc. These keys aspects have been compared to form a review for detailing the factors pertaining to the robotic autonomy for the improvement of mobility in unfamiliar terrain. Multiple applications can be converged to form a single task in each case such as safety, power resource management, etc. laying the foundation of any ML/DL model deployed on the rover.

Chapter 9 The recent agricultural revolution integrates modern-industry technology to increase precision agriculture output and efficiency. "Smart farming" aims to improve both the quantity and quality of agricultural products. The Internet of Things (IoT) is being used to improve Agro-waste management, as well as crop classification and disease detection. Garbage collection and wireless sensor networks on smart bins are employed in the IoT to progress waste management. Vermicomposting is a process that uses earthworms and other associated bacteria to create incredibly fertile compost. Waste management methods for flower waste, bagasse, banana agro-waste, and agro-industrial waste water have been illustrated.

Chapter 10 the integration of robot technology with additive manufacturing processes is discussed to improve the processes for making complex objects. A case study has been used to explain two-phase and multi-plane three-dimensional printing using industrial robots. Using previous research activities, a brief literature on 3D printing technology, parameters, and applications has been developed. The case study on the aerofoil wing made by industrial robotics programming has been explained. The flow chart that can exhibit both three phase and multi-plane processes has been illustrated.

## CONCLUSION

In summary, Chapter 1 highlights the importance of technology in enhancing the learning process and explores the role of AI algorithms in interpreting learner responses. Chapter 2 discusses the potential benefits of AI supporting robots in elder-care tasks. Chapter 3 focuses on implementing various learning frameworks for detecting compound emotions in facial expression recognition systems. Chapter 4 discusses the applications of robotics in various fields, such as industry, health, and space exploration. Chapter 5 explores the potential applications of social robots in healthcare, specifically in the context of treatment and rehabilitation. Chapter 6 explains how machine learning can be useful in achieving legged locomotion in robots. Chapter 7 discusses the use of task offloading in cloud computing for extending the capabilities of robots in smart city and manufacturing control applications. Chapter 8 analyzes the use cases of ML/DL models for autonomous navigation of planetary rovers. Chapter 9 explores the integration of IoT in agro-waste management and crop classification. Finally, Chapter 10 discusses the integration of robot technology with additive manufacturing processes to improve the manufacturing of complex objects.

# Acknowledgment

Writing a book is not an individual effort, and there are many people who deserve to be recognized and appreciated for their contributions. We would like to take this opportunity to express our sincere gratitude to all those who have helped and supported us throughout the process.

Firstly, we would like to thank our family members and friends for their encouragement, patience, and understanding during the time we were working on this book. Their support gave us the energy and motivation to keep going, even when the task seemed daunting.

We are also grateful to the team at the publisher, who have been so supportive and helpful throughout the publication process. Their professionalism and expertise have been invaluable, and we could not have completed this project without their guidance.

We would also like to express my appreciation to the researchers, experts, and practitioners who provided us with their valuable insights and knowledge. Your contributions helped us to shape our ideas and expand our understanding of the field of AI-enabled social robotics in human care services.

Finally, we would like to thank the readers of this book. We hope that this work will be of value to you and that it will contribute to the ongoing discussion about the role of AI-enabled social robotics in human care services.

Thank you all for your support and encouragement.

Editors

# Chapter 1
# Computational Psychometrics Social Analysis of Learners in Their Learning Behaviour Using AI Algorithms

**Ashima Bhatnagar Bhatia**
*Vivekananda Institute of Professional Studies, Delhi, India*

**Kavita Mittal**
*JIMS NCR, India*

**Radhika Mahajan**
*Vivekananda Institute of Professional Studies, Delhi, India*

**Pawan Whig**
*Vivekananda Institute of Professional Studies, Delhi, India*

## ABSTRACT

*Learning is a complex and continuous process or phenomena. Learning processes or phenomena include the acquisition of new calculative or declarative knowledge, the development of motor and cognitive skills through instruction or practice, the organization of new knowledge into general, effective representations, and the discovery of new facts and theories through observation and experimentation. When interpreting learner responses, AI algorithm is useful for psychometrics analysis. The success of today's technology-enhanced learning may be boosted by socializing the material and learning resources to every learner, hence maximizing the learning process. Psychometric analysis entails insinuating what a learner understands and can perform in the real world based on minimal observation is supported in a state testing setting. Evaluation, from the perspective of learning analytics, comprises evaluating reports shows in digital educational experiences to evaluate learners' behaviour with the purpose of favorably affecting the learning process.*

DOI: 10.4018/978-1-6684-8171-4.ch001

## INTRODUCTION

Using numerous characteristics, such as gender, age, and location of learners, computational psychometric social analysis of learning is quantified (Alavi, M., & Leidner, D. E. 2001). This study's comprehensive holistic learning and evaluation methodology and the way the computational psychometrics paradigm combines all these challenging elements. Several publications are being researched to offer specifics on these strategies and their evaluations (Fritz & Klingler, 2023). While there have been significant advances in the study and creation of combined nursing and evaluation schemes, more work is needed to improve the methods, and gage them awake. The goal is to be talented to provide in height superiority instructional resources and feedback to all learners, irrespective of race, place, and gender Chen, G., & Lee, T. H. (2019).

By carefully while examining the test data and making sure the standard-compliance of the test being utilised, psychometric social analysis is the organised method used to determine an individual's psychometric characteristics. To assure the integrity of the test, psychometric analysis examines both the test questions and the responses received. To assess the test's validity, each test and its accompanying items go through several stages of psychometric analysis.

Decisions involving high stakes talent are made using psychometric assessment. Skilled tests analyse the results of the social psychometric tests in-depth and produce reports that used to calculate the social behaviour of learners Chou, C. C., & Tsai, C. C. (2018).

A learner's personality traits, cognitive skills, and behavioral tendencies can all be accurately measured using psychometric assessments, which are measurement tools. In order to make crucial judgments about hiring, role fit, work styles, etc. in the education and employment settings, social psychometric exams have found use. Psychometric tests are dependable, thoroughly researched, and scientifically validated tools that provide a comprehensive understanding of a learner's psychological characteristics.

While psychometrics is a top-down and theory-driven area, machine learning is a lowest part data-driven study topic. Whereas psychometrics and cognitive analytics target similar goals, such as formative assessment, they are based on different approaches and ideas and significantly divided. Integration of investigate fields creates investigate topic called as 'computational social psychometrics' at the junction of both professions Fong, C. J., & Lam, H. C. (2020). In the future years, computational social psychometrics will be a major driving force in assessment research and practice. Computational social psychometrics may be used in areas such as emotional, motivational, (meta)cognitive, collaborative, and psychomotor learning Hsu, Y. C., & Lin, J. F. (2018). By utilizing multimodal electronic trace data, it offers up radically new evaluation avenues that might lead to a new input society with non-invasive evaluation.

## LUDWIG - DECLARATIVE MACHINE LEARNING

With Ludwig's labelling to deep learning, you have complete control over the parts of the pipelines that are important to you while letting Ludwig decide what's acceptable for the remainder.

*Figure 1. Ludwig - declarative machine learning*

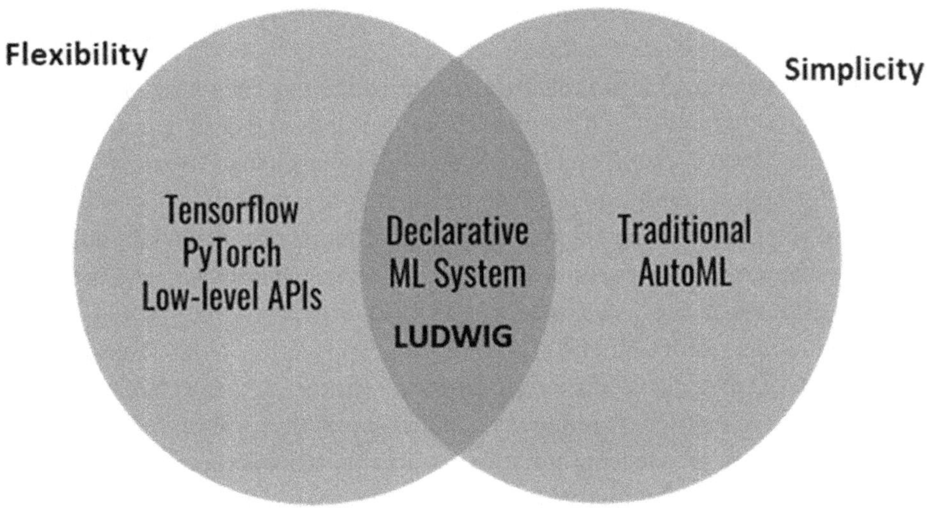


Ludwig is used by researchers, scientists, engineers, and analysts to investigate cutting-edge model structural design, perform hyperparameter searches, measure up to data larger than accessible RAM and multi-nodule cluster, and ultimately serve the best version in operation Huang, R. H., Liu, D. J., & Tlili, A. (2019). In figure 1, Ludwig is union of Tensorflow PyTorch low-level APIs and Traditional AutoML, which is a combination of flexibility and simplicity.

Furthermore, Ludwig may be easily extended by users by introducing new models, metrics, failures, and processing procedures that can be registered to make them immediately usable in the same combined configuration system. This is made possible by the usage of generic APIS all through codebase Kim, M. J., Kim, M., & Kim, H. (2019).

## Ludwig Benifits

The engineering complexity of machine learning is handled by Ludwig with minimal boilerplate, liberating research scientists to concentrate on creating models with the highest level of extraction. Preprocessing of the data, hyperparameter optimization, device organization, and delivered model training are all free services.

1. **Build your benchmarks quickly** - It only takes a little configuration change to set up a state-of-the-art standard and compare it to a new model.
2. **Apply new architectures to various issues and datasets with ease -** Using the vast array of tasks and datasets that Ludwig provides, apply new models. Every user has access to Ludwig's comprehensive benchmarking toolbox, which enables them to conduct experiments with a variety of models and datasets with just a few settings.
3. **Highly customizable models, metrics, and data preprocessing -** To tailor the pipeline to your needs, all facets of the model structural design, training loop, hyperparameter search, and backend base may be changed as extra fields in the declarative configuration. Visit the Ludwig Configuration documentation for further information on what may be configured. Multi-modal, multi-task learning out- without writing any code: Combine tabular data, text, graphics, and even music to create complicated model setups.
4. **Rich model exporting and tracking -** Use applied tools like "Tensorboard, Comet ML, Weights & Biases, MLFlow, and Aim Stack" to mechanically track all trials and metrics. Training may be transferred from your local machine to the cloud without requiring any code modifications thanks to automatic scaling to multi-GPU, multi-node clusters. Additionally, Huggingface Transformers' pre-trained models are natively integrated with Transformers - Ludwig. Without writing any code, users can select from a large selection of cutting-edge pre-trained PyTorch models.
5. **AutoML low-code interface -** Ludwig with merely a dataset, the target column, and a time budget, AutoML enables users to generate trained models. Ludwig creates it simple to serve deep learning models, as well as GPUs, for easy productionisation. Launch a REST API for the Ludwig model you've trained.

For each possible use case, Ludwig offers a variety of model structural design that may be coupled to produce an end-to-end standard Lee, K. C., Chai, C. S., & Tsai, C. C. (2018). Ludwig gives the buildings to create your city, like how deep learning libraries supply the building blocks to create your structure. You can choose from the available buildings or add your own structure to the list of ones that are already there. The guiding design ideas we incorporated into the toolbox are:

- It is not necessary to know how to code to *train a model* and utilize it to make projections.
- *Generality*: a novel data type-based method for designing deep learning models that makes the tool useable in a wide range of scenarios.
- *Flexibility:* While novices will find it simple to use, experienced users have great control over model development and training.
- *Extensibility*: simple addition of new feature data kinds and model architectures.
- *Understanding*: Although deep learning prototype internals are sometimes regarded as "black boxes," we offer common representations to comprehend their operation and evaluate their projections.

## Ludwig Working Procedure

A tabular file (like a CSV) having the data and a "YAML configuration file" indicating which columns of the tabular file are input features and which columns are output objective variables are all that Ludwig's users need to train a deep learning model Maris, G., & Bechger, T. (2019). Due to the configuration file's simplicity, prototyping can go more quickly, perhaps cutting hours of code down to a few minutes. Ludwig will use multi-task learning, which involves understanding to anticipate all the outputs concurrently, if more than one output target variable is supplied.

Additional information, such as preprocessing details for each element in the dataset, the encoder or decoder to use for each aspect, architectural constraints for each encoder and decoder, and training parameters, may be included in the model definition. In order to make it simple for beginners to train complex models, default standards of pre-processing, training, and different model architectural parameters are either chosen based on our understanding or are adopted from the academic literature. Experts have complete flexibility because they can each set them separately in the model configuration file. Every model that has been trained using Ludwig is preserved and can be imported again to make forecasts on new data. Models might be loaded in a serving environment, for instance, to provide forecasts in software programs.

## PYCARET – DEMOCRATIZING MACHINE LEARNING

PyCaret empowers anyone to build low-code, powerful, end-to-end machine learning solutions. There are tons of ways for you to get started. PyCaret is a mature AutoML framework in Python that helps us train on our dataset with ease and automatically come up with the best model for prediction Wu, C. H., Chen, C. H., & Huang, Y.

M. (2018). By means of a smart brute force tactic, it runs the data and trains it on all major regression models and comes up with the best fit (meaning the model with the best accuracy for us). To make life much easier and standardized for data preparation, PyCaret supports *Pandas dataframe* as an input method. In simple words we can have any pandas *dataframe* object and pass it to PyCaret, and it will automatically perform all the transformations for the data preparation for us. Below Figure 2 shows the PyCaret flow diagram, in which data set and transformation function input in the model building and Configs multiple models, then input for model testing for the creation of best model(Whig, Velu, & Naddikatu, 2022).

*Figure 2. Flow diagram of PyCaret*

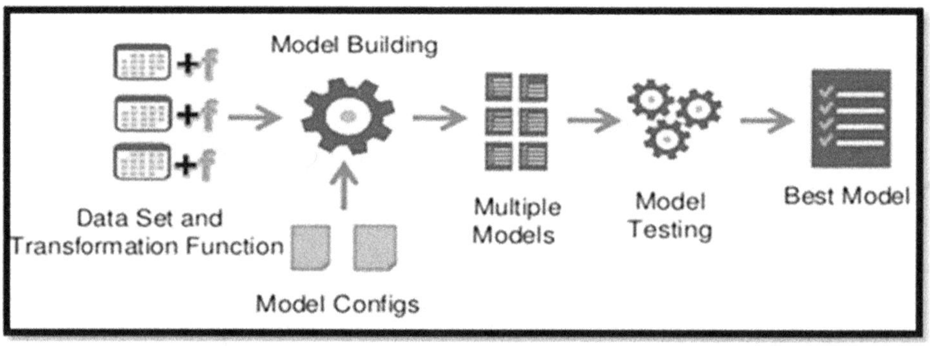


## Analysis Using PyCaret

Models may be trained, visualized, evaluated, interpreted, and deployed with very little code. When approaching supervised machine learning challenges, it might be tempting to just observe how a random forest or gradient boosting model performs and then stop testing if we are pleased with the results. In Figure 3, flow process of PyCaret components starting from the data preparation to analysis & interpretability. In which model training and selection are the part of process followed by hyperparameter tuning and experiment logging(Alkali et al., 2022).

PyCaret is an open-source, low-code machine learning library in Python that aims to reduce the cycle time from hypothesis to insights. It is well suited for seasoned data scientists who want to increase the productivity of their ML experiments by using PyCaret in their workflows or for citizen data scientists and those new to data science with little or no background in coding. A simple supervised machine learning job is regression, which calculates the connection between the dependent variable y and other factors(Whig, Kouser, Velu, et al., 2022; Whig, Velu, & Sharma,

2022). Rather than classification, which is used to forecast single value called as classes, regression may be used to forecast attribute variables such as the worth of a home. The PyCaret regression module, which is powered by sklearn, allows you to develop and test linear regression with just a few lines of coding. It comes with several algorithms, in addition to the capability to plot and tune hyperparameters.

*Figure 3. Analysis using PYCARET components*

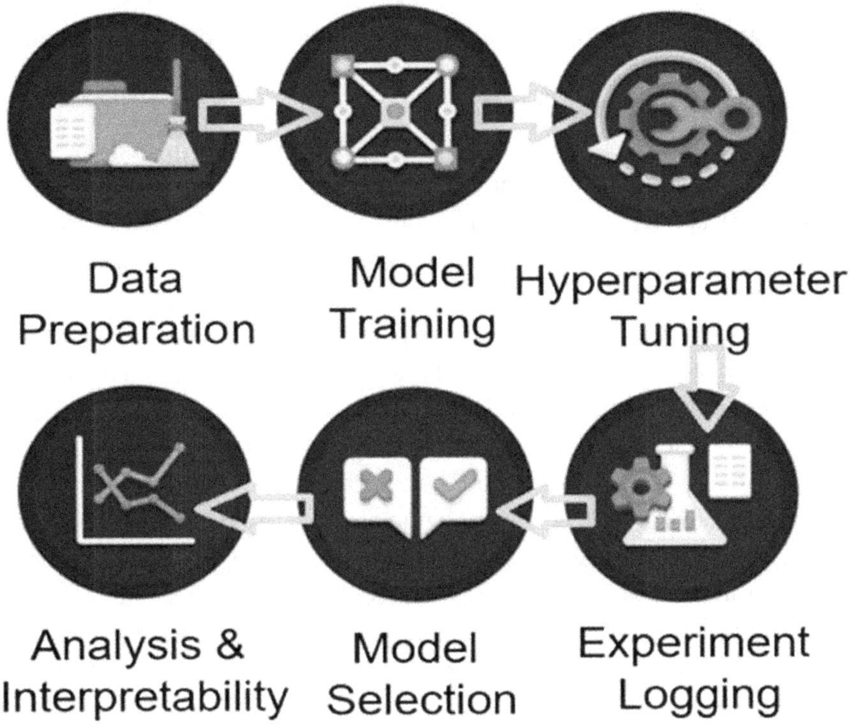


## PyCaret Components

PyCaret is a machine learning platform component that operates in number of fields. Some of these are:

1. **Analyses of model performance -** Pycaret has 60 graphs that may evaluate and explain model performance and provide results instantly without creating any complicated code, which is important because analyzing the performance of a trained ML model.

2. **PyCaret Data Preparation -** It operates in many data preparation segments with a high level of automation.

Preparing Data for Missing Values Ordinal encoding, cardinal encoding, one hot encoding, normalization, and transformation. Feature interaction, polynomial features, trigonometry features, group features, bin numerical features, and combine rare levels are all examples of feature engineering. Feature selection: Feature Importance, Ignore Low Variance, Remove Multicollinearity, Principal Component Analysis

*Benefits of Pycaret* - PyCaret is a low-code library, and it too has several benefits. Among them are - Enhanced output; simple deployment; company solution

## Model Block Diagram Used

Psychometric social analysis on learners bases on various parameters like age, gender, and geographical area, on primary data almost more than 500 responses received. To calculate the Psychometric social analysis of learners' while learning having motivational level, satisfaction level, stress level and time duration or breaks, the block diagram of the model used to show in Figure 4 Model Block Diagram Used. A thorough main facts has been composed founded on enquiries given below in Table 1 Questionnaire and responses received from strongly agree to strongly disagree and after considering them Psychometric social analysis on learners calculated. In addition to the help of deep learning algorithm is understood as result by seeing the result of all stated limits in Table 2 Mentioned parameter and responses received(Jupalle et al., 2022; Whig, Velu, & Bhatia, 2022). The selected sample is then sent back into the machine and saved in the database to improve the system's accuracy. At training and validation, it is discovered that the damage is diminishing, and, in other respects, the system's precision continues to increase, more historical periods, more precision.

***Inference of figure 5***, Following are many chart kinds based on information from table 1 and table 2. In Figure 5, shows pie chart on responses received for the entire analysis of all levels. Range from strongly agree to strongly disagree. According to responses received for all levels agree percentage is 49%, neutral 25% and disagree is 26%. Agree percentage is almost double than disagree percentage.

Conclusion from Figure 6 According to the age-based bar charts above, around 66% of respondents are satisfied with their early learning experiences, 21% are undecided or neutral, and 13% disapprove. Although the amount of agreement increases by 12% and reaches 77%, the neutral percentage keeps nearly constant at 21%, and the number of dissent decreases by about half to 6% as learners develop and learn skills.

*Figure 4. Model block diagram used*

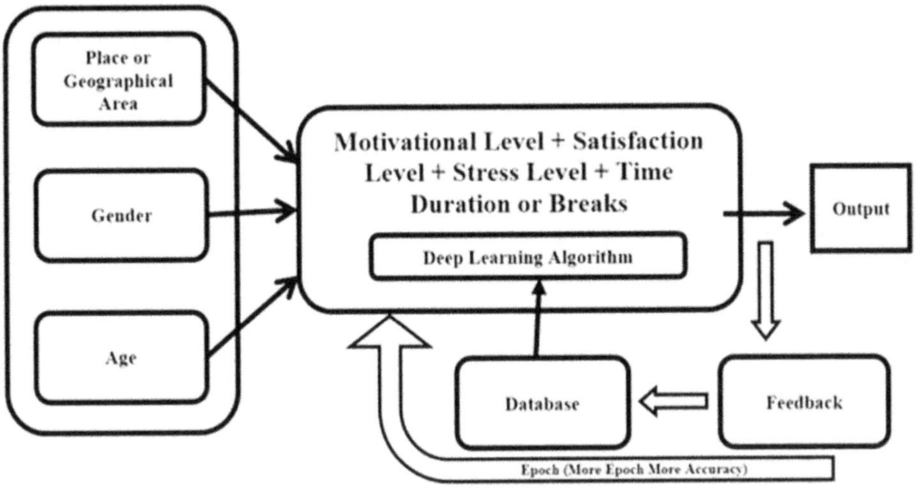


*Table 1. Questionnaire and responses received*

| Questions | Strongly Agree | Agree | Neutral | Disagree | Strongly Disagree |
|---|---|---|---|---|---|
| ***Motivational Level*** | | | | | |
| 1.1 Always feel motivated during learning | 179 | 181 | 132 | 27 | 8 |
| 1.2 Always spend maximum time for learning | 103 | 172 | 187 | 57 | 8 |
| 1.3 Can spend time for learning instead of hectic schedule | 112 | 158 | 173 | 54 | 30 |
| 1.4 Must spend time for learning instead of physical stress | 118 | 177 | 146 | 60 | 26 |
| 1.5 Must spend time for learning instead of emotional stress | 135 | 149 | 139 | 78 | 26 |
| 1.6 Always feel motivated in learning after praying or worshiping | 204 | 139 | 123 | 37 | 24 |
| ***Satisfaction Level*** | | | | | |
| 2.1 Always Feel Mentally Satisfied while Learning | 203 | 191 | 93 | 31 | 9 |
| 2.2 Satisfaction level increases while learning even if topic is from area of your interest | 342 | 120 | 48 | 14 | 3 |
| 2.3 Satisfaction level increases while learning if topic is interesting but not from area of your interest | 170 | 170 | 128 | 38 | 21 |
| 2.4 Can Feel Satisfied while Learning if topic is not interesting | 59 | 86 | 167 | 140 | 75 |
| 2.5 Interesting Learning or topic reduce Physical Stress | 198 | 159 | 114 | 40 | 16 |
| 2.6 Interesting Learning or topic reduce Mental Stress | 245 | 161 | 99 | 15 | 7 |
| 2.7 Praying reduces Mental Stress | 213 | 132 | 122 | 35 | 25 |
| 2.8 If Learning is Interesting than Time flies | 266 | 159 | 80 | 17 | 5 |

*continues on following page*

*Table 1. Continued*

| Questions | Strongly Agree | Agree | Neutral | Disagree | Strongly Disagree |
|---|---|---|---|---|---|
| ***Stress Level*** | | | | | |
| 3.1 Always Feel Mental Stress while Learning because you are not interested in Learning | 82 | 94 | 142 | 115 | 94 |
| 3.2 Always Feel Mental Stress while Learning even if topic is from area of your interest because you are not interested in Learning | 70 | 91 | 125 | 126 | 115 |
| 3.3 Always Feel Mental Stress while Learning if topic is interesting but not from area of your interest | 69 | 79 | 160 | 127 | 92 |
| 3.4 Always Feel Physical Stress while Learning even if topic is from your area of interest | 55 | 76 | 126 | 120 | 150 |
| 3.5 Always Feel Physical Stress while Learning even if topic is not from your area of interest | 76 | 107 | 141 | 112 | 91 |
| 3.6 Always Feel Physical Stress while Learning even if topic is interesting but you are not interested in learning | 78 | 100 | 148 | 102 | 99 |
| 3.7 Always Feel Emotional Stress while Learning because you are not interested in learning | 70 | 101 | 143 | 110 | 103 |
| 3.8 Always Feel Emotional Stress while Learning even if topic is from your area of interest | 45 | 87 | 125 | 133 | 137 |
| 3.9 Always Feel Emotional Stress while Learning even if topic is not from your area of interest | 77 | 109 | 137 | 109 | 95 |
| 3.10 When Stress level increases you start praying or worshiping | 105 | 126 | 129 | 83 | 84 |
| ***Stress Vs Break or Time Duration*** | | | | | |
| 4.1 Under Mental Stress Learners Learn better | 52 | 98 | 123 | 97 | 157 |
| 4.2 Under Physical Stress Learners Learn better | 42 | 74 | 118 | 129 | 164 |
| 4.3 Under Emotional Stress Learners Learn better | 43 | 82 | 100 | 121 | 181 |
| 4.4 If Learner's is Emotional then they perform better | 63 | 117 | 176 | 94 | 77 |
| 4.5 If Learner's is Physically fit then they perform better | 231 | 146 | 107 | 28 | 15 |
| 4.7 Time duration decides stress level | 126 | 167 | 156 | 53 | 25 |
| 4.6 If Learner's is in mental stress but physically fit then they perform better | 49 | 79 | 154 | 141 | 104 |
| 4.8 Short Interval or breaks reduces mental stress level effect on performance | 226 | 166 | 100 | 24 | 11 |
| 4.9 Short Interval or breaks reduces physical stress level effect on performance | 207 | 180 | 103 | 28 | 9 |
| 4.10 Short Interval or breaks reduces emotional stress level effect on performance | 180 | 181 | 125 | 26 | 15 |
| 4.11 Single Long Interval is better than many short intervals | 84 | 104 | 131 | 105 | 103 |

*Table 2. Mentioned parameter and responses received*

| Parameters | Strongly Agree | Agree | Neutral | Disagree | Strongly Disagree |
|---|---|---|---|---|---|
| Entire learner's responses | 4577 | 4518 | 4520 | 2626 | 2204 |
| Female learner's responses | 2029 | 1966 | 1859 | 1158 | 1108 |
| Male learner's responses | 2548 | 2552 | 2661 | 1468 | 1096 |
| Delhi & NCR learner's responses | 2992 | 3071 | 3057 | 1797 | 1543 |
| Outside Delhi & NCR learner's responses | 1585 | 1447 | 1463 | 829 | 661 |
| Less than equal to 18 Years learner's responses | 1101 | 804 | 894 | 494 | 487 |
| Between 19 years to 22 years learner's responses | 1897 | 2180 | 2152 | 1218 | 918 |
| Between 23 years to 30 years learner's responses | 549 | 543 | 543 | 379 | 296 |
| Between 31 years to 40 years learner's responses | 552 | 534 | 458 | 244 | 312 |
| Greater than equal to 41 years learner's responses | 478 | 457 | 473 | 291 | 191 |

*Figure 5. Entire pie chart*

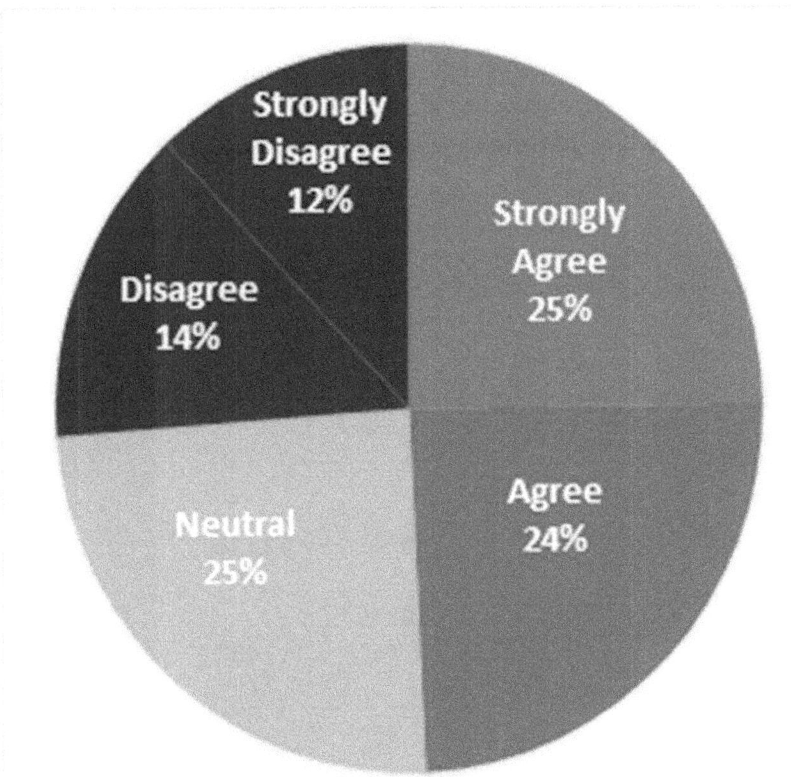

*Figure 6. All levels bar chart*

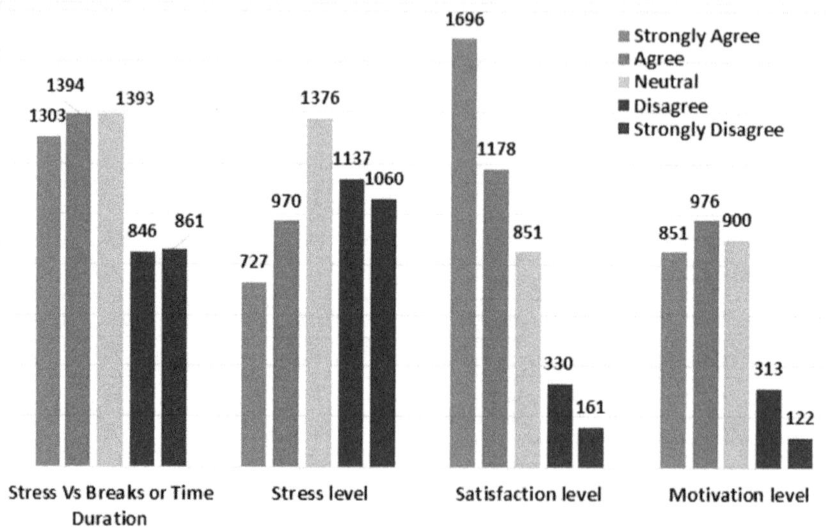


In above figure 6 is a bar chart of all levels, strongly agree percentage in satisfaction level is high as compared to other levels, neutral percentage is high in stress level as compared to other levels, whereas motivation level and satisfaction level in learners are high rather than other levels.

*Figure 7. All levels line chart gender-wise*

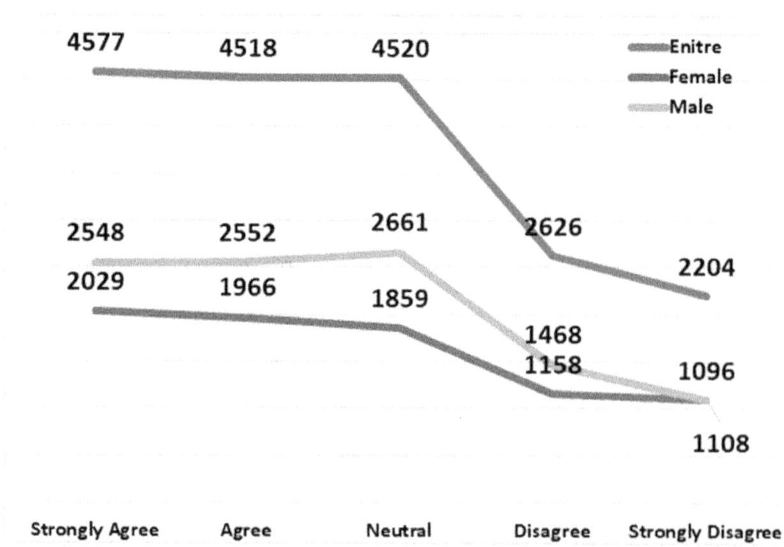

In above figure 7 is a line chart gender-wise of all levels, agree percentage is almost same irrespective of gender. Disagree percentage is less as compared to agree percentage in all levels.

*Figure 8. All levels line chart geographical location-wise*

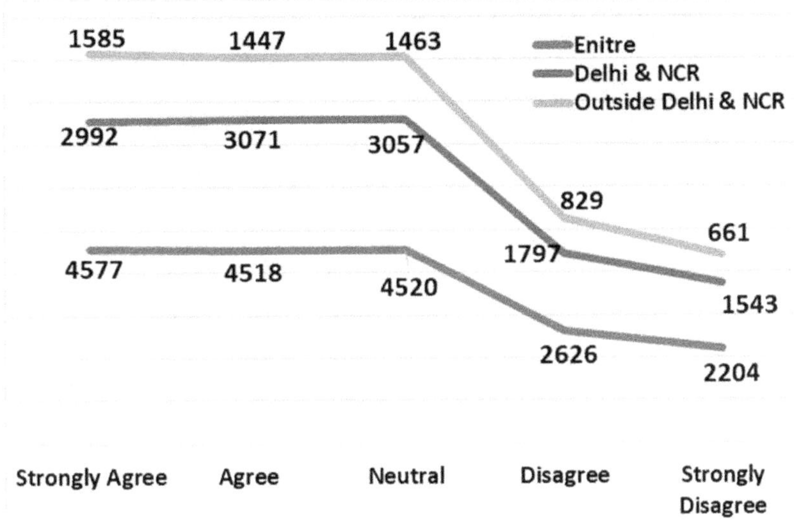


Similarly, in above figure 8 is a line chart area-wise of all levels, agree percentage is almost same irrespective of gender. Disagree percentage is less as compared to agree percentage in all levels.

In below figures from 9 – 13 is a pie chart age-wise strongly agree is more in 18 years age rather than 19 – 20 years age group, they are more agree. In rest age group agree is almost same. Which implies in all age group agree percentage is more as compared to disagree, i.e., social analysis of learner's behaviour agrees. In figure 14 represent age-wise in the form of bar chart.

## Algorithmic Model

ML is a branch of AI and computer discipline that emphases on employing complex procedures to reproduce how people acquire to continuously increase correctness. Coding according to dataset in Ludwig, which is a part of Deep Learning (DL). DL is division of AI and processer programming that emphases on employing complex procedures to reproduce how people absorb to continuously increase correctness.

*Figure 9. Age-wise charts – entire (<= 18 years)*

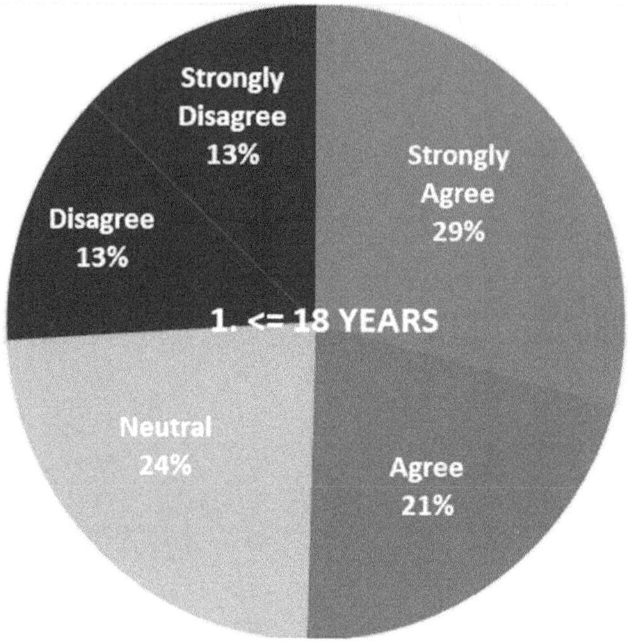


*Figure 10. Age-wise charts – entire (between 19 years – 22 years)*

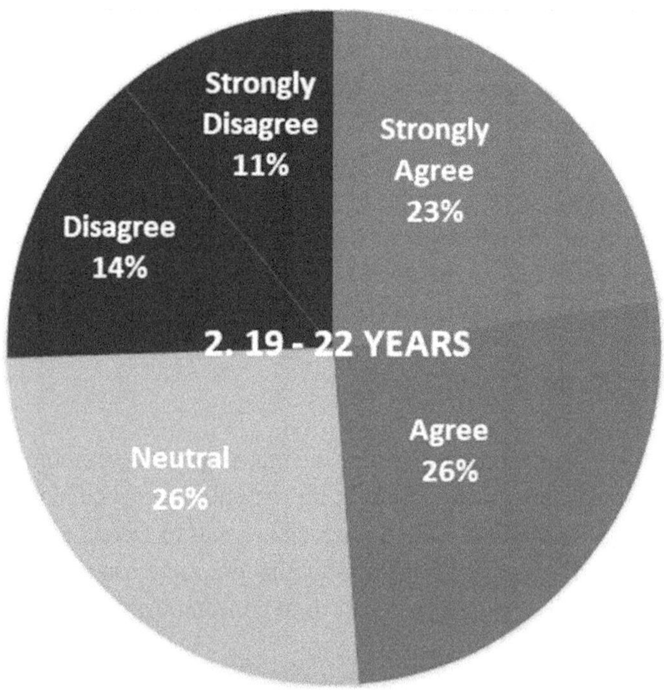

*Figure 11. Age-wise charts – entire (between 23 years – 30 years)*

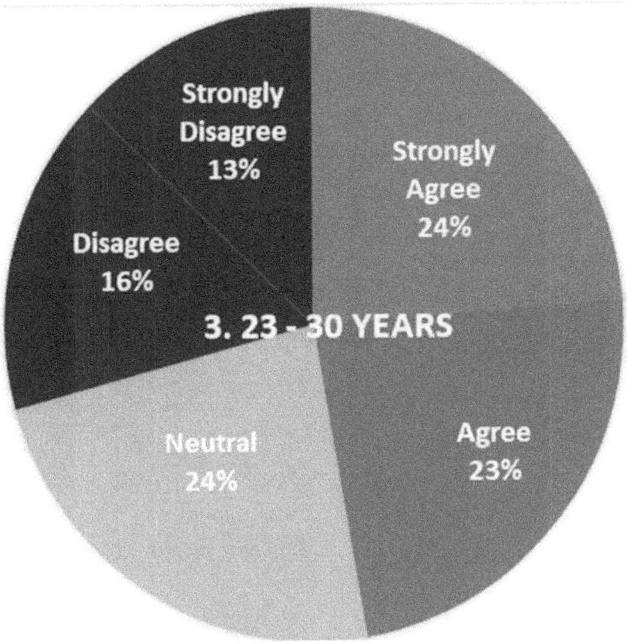


*Figure 12. Age-wise charts – entire (between 31 years – 40 years)*

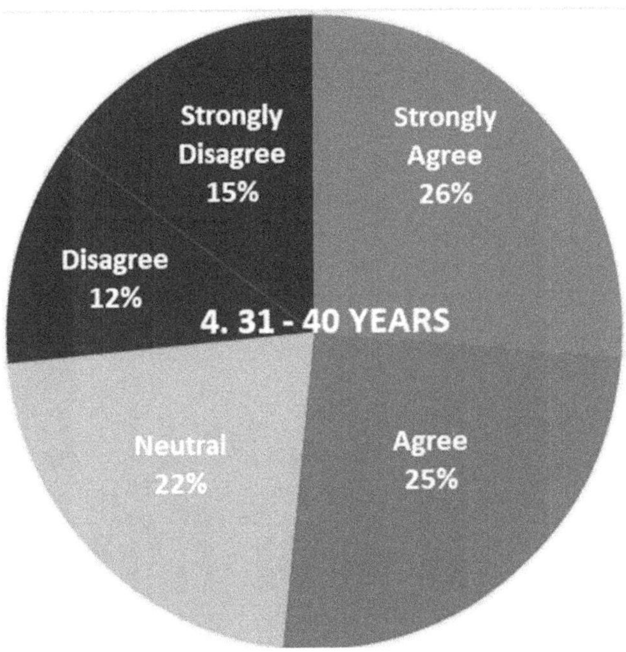

*Figure 13. Age-wise charts – entire (>= 41 years)*

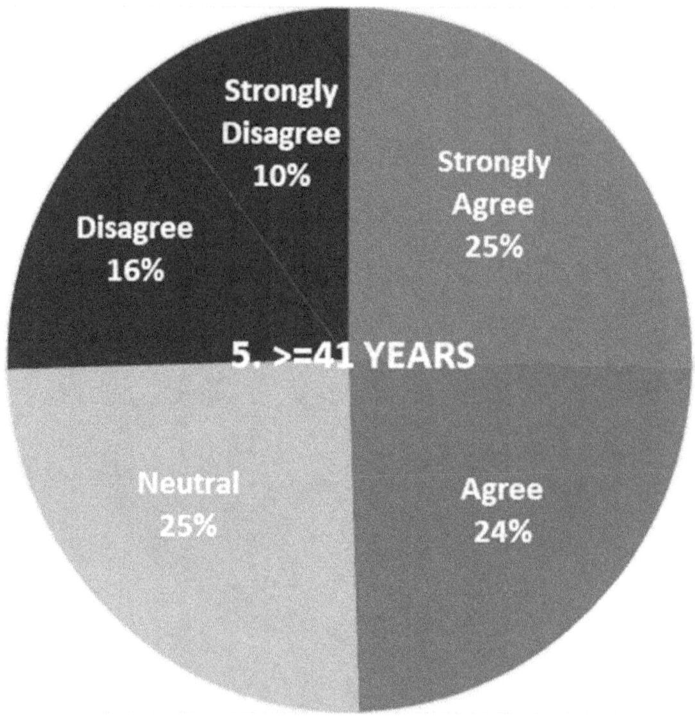


*Figure 14. Entire all levels – age wise bar chart*

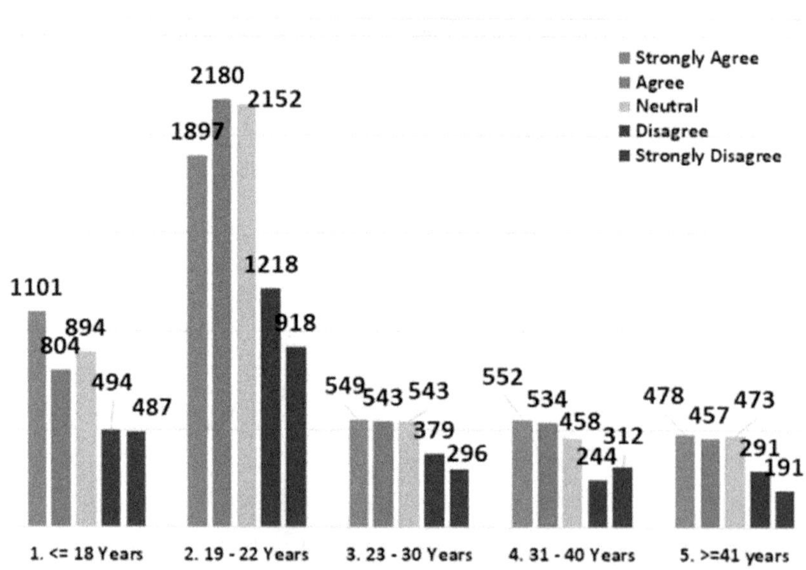

*Figure 15. All levels Data Frame(df) Headings*

```
%%writefile config.yaml
input_features:
    -
        name: Learner's Age
        type: numerical
    -
        name: Learner's Educational Status
        type: numerical
    -
        name: Learner's Gender
        type: numerical
    -
        name: Learner's State
        type: numerical
    -
        name: Learning Experiences
        type: numerical
    -
        name: Learner's Highest Education / Learning State
        type: numerical

output_features:
    -
        name: all levels
        type: numerical
```

In Above figure 15, Numerous DataFrame(df) Headings of all levels mentioned in the dataset along through the queries are gotten using df.Column (x and y coordinates) as exposed under. In figure 16 and 17 x-axis coordinates and y-axis coordinates represent respectively. Pandas DataFrame is a two-dimensional tabularized data arrangement with labeled axes that is the shape and possibly heterogeneous.

Declarative learning in machine learning. Ludwig, a suite of tools based on deep learning, was introduced to us. Because of the toolbox's versatility, extensibility, and user-friendliness, together authorities and beginners may train DL mockups, use them to generate forecasts, and test with various buildings deprived of having to engrave any cypher. We had a straightforward use case for this toolkit through this object.

## DATASET WITH DIFFERENT PARAMETERS

The collection of an adequate dataset with different parameters of all levels of learning effort. PyCaret contains a range of instance datasets aimed at various types of ML tasks, the learner's dataset with different parameters having sample size 527 passes through machine shown in figure 18. On that bases model created.

*Figure 16. X-axis df coordinates*

```
df.columns

Index(['Timestamp', 'Email Address', 'Learner's Age',
       'Learner's Educational Status', 'Learner's Gender', 'Learner's State',
       'Learning Experiences', 'Learner's Highest Education / Learning State',
       '1.1 Always feel motivated during learning',
       '1.2 Always spend maximum time for learning',
       '1.3 Can spend time for learning instead of hectic schedule',
       '1.4 Must spend time for learning instead of physical stress',
       '1.5 Must spend time for learning instead of emotional stress',
       '1.6 Always feel motivated in learning after praying or worshiping ',
       '2.1 Always Feel Mentally Satisfied while Learning',
       '2.2 Satisfaction level increases while Learning even if topic is from area of your interest',
       '2.3 Satisfaction level increases while Learning if topic is interesting but not from area of your interest ',
       '2.4 Can Feel Satisfied while Learning if topic is not interesting',
       '2.5 Interesting Learning or topic reduce Physical Stress ',
       '2.6 Interesting Learning or topic reduce Mental Stress ',
       '2.7 Praying reduces Mental Stress ',
       '2.8 If Learning is Interesting than Time flies',
       '3.1 Always Feel Mental Stress while Learning because you are not interested in Learning',
       '3.2 Always Feel Mental Stress while Learning even if topic is from area of your interest because you are not interested in Learning',
       '3.3 Always Feel Mental Stress while Learning if topic is interesting but not from area of your interest ',
       '3.4 Always Feel Physical Stress while Learning even if topic is from your area of interest',
       '3.5 Always Feel Physical Stress while Learning even if topic is not from your area of interest',
       '3.6 Always Feel Physical Stress while Learning even if topic is interesting but you are not interested in learning',
       '3.7 Always Feel Emotional Stress while Learning because you are not interested in learning',
       '3.8 Always Feel Emotional Stress while Learning even if topic is from your area of interest',
       '3.9 Always Feel Emotional Stress while Learning even if topic is not from your area of interest',
       '3.10 When Stress level increases you start praying or worshiping ',
       '4.1 Under Mental Stress Learners Learn better',
       '4.2 Under Physical Stress Learners Learn better',
       '4.3 Under Emotional Stress Learners Learn better',
       '4.4 If Learners is Emotional then they perform better',
       '4.5 If Learners is Physically fit then they perform better',
       '4.6 If Learners is in mental stress but physically fit then they perform better',
       '4.7 Time duration decides stress level ',
       '4.8 Short Interval or breaks reduces mental stress level effect on performance',
       '4.9 Short Interval or breaks reduces physical stress level effect on performance',
       '4.10 Short Interval or breaks reduces emotional stress level effect on performance',
       '4.11 Single Long Interval is better than many short intervals '],
      dtype='object')
```

## Modelling Create using PyCaret

The create model () method allows you to create a regression model depending on your preferred algorithm. Gradient Boosting Regressor used it performed the best in compare models (). The create model () method evaluates model correctness using k-fold cross-validation. The dataset is initially divided into k subsamples in this procedure shown in figure 19, one of which is kept for verification and the remainder is utilised to sleeper the perfect. This method is done k aeras, and apiece subset is utilised as authentication information just once.

*Figure 17. Y-axis df.columns coordinates*

```
[ ] y_columns = ['1.1 Always feel motivated during learning',
         '1.2 Always spend maximum time for learning',
         '1.3 Can spend time for learning instead of hectic schedule',
         '1.4 Must spend time for learning instead of physical stress',
         '1.5 Must spend time for learning instead of emotional stress',
         '1.6 Always feel motivated in learning after praying or worshiping ',
         '2.1 Always Feel Mentally Satisfied while Learning',
         '2.2 Satisfaction level increases while Learning even if topic is from area of your interest',
         '2.3 Satisfaction level increases while Learning if topic is interesting but not from area of your interest ',
         '2.4 Can Feel Satisfied while Learning if topic is not interesting',
         '2.5 Interesting Learning or topic reduce Physical Stress ',
         '2.6 Interesting Learning or topic reduce Mental Stress ',
         '2.7 Praying reduces Mental Stress ',
         '2.8 If Learning is Interesting than Time flies',
         '3.1 Always Feel Mental Stress while Learning because you are not interested in Learning',
         '3.2 Always Feel Mental Stress while Learning even if topic is from area of your interest because you are not interested in Learning',
         '3.3 Always Feel Mental Stress while Learning if topic is interesting but not from area of your interest ',
         '3.4 Always Feel Physical Stress while Learning even if topic is from your area of interest',
         '3.5 Always Feel Physical Stress while Learning even if topic is not from your area of interest',
         '3.6 Always Feel Physical Stress while Learning even if topic is interesting but you are not interested in learning',
         '3.7 Always Feel Emotional Stress while Learning because you are not interested in learning',
         '3.8 Always Feel Emotional Stress while Learning even if topic is from your area of interest',
         '3.9 Always Feel Emotional Stress while Learning even if topic is not from your area of interest',
         '3.10 When Stress level increases you start praying or worshiping ',
         '4.1 Under Mental Stress Learners Learn better',
         '4.2 Under Physical Stress Learners Learn better',
         '4.3 Under Emotional Stress Learners Learn better',
         '4.4 If Learners is Emotional then they perform better',
         '4.5 If Learners is Physically fit then they perform better',
         '4.6 If Learners is in mental stress but physically fit then they perform better',
         '4.7 Time duration decides stress level ',
         '4.8 Short Interval or breaks reduces mental stress level effect on performance',
         '4.9 Short Interval or breaks reduces physical stress level effect on performance',
         '4.10 Short Interval or breaks reduces emotional stress level effect on performance',
         '4.11 Single Long Interval is better than many short intervals ']

  y = df.loc[:, y_columns].copy()
```

Setting up the environment in which machine learning experiments will run is the first thing need to accomplish. The setup () function from the pycaret.classification module has to be initialized.

The variable for all levels with different parameters and must define the dataframe (df) of dataset in the setup function. The experiment name and other variables are also adjustable. In figure 19 below shows setup () function with all level parameters and targets. After complete run, Setup() successfully Completed message appears.

*Figure 18. Dataset with different parameters*

| | Learner's Age | Learner's Educational Status | Learner's Gender | Learner's State | Learning Experiences | Learner's Highest Education / Learning State | all levels |
|---|---|---|---|---|---|---|---|
| 0 | 4 | 16 | 0 | 9 | 0 | 8 | 15.125 |
| 1 | 4 | 16 | 1 | 9 | 1 | 8 | 15.000 |
| 2 | 0 | 20 | 1 | 9 | 1 | 8 | 18.875 |
| 3 | 0 | 20 | 1 | 9 | 2 | 8 | 12.750 |
| 4 | 0 | 20 | 1 | 9 | 2 | 8 | 13.375 |
| ... | ... | ... | ... | ... | ... | ... | ... |
| 522 | 4 | 16 | 0 | 7 | 5 | 14 | 7.750 |
| 523 | 0 | 23 | 0 | 7 | 2 | 6 | 12.750 |
| 524 | 4 | 16 | 0 | 21 | 0 | 20 | 15.000 |
| 525 | 1 | 20 | 1 | 7 | 1 | 6 | 14.125 |
| 526 | 3 | 17 | 0 | 7 | 5 | 14 | 16.000 |

527 rows × 7 columns

*Figure 19. Modelling using PYCARET*

```
clf1 = setup(data = data, target = 'all levels')
```

| | Description | Value |
|---|---|---|
| 0 | session_id | 2432 |
| 1 | Target | all levels |
| 2 | Original Data | (527, 7) |
| 3 | Missing Values | False |
| 4 | Numeric Features | 3 |
| 5 | Categorical Features | 3 |
| 6 | Ordinal Features | False |
| 7 | High Cardinality Features | False |
| 8 | High Cardinality Method | None |
| 9 | Transformed Train Set | (368, 21) |
| 10 | Transformed Test Set | (159, 21) |
| 11 | Shuffle Train-Test | True |
| 12 | Stratify Train-Test | False |
| 13 | Fold Generator | KFold |

For regression problems, Pycaret uses the following evaluation metrics,

```
# return best model
best = compare_models() run for the best results.
```

According to above return best model, below in figure 20 is regression output received in PyCaret, for Regression: MAE, MSE, RMSE, R2, RMSLE, MAPE

*Figure 20. Regression output of Model Designed*

| | Model | MAE | MSE | RMSE | R2 | RMSLE | MAPE | TT (Sec) |
|---|---|---|---|---|---|---|---|---|
| br | Bayesian Ridge | 1.7398 | 5.377500e+00 | 2.2933 | -2.630000e-02 | 0.1412 | 0.1179 | 0.026 |
| en | Elastic Net | 1.7393 | 5.379800e+00 | 2.2938 | -2.670000e-02 | 0.1412 | 0.1179 | 0.034 |
| dummy | Dummy Regressor | 1.7450 | 5.390700e+00 | 2.2970 | -3.070000e-02 | 0.1415 | 0.1184 | 0.018 |
| llar | Lasso Least Angle Regression | 1.7450 | 5.390700e+00 | 2.2970 | -3.070000e-02 | 0.1415 | 0.1184 | 0.041 |
| lasso | Lasso Regression | 1.7448 | 5.392300e+00 | 2.2973 | -3.090000e-02 | 0.1415 | 0.1184 | 0.026 |
| omp | Orthogonal Matching Pursuit | 1.7720 | 5.534700e+00 | 2.3276 | -5.860000e-02 | 0.1434 | 0.1203 | 0.026 |
| huber | Huber Regressor | 1.7693 | 5.749600e+00 | 2.3612 | -7.940000e-02 | 0.1445 | 0.1174 | 0.088 |
| ridge | Ridge Regression | 1.8067 | 5.733500e+00 | 2.3665 | -9.210000e-02 | 0.1460 | 0.1227 | 0.027 |
| lr | Linear Regression | 1.8121 | 5.767300e+00 | 2.3732 | -9.830000e-02 | 0.1464 | 0.1231 | 0.494 |
| ada | AdaBoost Regressor | 1.9151 | 6.027900e+00 | 2.4324 | -1.659000e-01 | 0.1516 | 0.1335 | 0.035 |
| lightgbm | Light Gradient Boosting Machine | 1.9015 | 6.357200e+00 | 2.4993 | -2.276000e-01 | 0.1546 | 0.1294 | 0.099 |
| gbr | Gradient Boosting Regressor | 1.8851 | 6.370000e+00 | 2.4977 | -2.354000e-01 | 0.1538 | 0.1290 | 0.057 |
| knn | K Neighbors Regressor | 1.9139 | 6.550800e+00 | 2.5405 | -2.840000e-01 | 0.1568 | 0.1300 | 0.086 |
| rf | Random Forest Regressor | 1.9607 | 6.889900e+00 | 2.6024 | -3.362000e-01 | 0.1607 | 0.1338 | 0.625 |
| par | Passive Aggressive Regressor | 2.2869 | 8.660900e+00 | 2.8397 | -5.819000e-01 | 0.1776 | 0.1561 | 0.028 |
| et | Extra Trees Regressor | 2.1726 | 8.519100e+00 | 2.8908 | -6.670000e-01 | 0.1778 | 0.1482 | 0.534 |
| dt | Decision Tree Regressor | 2.2881 | 9.635000e+00 | 3.0732 | -8.653000e-01 | 0.1902 | 0.1559 | 0.031 |
| lar | Least Angle Regression | 1241.9611 | 2.519789e+07 | 1589.5476 | -4.248722e+06 | 0.7845 | 89.4401 | 0.033 |

## Model Making in PyCaret

In Pycaret, building a model is simple and easy. The object name is the only argument you must add to the create model () method. In figure 21 shows mean and std value when passes through model ().

*Figure 21. Model calculate Mean and Std values*

| | MAE | MSE | RMSE | R2 | RMSLE | MAPE |
|---|---|---|---|---|---|---|
| **Fold** | | | | | | |
| **0** | 1.9450 | 6.7636 | 2.6007 | -0.2902 | 0.1563 | 0.1260 |
| **1** | 2.2070 | 9.1722 | 3.0286 | -0.1724 | 0.1768 | 0.1391 |
| **2** | 1.8699 | 6.6768 | 2.5839 | -0.1258 | 0.1563 | 0.1244 |
| **3** | 1.5848 | 3.9417 | 1.9854 | -0.3625 | 0.1301 | 0.1164 |
| **4** | 1.8902 | 6.2670 | 2.5034 | -0.3050 | 0.1553 | 0.1295 |
| **5** | 1.3772 | 3.3542 | 1.8315 | -0.2212 | 0.1131 | 0.0925 |
| **6** | 2.0856 | 7.7504 | 2.7839 | -0.0375 | 0.1779 | 0.1443 |
| **7** | 1.9833 | 6.6996 | 2.5884 | -0.2390 | 0.1626 | 0.1374 |
| **8** | 1.9159 | 6.1675 | 2.4835 | -0.1869 | 0.1520 | 0.1278 |
| **9** | 2.1558 | 6.7790 | 2.6037 | -0.3358 | 0.1659 | 0.1560 |
| **Mean** | 1.9015 | 6.3572 | 2.4993 | -0.2276 | 0.1546 | 0.1294 |
| **Std** | 0.2402 | 1.5914 | 0.3328 | 0.0955 | 0.0189 | 0.0163 |

## Model Performance's Visualization

The interesting part of PyCaret is that it allows to evaluate model performance easily and generate different graphs for clear explanation. It's a bit tedious to create each of these graphs, so having a quick way to evaluate model. In figure 22 shows value of Train $R^2$ and test $R^2$ in residuals regression dot plot graph.

In figure 23 shows value of $R^2$ lies on best fit and identity line in Prediction Error Dot Plot Graph.

Whether and to what extent the feature importance plot used for training assisted in predicting the target is a key component of our models' post-training analysis. The feature importance attribute of tree-based machine learning algorithms like Random Forest and XGBoost produces an array with a value between 0 and 100 for each feature, representing how helpful the model found each feature in trying to predict the target. This provides the chance to determine which features were merely noise and which ones helped the model to be accurate. Check that the model is functioning as expected using this data, eliminate features that are not useful, and use it to make assumptions about new features that could be developed for a subsequent iteration of the model as shown in figure 24.

*Figure 22. Residuals Regression Dot Plot Graph*

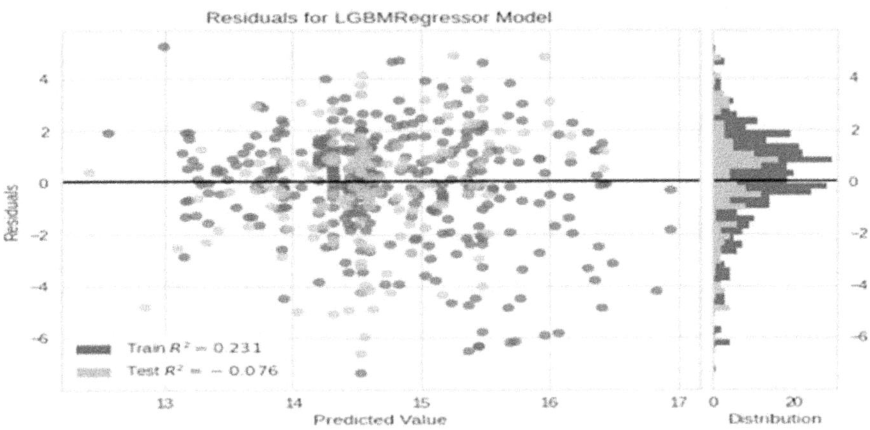


*Figure 23. Prediction Error Regression Dot Plot Graph*

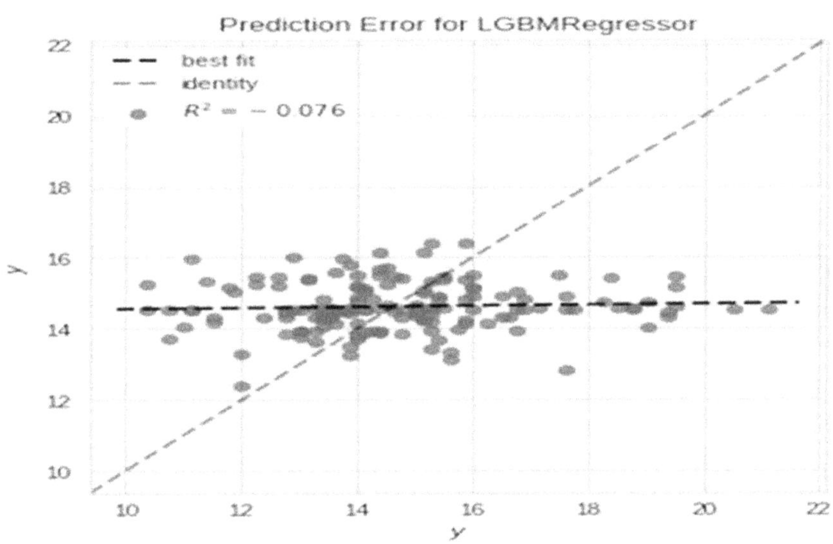


## Ludwig Classifier Analysis

Over the past ten years, deep learning models need shown to be incredibly successful in a variety of machine learning missions in vision, speech, and language. Ludwig is exceptional in that it helps non-experts understand deep learning while enabling quicker iteration cycles for model improvement for specialists in machine learning research and development. Ludwig enables professionals and academics to streamline the prototype procedure and accelerate data processing, freeing them up to concentrate on creating deep learning systems somewhat than data management. The Learning Curve As shown in Figure 25, the outcome acquired using machine learning algorithm (Ludwig) at the 100th epoch for estimating all levels, it is discovered that the loss is diminishing and, in other terms, the machine's accuracy continues to increase.

*Figure 24. Feature Importance Plot graph*

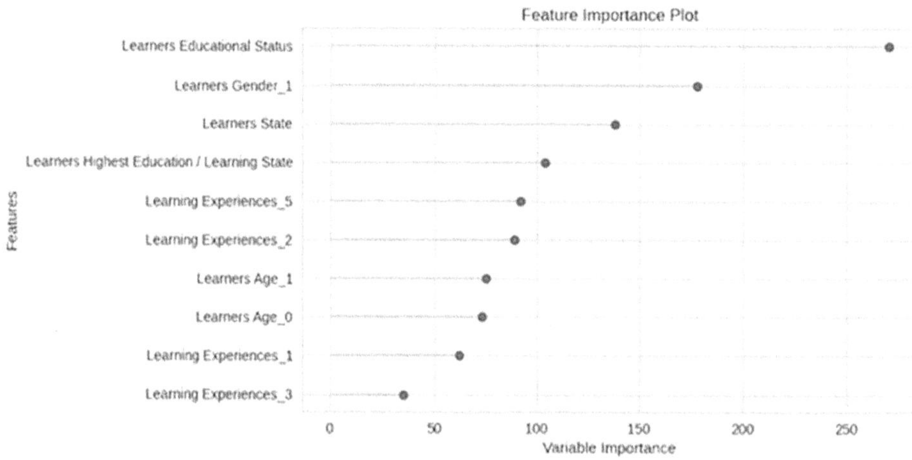


Ludwig is a deep learning toolkit designed to take use of inheritance and code modularity. It was constructed at the degree of abstraction mentioned above. Ludwig makes it simpler for practitioners to reuse, expand, and favor best practices in addition to creating deep learning models by only expressing their data and tasks. These equivalence classes are named using the data types of the projected outputs by the decoding functions and the data types of the inputs encoded by the encoding functions (image, text, series, category, etc.). Unlike the deep learning frameworks that are currently available, which abstract at the tensor operation or layer level, this kind of abstraction enables a higher degree of interface. This is accomplished by offering abstract APIS for every data type, enabling an addition by enabling the creation of any new implementation of the interface.

Ludwig is founded on the idea of a declarative model definition, which democratizes deep learning models by making them accessible to a much larger audience. In the past 10 years, a range of language, audio, and visual machine-learning applications have shown deep-learning replicas to be very successful. Ludwig is unique in that it enables faster model development iteration loops for machine learning professionals while also assisting non-experts in understanding deep learning. Ludwig gives professionals and researchers the ability to expedite the sample process and accelerate data processing, allowing them time to concentrate on developing deep learning programs instead than data administration. The learning curve as shown in figure 25 and the outcome as shown in figure 26 obtained using a classifier (Ludwig) at the 1st epoch for estimating engagement at work.

*Figure 25. Loss with Number of epochs learning Curve*

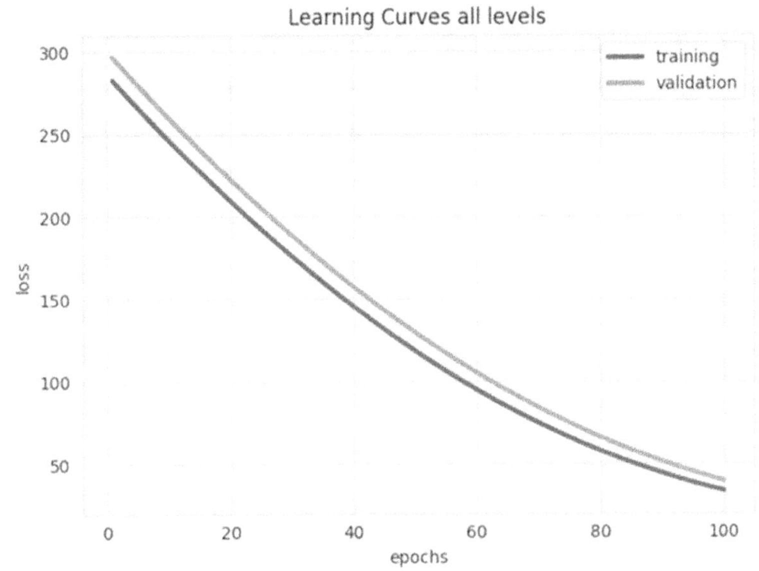


*Figure 26. Result obtained using deep learning algorithm at 1st epoch*

```
Epoch   1
Training: 100% 3/3 [00:00<00:00, 194.51it/s]
Evaluation train: 100% 3/3 [00:00<00:00, 538.68it/s]
Evaluation vali: 100% 1/1 [00:00<00:00, 164.89it/s]
Evaluation test: 100% 1/1 [00:00<00:00, 498.67it/s]
Took 0.8425s
```

| all levels | loss | error | mean_squared_error | mean_absolute_error | r2 |
|---|---|---|---|---|---|
| train | 278.7071 | 16.4338 | 278.7071 | 16.4351 | -54.1318 |
| vali | 292.5583 | 16.8783 | 292.5583 | 16.8783 | -61.9167 |
| test | 277.4130 | 16.3349 | 277.4130 | 16.3349 | -51.4740 |

| combined | loss |
|---|---|
| train | 278.7071 |
| vali | 292.5583 |
| test | 277.4130 |

```
Validation loss on combined improved, model saved
```

And it was determined that the loss was decreasing, and that the machine's accuracy continued to rise with a greater number of epochs, indicating an increase in consistency as shown in figure 27 classifier (Ludwig) at the 100th epoch.

*Figure 27. Result obtained using deep learning algorithm at 100th epoch*

```
Epoch 100
Training: 100% 3/3 [00:00<00:00, 273.38it/s]
Evaluation train: 100% 3/3 [00:00<00:00, 776.58it/s]
Evaluation vali : 100% 1/1 [00:00<00:00, 692.59it/s]
Evaluation test : 100% 1/1 [00:00<00:00, 552.46it/s]
Took 0.5465s
```

| all levels | loss | error | mean squared error | mean absolute error | r2 |
|---|---|---|---|---|---|
| train | 33.3281 | 4.1638 | 33.3281 | 5.1010 | -5.5874 |
| vali | 39.2998 | 4.8270 | 39.2998 | 5.7197 | -7.4517 |
| test | 35.8938 | 4.3889 | 35.8938 | 5.2875 | -5.7895 |

| combined | loss |
|---|---|
| train | 33.3281 |
| vali | 39.2998 |
| test | 35.8938 |

```
Validation loss on combined improved, model saved

Best validation model epoch: 100
Best validation model loss on validation set combined: 39.29975128173828
Best validation model loss on test set combined: 35.89380645751953

Finished: experiment run
Saved to: results/experiment run
Evaluation: 100% 1/1 [00:00<00:00, 464.38it/s]
```

Declarative learning in machine learning was discussed in this chapter. Ludwig, a suite of tools based on deep learning, was introduced to us. Because of the toolbox's versatility, extensibility, and user-friendliness, both experts and novices may train deep learning models, use them to make predictions, and experiment with various architectures without having to write any code. A straightforward use case for this toolbox through this chapter. I urge you to read the official materials and the study report to learn how to approach challenging ML challenges (Kautish et. al 2022) in the manner we have covered in this book chapter.

## CONCLUSION

The learners' gender, age, and location are only parameters of the variables used to determine how satisfied they are with their social learning. Like how social learning satisfaction changes with age, females report higher levels of satisfaction than males do. For instance, learning interest rises early as we approach earning age, progressively drops as we become older, and then rises again. Like this, outside Delhi & NCR, satisfaction ratings are greater than those within Delhi & NCR. The comprehensive knowledge and assessment framework of this study also investigates how the computational psychometry example considers all these difficult components. All level was designed with the hypothesis that enhancing learning, assessment, and guiding will boost apprentices' chances of absorbing a beneficial, thorough education and information. The Ludwig Classifier used to calculate all level with different parameters was gained using a deep learning model during the 100th epoch and model build with PyCaret, and it was found that the loss was decreasing or that the machine's accuracy was increasing. Every subject discussed in this chapter has new capabilities, or those skills have at the very least been enhanced.

Chose to open-source Ludwig and Pycaret because that both novice machine learning practitioners and seasoned deep learning developers and academics can benefit from using it. Without writing any code, novices may easily train and test deep learning models. By carrying out common data preprocessing and visualization, experts can establish solid criteria to evaluate their models alongside and have an experimentation setup that creates it simple to test new hypotheses and evaluate models.

## FUTURE SCOPE

Several publications are being analyzed while offering the specifics of these tactics and their assessments. The approaches need to be improved, scaled up, and frequently evaluated for fairness, efficacy, and validity even though integrated monitoring and evaluation systems have undergone extensive research and innovation. No matter their race or location, it is the intention to be able to offer all students the best educational materials and feedback. Future research will use computational psychometrics to analyse learners' varied learning factors. the creation of fresh learning-friendly flexible study diagnostic models, as well as the use of multimodal analytic approaches, machine learning, and artificial intelligence (AI) to enhance current psychometric models.

## REFERENCES

Alavi, M., & Leidner, D. E. (2001). Review: Knowledge management and knowledge management systems: Conceptual foundations and research issues. *Management Information Systems Quarterly*, *25*(1), 107–136. doi:10.2307/3250961

Y. Alkali, I. Routray, & Whig, P. (2022). Strategy for Reliable, Efficient and Secure IoT Using Artificial Intelligence. *IUP Journal of Computer Sciences, 16*(2).

AlkaliY.RoutrayI.WhigP. (2022). Study of various methods for reliable, efficient and Secured IoT using Artificial Intelligence. SSRN 4020364. doi:10.2139/ssrn.4020364

Anand, M., Velu, A., & Whig, P. (2022). Prediction of Loan Behaviour with Machine Learning Models for Secure Banking [JCSE]. *Journal of Computing Science and Engineering : JCSE*, *3*(1), 1–13.

Bhatia, A. B., & Mittal, K. (2022). Measuring Computational Psychometrics Analysis Motivational Level in Learner's using Different Parameters through Deep Learning Algorithm. [JCSE]. *Journal of Computing Science and Engineering : JCSE*, *3*(2), 120–136.

Bhatnagar, A., & Mittal, K. (2022). Computational Psychometrics Analysis of Learners' *Motivational Level Using Different Parameters. In International Conference on Innovative Computing and Communications* (pp. 493-507). Springer, Singapore.

Chen, G., & Lee, T. H. (2019). Predictive validity of self-efficacy and academic goal orientation in online learning. *Computers & Education*, *128*, 296–309.

Chou, C. C., & Tsai, C. C. (2018). Effects of personalized game-based instruction on students' learning performance and motivation in mathematics. *Journal of Educational Technology & Society*, *21*(3), 42–54.

Drachsler, H., & Goldhammer, F. (2020). Learning Analytics and e-assessment—towards computational psychometrics by combining psychometrics with learning analytics. In *Radical Solutions and Learning Analytics* (pp. 67–80). Springer. doi:10.1007/978-981-15-4526-9_5

Fong, C. J., & Lam, H. C. (2020). Applying artificial intelligence in education: A review of emerging trends and issues. *Educational Technology Research and Development*, *68*(6), 2709–2730.

Hsu, Y. C., & Lin, J. F. (2018). Using educational data mining to improve student learning: A literature review. *Journal of Educational Technology & Society*, *21*(4), 74–86.

Huang, R. H., Liu, D. J., & Tlili, A. (2019). Effects of integrating learning analytics into a flipped classroom on students' learning outcomes and satisfaction. *Internet and Higher Education*, *43*, 100705.

Jeon, H., Oh, H., & Lee, J. (2018). *Machine learning-based fast reading algorithm for future ICT-based education*. In: *2018 International Conference on Information and Communication Technology Convergence (ICTC)* (pp 771–775). . IEEE. 10.1109/ICTC.2018.8539447

Jupalle, H., Kouser, S., Bhatia, A. B., Alam, N., Nadikattu, R. R., & Whig, P. (2022). Automation of human behaviors and its prediction using machine learning. *Microsystem Technologies*, *28*(8), 1–9. doi:10.100700542-022-05326-4

Kaur, R. & Kautish, S. (2022). Multimodal sentiment analysis: A survey and comparison. *Research Anthology on Implementing Sentiment Analysis Across Multiple Disciplines*, 1846-1870.

Kautish, S., Peng, S. L., & Obaid, A. J. (Eds.). (2021). *Computational intelligence techniques for combating COVID-19*. Springer International Publishing. doi:10.1007/978-3-030-68936-0

Kautish, S., Reyana, A., & Vidyarthi, A. (2022). SDMTA: Attack detection and mitigation mechanism for DDoS vulnerabilities in hybrid cloud environment. *IEEE Transactions on Industrial Informatics*, *18*(9), 6455–6463. doi:10.1109/TII.2022.3146290

Kim, M. J., Kim, M., & Kim, H. (2019). An overview of recent applications of artificial intelligence in education: A systematic review. *Journal of Educational Technology & Society*, *22*(2), 1–16.

Kučak, D., Juričić, V., & Đambić, G. (2018). Machine Learning In Education-A Survey Of Current Research Trends. *Annals of DAAAM & Proceedings*, (pp. 29). Springer.

Lee, K. C., Chai, C. S., & Tsai, C. C. (2018). Computational thinking research in education. *Educational Technology Research and Development*, *66*(4), 765–768.

Madhu, G., Govardhan, A., Ravi, V., Kautish, S., Srinivas, B. S., Chaudhary, T., & Kumar, M. (2022). DSCN-net: A deep Siamese capsule neural network model for automatic diagnosis of malaria parasites detection. *Multimedia Tools and Applications*, *81*(23), 34105–34127. doi:10.100711042-022-13008-6

Maris, G., & Bechger, T. (2019). Bayesian network modeling for psychometric applications. *Frontiers in Psychology*, *10*, 438.

Pelánek, R. (2016). Applications of the Elo rating system in adaptive educational systems. *Computers & Education*, *98*, 169–179. doi:10.1016/j.compedu.2016.03.017

Studer, S., Bui, T. B., Drescher, C., Hanuschkin, A., Winkler, L., Peters, S., & Müller, K. R. (2021). Towards CRISP-ML (Q): a machine learning process model with quality assurance methodology. *Machine Learning and Knowledge Extraction*, *3*(2), 392-413. https://blog.devgenius.io/predicting-tesla-stocks-tsla-using-python-pycaret-45af9ed47de9

Tomar, U. Chakroborty, N. Sharma, H., & Whig, P. (2021). AI based Smart Agricuture System. *Transactions on Latest Trends in Artificial Intelligence*, 2(2).

Von Davier, A. A., Deonovic, B., Yudelson, M., Polyak, S. T., & Woo, A. (2019). Computational psychometrics approach to holistic learning and assessment systems. In *Frontiers in Education* (p. 69). Frontiers. doi:10.3389/feduc.2019.00069

Whig, P., Kouser, S., Velu, A., & Nadikattu, R. R. (2022). Fog-IoT-Assisted-Based Smart Agriculture Application. In Demystifying Federated Learning for Blockchain and Industrial Internet of Things. IGI Global.

Whig, P., Nadikattu, R. R., & Velu, A. (2022). COVID-19 pandemic analysis using application of AI. Healthcare Monitoring and Data Analysis Using IoT: Technologies and Applications, (p. 1). IET. doi:10.1049/PBHE038E_ch1

Whig, P., Velu, A., & Bhatia, A. B. (2022). Protect Nature and Reduce the Carbon Footprint with an Application of Blockchain for IIoT. In Demystifying Federated Learning for Blockchain and Industrial Internet of Things. IGI Global.

Whig, P., Velu, A., & Naddikatu, R. R. (2022). The Economic Impact of AI-Enabled Blockchain in 6G-Based Industry. In AI and Blockchain Technology in 6G Wireless Network. Springer.

Whig, P., Velu, A., & Nadikattu, R. R. (2022). Blockchain Platform to Resolve Security Issues in IoT and Smart Networks. In AI-Enabled Agile Internet of Things for Sustainable FinTech Ecosystems. IGI Global.

Whig, P., Velu, A., & Ready, R. (2022). Demystifying Federated Learning in Artificial Intelligence with Human-Computer Interaction. In Demystifying Federated Learning for Blockchain and Industrial Internet of Things. IGI Global.

Whig, P., Velu, A., & Sharma, P. (2022). Demystifying Federated Learning for Blockchain: A Case Study. In Demystifying Federated Learning for Blockchain and Industrial Internet of Things. IGI Global. doi:10.4018/978-1-6684-3733-9.ch008

Wu, C. H., Chen, C. H., & Huang, Y. M. (2018). Using artificial intelligence to analyze and evaluate students' online learning behaviors. *Journal of Educational Technology & Society*, *21*(4), 58–69. https://www.learndatasci.com/tutorials/introduction-pycaret-machine-learning/

Wu, J. H., Tennyson, R. D., & Hsia, T. L. (2010). A study of student satisfaction in a blended e-learning system environment. *Computers & Education*, *55*(1), 155–164. doi:10.1016/j.compedu.2009.12.012

## ADDITIONAL READINGS:

Buss, A. H., & Plomin, R. (1984). *Temperament: Early developing personality traits*. Lawrence Erlbaum Associates.

DiStefano, C., Zhu, M., & Mîndrilă, D. (2018). Understanding and Using Factor Scores: Considerations for the Applied Researcher. *Practical Assessment, Research & Evaluation*, *23*(1), 1–13.

Huang, Y. M., & Wu, W. C. (2020). Development and evaluation of an adaptive e-learning system with personalized learning paths based on learning behavior analysis and a hybrid recommendation algorithm. *Interactive Learning Environments*, *28*(5), 665–680.

Korkmaz, Ş., Gürkan, G., & Özturan, M. (2021). Predicting the Effect of Learning Analytics on Learning Behaviors and Learning Outcome in Distance Education. *Journal of Educational Technology & Society*, *24*(2), 139–151.

Kyriakides, L., Creemers, B. P., & Antoniou, P. (2009). Using multilevel analysis to measure the impact of classroom factors on student achievement in primary education. *Journal of Classroom Interaction*, *44*(1), 14–23.

Lee, M. C., & Weng, P. (2019). Comparing classification accuracy of three AI algorithms in predicting university students' academic achievement. *Journal of Educational Computing Research*, *57*(4), 1065–1080.

Liu, Y., & Yao, X. (2019). A new perspective on the prediction of academic performance: A computational psychometrics approach. *Educational Psychology Review*, *31*(4), 1059–1079.

Loh, W. Y. (2014). Fifty years of classification and regression trees. *International Statistical Review*, *82*(3), 329–348. doi:10.1111/insr.12016 PMID:25844011

Rummel, N., Spada, H., & Hauser, S. (2018). Designing Adaptive Collaborative Learning Support: The Role of a Computational Model of Collaboration. In Adaptive and Adaptable Learning (pp. 211-226). Springer.

Volet, S. E., & Ang, G. (2012). Culturally mixed groups on international campuses: An opportunity for inter-cultural learning. *Higher Education*, *63*(5), 627–644.

## KEY TERMS AND DEFINITIONS

**Artificial Intelligence:** Artificial Intelligence (AI) refers to the simulation of human intelligence in machines that are programmed to think and act like humans.

**Computational Psychometrics:** It is a field of study that combines psychology, statistics, and computer science to develop and apply measurement models to human behavior data.

**Deep Learning:** It is a subset of machine learning that utilizes neural networks to model complex patterns and relationships in data.

**Epochs:** In machine learning, an epoch refers to a complete iteration of the training data through a neural network.

**Learners:** In the context of education, learners refer to students or individuals who are engaged in the process of learning new skills or knowledge.

**Ludwig Classifier:** Ludwig Classifier is an open-source tool that enables the creation of machine learning models without requiring any specialized programming skills or knowledge.

**PyCaret:** PyCaret is a machine learning library in Python that simplifies the process of training and deploying machine learning models.

# Chapter 2
# Decentralized Interaction With Elder-Care Tasks:
## AI Supporting Robot

**Umma Khatuna Jannat**
*Karpagam Academy of Higher Education, India*

**M. Mohan Kumar**
*Karpagam Academy of Higher Education, India*

**Syed Arif Islam**
*Karpagam Academy of Higher Education, India*

## ABSTRACT

*The rate of ageing in the world is extremely fast. Elderly people are becoming more and more prevalent all over the world. The need for services like hospitals, retirement communities, and assisted living facilities has increased as the proportion of older persons in the population rises. For legislators, medical experts, and society as a whole, recovering command over this populace's assumptions will introduce new difficulties. With the aid of smart robot technology, older people can live independent, satisfying lives while remaining safe and secure. As a result, the most notable aspect of this paper is an occy robot. An AI model is used to process facial recognition and speech data, and occy assists elderly people with daily tasks. Occy is more effective for home appliances and healthcare systems. Occy may do tasks like equipment control in the home, safety, and wellbeing sensing, travelling outside, emergency situations and everyday other tasks.*

DOI: 10.4018/978-1-6684-8171-4.ch002

## 1. INTRODUCTION

Everywhere in the world, technology is a trend that has an impact on how people live. Fascinating illustrations of the technology of the future are robots (Aceto, G., Persico, V., & Pescapé, A., 2019). Robotics has significantly impacted both society and industry as technology has advanced (Chiacchio, F., Petropoulos, G., & Pichler, D, 2018). Adults with autism are already common in various professions, such as those in the healthcare, military, and domestic support sectors. Even though the development of cognitive and physical capacities is highly individual, many older people eventually require assistance with daily living activities. As a result, elderly individuals become dependent on informal or formal help services, which are time-consuming and expensive. It also reduces their perception of their own independence. Most elderly people want to remain in their comfortable homes for as long as possible. Service robots may be able to offer support, enabling and extending independent lives in the home setting. The needs of this group are becoming an increasingly important concern for health professionals, government officials, caregivers, and families as the world's population continues to expand.

Due to these factors, robots that assist the elderly are sometimes perceived as serving an authoritarian purpose since they enable people to take care of themselves without assistance. Increasing the elderly population is frequently suggested as a solution to the growing senior citizen population. Actually, the utilization of robots in the senior consideration industry is expanding. There are many practical applications for AI methods. An AI model makes it feasible to develop digital smart systems, healthcare systems, and other applications (Carra, G., Argiolas, A., Bellissima, A., Niccolini, M., & Ragaglia, M, 2018). The most popular AI health care systems use deep learning to diagnose diseases, machine learning to create radiological models, automatic systems, and other techniques. The ageing of our current generation has created a greater need for new technology among older folks. The main arguments in favour of this are that there are insufficient hospital employees and that many people want to live as autonomously as possible rather than being institutionalised. Also require the utilisation of cutting-edge technologies and a sufficient number of healthcare personnel. Robotics now plays an important role in the care of the elderly. One level of household use robot that is specifically designed for use at home is a service robot (Demir, K. A., Döven, G., & Sezen, B, 2019) (Jannat, U. K., Kumar, M. M., & Islam, S. A.,2022). Both scientifically and commercially, the industry of building robots specifically for the house is growing.

Occy robot needs to be capable of a variety of tasks to be most useful in the home, including home monitoring, device control, personal help, and entertainment. The absence of healthcare and the level of living for the elderly will both improve as more robots are created to interact with humans and provide the kind of care that is frequently supplied by a licenced therapist. The approach taken is to develop and reassure the elderly person. The major objective is to create a robot that can assist like a friend an elderly person with needful work and household chores on a daily basis.

## 1.1 Overview of Robots in General

Modern robots can perform intricate motions, are extremely talented and adaptable, not to mention that they can communicate with humans and learn from them. These developments have considerably increased the possible uses for robots. These advancements in robotic technology have an impact on the service robotics industry and have led to a wide range of robotic applications (Ellitan, L., 2020). As a result of the current demographic shifts, reorganisation of the labour force, and economic crisis, the emphasis on the opportunities afforded by information, communication, and robotics technology, as well as the exploration activities and projects aimed at finding sustainable healthcare service solutions, has steadily risen (de Kervenoael, R., Hasan, R., Schwob, A., & Goh, E., 2020). The World Health Organization (WHO) predicts that between 2020 and 2050, the number of people 80 or older is projected to treble, reaching 426 million (Bossone, E., Ranieri, B., Coscioni, E., & Baliga, R. R., 2019). The world is currently dealing with significant societal issues as a result of an ageing population putting increasing strain on the healthcare system's ability to remain sustainable. First, there is a growing need for care and social services, but the supply is being limited by a shrinking labour force. Second, the expense of treatment skyrockets as people age, primarily as a result of the management of comorbidities and chronic diseases. A robot generally comprises of seven primary parts, no matter what its shape or size. This cooperation has a specific purpose.

Controller: The robot's movements should be composed by the leader. The region wherein a robot might work is alluded to as the functional climate. The microcontroller is additionally answerable for social occasion information from the climate utilizing its sensors.

Power Conversion Unit: The robot is controlled by the fusion reactor. Robots common power sources include flammable substances, electricity, and battery storage. The majority of the power that powers a building originates from the grid, and the system converts the grid's AC electricity into the building's DC electricity using an AC/DC electrical power converter.

Manipulator: The robots are fit for getting, changing, and exploding objects. The robot's control emulates the lengthy arm of what gives off an impression of being an individual. The joints that interface the arms to the upper arms, elbows, and wrists are normally alluded to as controllers. Joints frequently have a sliding or rotational movement. To characterize the potential stepper engine, the joints are set by response energy.

*Figure 1. Shows robot component parts*

End effector: The control connect is the framework's last part and connections to the info gadget. "Activity" alludes to the connection on the instrument handle. It mirrors the activities of the mechanical arm and the results.

Sensors: Robots may now be furnished with sensors that take exact estimations of their environmental elements, with the information from the sensors assisting with profoundly shaping the robot's way of behaving. This is as often as possible done to guarantee the robot's security. Automated innovation is utilized by gadgets to respond to ecological changes.

Actuator: Actuators are much of the time alluded to as the robot's muscles in most of advanced mechanics. The energy that drives robot's movement is utilized in the transformation.

Control and Errand Program: The work program is a progression of inquiries frequently given by the client, though the administration plan is an assortment of directions from the robot's regulator's producers. The controller should execute the developments prearranged by the errand programming to get done with a specific job.

## 1.2 Healthcare Facilities

Generally speaking, automation is not identical common in long-term care facilities. They should think about automating some of their care services. As a result of the COVID-19 reevaluation of systems and protocols, a comprehensive reevaluation of long-term automation in healthcare is currently underway. Health systems can use robots to improve care and efficiency, with a focus on reducing person-to-person contact by providing contact-free support that protects residents and staff from infectious diseases, such as robotic equipment for remote patient monitoring and home therapy (Getson, C., & Nejat, G. 2021). In hospitals and long-term care facilities, smart automation can be used in a variety of settings, ranging from front desks, where it can be used to screen all visitors and remind them of COVID-19 protocols and regulations, to outpatient settings, where it can be used to promote socialisation and keep isolated seniors in contact with their families and healthcare professionals, or by connecting to other sensors and devices to monitor a person's vital signs (Patil, R. V., Thote, A. M., & Chavan, S. T., 2022).

The adoption of AI-enabled robotics can be difficult and costly, and should be viewed as a long-term investment. Nevertheless, automation is being implemented in more industries due to social distancing laws and the health benefits it provides. These costs must be taken into account. In addition to the initial purchase price, many robot platforms require a monthly subscription fee; otherwise, the software risks becoming obsolete (Jin, B. E., & Shin, D. C., 2020). People are left with a robot they may have an emotional attachment to but no longer use if a business fails.

## 2. BACKGROUND

This section will cover related studies on acceptance of robots, robotic technology in eldercare, and determining influence more generally. Globally, life expectancies are increasing, which causes a rise in the proportion of retirees. The creation of conversational robots for experiments in eldercare is a topic of increasing interest for projects. As visual acuity declines in older persons, more services must be provided, which has a greater impact on healthcare expenses. More research is required to assist the elderly in maintaining emotional intelligence (Barreiro, C. A., & Treglown, L., 2020). Robotic technology could help social and rehabilitative robots relieve pressure on social care institutions. Large data processing, overlapping concerns, and feature set analysis challenges are the main drawbacks of traditional labour. While some initiatives concentrate on their creation or on gauging the responses to it by conducting trials with particular robots, some research focuses on the potential and requirements of these "eldercare companion robots," which will be used by the baby boomers, who will range in age from 66 to 84 in 2030. Thus, it appears that the burden of providing care is increasing—not just for families but also for society as a whole. This burden, in particular for elderly people who live alone, increases the risks to their safety and ability to maintain their lifestyles. The advantages of getting to know Paro, a simulated nursing care companion for seniors, include decreased irritability and depressive disorder in cognitive impairment, improved immune responses, less stress on caregivers, and improved affect and communication between senior citizens and their day care providers (Kang, H. S., Makimoto, K., Konno, R., & Koh, I. S., 2020). It is crucial to reduce the negative impacts of ageing, enhance the lives of older people, and enhance the general environment for good health. Smart ageing addresses these issues by effectively utilising modern biomedical, digital healthcare, computer, and communication technology (Song, I. Y., Song, M., Timakum, T., Ryu, S. R., & Lee, H. 2018). The main methodologies are examined by the authors and others, including designation and learning algorithms, which are crucial in the development of quantum computing and computational efficiency and are applicable in the field of developing artificial intelligence but complex in their application (Dunjko, V., & Briegel, H. J., 2018). As several assessment factors are taken into account, research methodologies used in data security, personal computing, and cloud services produce better conclusions. Previous research demonstrates that smart ageing enhances the health of the ageing population by integrating biotechnology, robotics, medical systems, and information technology in a seamless manner (Holroyd, C., 2022). In light of this, it is essential to recognise the traits and requirements of the elderly as service users and to use these to build sophisticated service technologies that assist elderly people's everyday living in a way that is compliant with their diverse demands and both their health and illnesses. Numerous studies have urged the adoption of quickly developing technologies to improve the lives of elderly people (Chang, V., Baudier, P., Zhang, H., Xu, Q., Zhang, J., & Arami, M. 2020). The older population's perspective and acceptance of such technologies must first be determined, as determining this population's primary needs is crucial for determining the path of product development when introducing communicative robots (Kautish et. al 2022). Recently, researchers from other countries studied elderly residents in nursing homes to learn about their attitudes toward robot services, level of technological acceptability, and experiences using them. In particular, the creation of robots to care for the elderly is advancing in advance of our transition into a post-aging society in early 2026. From 2016 to 2020, there were an average of 72 new applications for elderly care robots. The features and health circumstances of senior people must be better understood in order to create robot technology services that are efficient. This is because it is necessary to determine how robot technology will fit into the daily lives of older people as well as their market acceptance.

# 3. OVERVIEW OF THE OCCY

## 3.1 Proactive Behaviour

### Integrity, Security, and Safety

The informants placed a great focus on safety and reducing hazards. In order to prevent injuries, occy is safe and honest support system. Additionally, the preservation of integrity and privacy is always of paramount concern. It is crucial to evaluate what data should be stored or not and how it should be stored, as the more occy robots are employed, the more data is recorded and the more behaviours are mapped. When it comes to the exchange of digital data, ignorance and naiveté can occasionally reign supreme.

### Consistency

Occy never needs to divide their attention between multiple tasks. His work is independent of the work of others at all times. He will not encounter unforeseen emergencies or be required to relocate to complete a separate urgent task. He is always present and performing his duties. In general, automation is considerably more reliable than human labour.

### Perfection

Occy will always offer superior service. He is less likely to commit errors because he is programmed for precise, repetitive motion. Robots serve as both a quality assurance system and an employee in a sense. The absence of preferences and idiosyncrasies, coupled with the eradication of human error, will always result in a product that is dependably flawless.

### Speed

Occy neither require nor are easily distracted by breaks. He did not request a day off or an earlier departure. Occy will never become anxious and move more slowly. Moreover, he does not need an invitation to attend employee meetings or training sessions. Occy is capable of continuous operation, which accelerates production.

### Happier Workers

Employees are more likely to be satisfied because they are frequently assigned tasks that they do not particularly enjoy, such as menial labour, repetitive motion, and dangerous occupations. Occy will focus on more engaging tasks that are less likely to irritate them. People are exhausted from performing laborious tasks, but an occy robot will gladly delegate them.

### Longevity and Dependability

Due to the testing equipment's anticipated interaction with vulnerable and elderly individuals, occy must be reliable.

### Individualization, Modularity, and Adaptability

Over time, the needs of the users will evolve. After providing component services, occy is able to provide essential services that are adaptable to the varying needs and competencies of customers.

### Employability

Occy doesn't eliminate jobs. Only modifies the current jobs. More personnel are required to construct robots as there is an increase in demand. Offering employees a reason to remain engaged by preparing them to work with robots. Because of the developments, they also get the exceptional chance to learn brand-new technical or engineering-related abilities.

## 4. ARCHITECTURE

### 4.1 Design of the Prototype Robot

The working diagram given in describes the entire system (Fig. 3). The entire system is shown in Figure 2. There are digital programmes that include sensors and algorithms that are ready to be executed. When the robot has been completely prepared physically, the algorithm will begin initialization. Robot occy has many sensors, including a voice sensor, a face recognition sensor, and an environment sensor, and it will capture input and recognition from these sensors and send it to its data analysis section.

*Figure 2. Working flow chart of the robot*

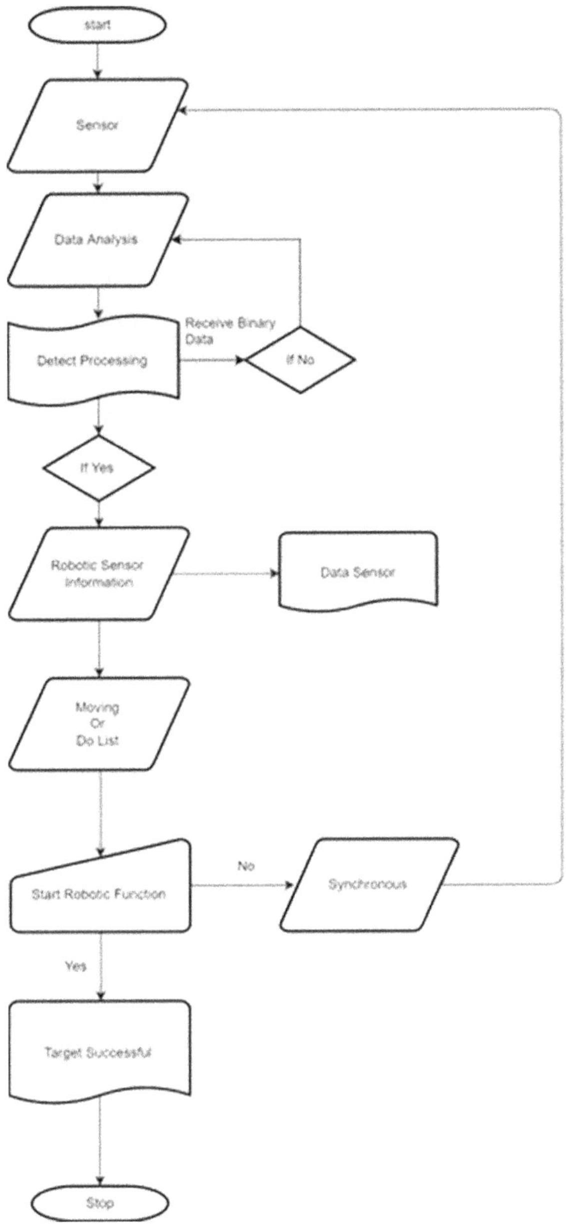


When it starts analysing the data with the information from the sensor, if its input is correctly detected, detect processing will start, and if the processing is active then it will proceed to its next test through the "yes" function.

And if the detection processing fails, either due to a failure or because it cannot catch the processing, it will return to data analysis using the received binary data to extract the information from the previous sensor.

So when the robotic sensor information is properly validated again from the data server and its previous status then the robot will perform its workings from the moving or its do list.

When the robotic function is set to "no," it will synchronise to show the sensor again, and when the robotic function is active, it will target through the "yes" function. When the target is successfully completed, it will stop the run function.

When human detects and assists, occy, begins modelling here, its modelling word is "starts, motor and sensor," which means assume it will begin working through its sensor and motors through activity.

Then occy considers the next step based on the analysis: the receive control and send, report options will come through, which the occy robot will perform its functions for the next steps, will perform its functions for the next steps, such as one hand will compile and on the other hand edit, debug, viewing and output through the performance of these tasks.

So when the receive control signal report function is already compiled, occy can check its performance through the output visualization, GUI application, vision. When that performance is read as output, occy can move on to the model database.

The model database then analyzes everything and clears everything as output, then proceeds to the next final input processing, and from final input processing to output processing, all the work efficiency tasks are completed to meet its targets.

## 5. METHODOLOGY

The mathematical model for calculating some of the cycle time of an occy robot is created in this section. Take into account the following fundamental presumptions while forming the model.

### 5.1 Occy Robot Modeling and Formulations

To approach big masses, one can investigate bending moments. The positions, orientations, and greater incidence may be seen as having changed throughout time based on the current motion. The fundamental equations of controls in robotics are constructed using dynamics, and the justification for motion is the deceiver's dynamic transfer function. Torque is the mechanism responsible for the dynamic movement of both manipulating arms of a robot's arm. Dynamic modelling involves the creation of differential equations for its manipulation as a function of factors

*Figure 3. Working diagram of occy robot*

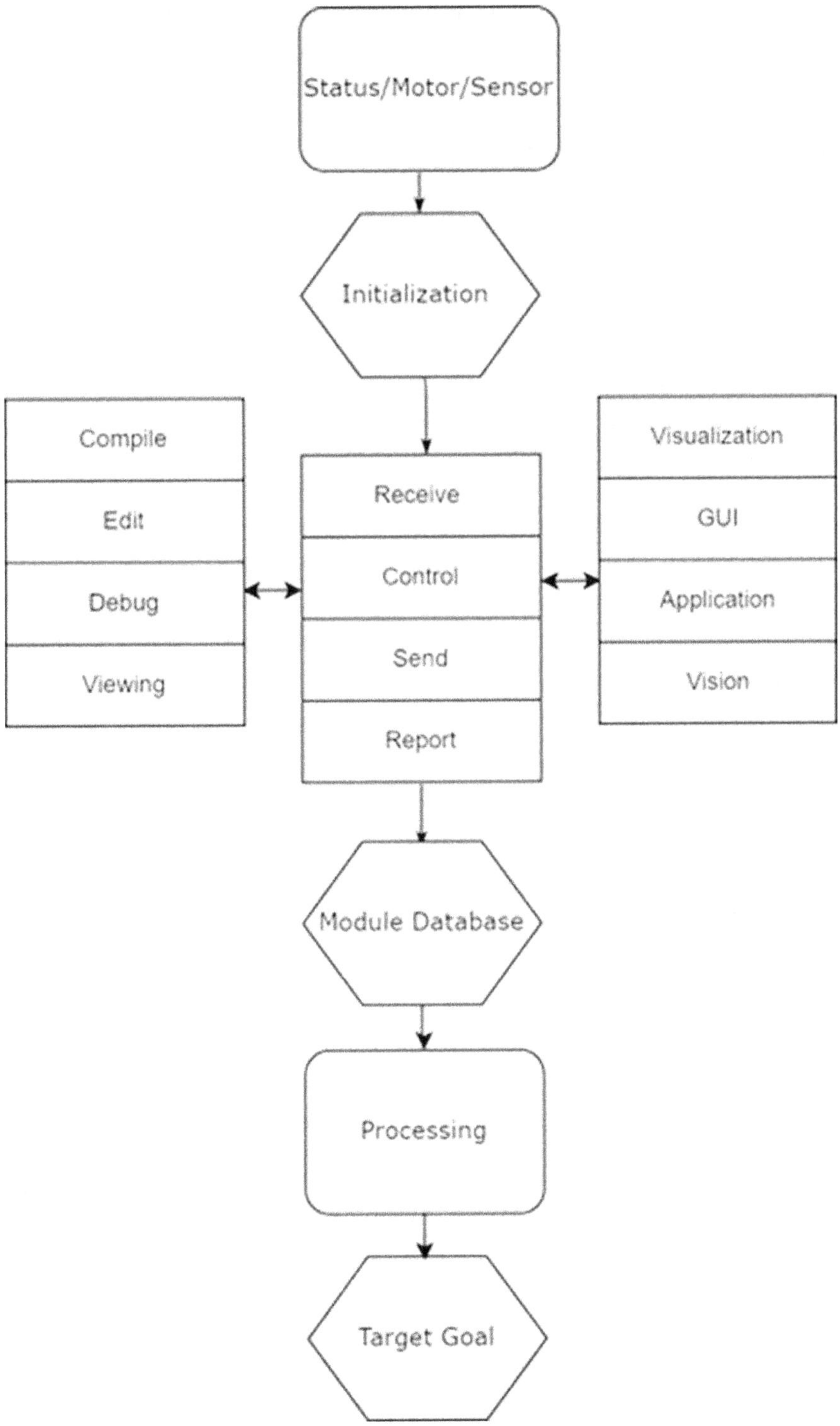

such as displacements. The locations, velocities, and deceleration of the robotic manipulator's components are calculated in real-time by the mechanism algorithm, image recognition, contact detection, bins, and time to change when using dynamic modelling, where the components are bound by a set of forces.

### 5.1.1 Real-Time Analysis Technique

The primary algorithms and related equations are presented in this section. Equations are presented in spatial notation for the sake of conciseness, but readers who are unfamiliar with this notation should still be able to understand the material. The linear and angular components of physical variables like velocity, acceleration, and force are both contained in the four-by-one spatial vectors. They are divided into two vector spaces for technical reasons: motion-type vectors in $P^4$ and force-type vectors in $B^4$. Four by four matrices are used to express tensor quantities such as inertia. Vector spaces in a dual system make up the underlying algebra.

A system of n independent rigid bodies' motion equation can be expressed in the form and several similar forms. These kinds of composite vectors and matrices serve as the foundation for the spatial operator. The two vector spaces $P^{4n}$ and $B^{4n}$, which are the cartesian products of n lots of $P^4$ and $B^4$, respectively, are introduced for the spatial algebra treatment.

$$b = yA + v \times Yv \tag{1}$$

where

$$b = \left[ b_1^t, \ldots, b_n^t \right]^t \in B^{4n},$$

$$A = \left[ A_1^t, \ldots, A_n^t \right]^t \in P^{4n},$$

$$Y = \mathrm{diag}\left( Y_1, \ldots, Y_n \right) : P^{4n} \mapsto B^{4n}$$

Assume that the kinematic restrictions placed on the aforementioned system limit the (instantaneous) velocity to an n-dimensional subspace $K \subset P^{4n}$. Therefore, the acceleration is restricted by

$$A - A_0 \in S \subset P^{4n},$$

where $A_0$ is a vector of velocity. Range (k) = K indicates that if K is a 4n ×N matrix, where α vector of unknowns, exists then

$$A = A_0 + k\alpha \tag{2}$$

The system will experience certain forces as a result of these kinematic restrictions. Let's use the formula $b = b_{ka} + b_{uk}$ to represent the overall applied force, where $b_{ka}$ represents known active forces and $b_{uk}$ represents unidentified constraining forces. Consequently, the confined system's equation of motion is

$$b_{k\alpha} + b_{uk} = YA + v \times Yv \tag{3}$$

According to the virtual work concept, $k^t b_{uk}=0$, if we replace with in Eq. 2 and multiply the outcome by $k^t$ before adding, we get

$$k^t Y k\alpha = k^t(b_{k\alpha} - YA_0 - v \times Yv) \tag{4}$$

Since both apply constraints in the same way and only differ in the starting system, this equation has the same structure as Eq. 2. Starting with a system of unconstrained rigid bodies in Eq. 4 and a kinematic tree in Eq. 2, respectively. Despite being larger than those in Eq. 2, the matrices in Eq. 4 have a sparse structure that is helpful for characterising low-order algorithms.

### 5.1.2 The Mechanism Algorithm

A set of n movable links (rigid bodies), numbered 1 through n, a fixed base link, numbered 0, and a set of n joints connecting the links in such a way that joint $\ell$ connects from link $\lambda(\ell)$ to link $\ell$ where $\lambda(\ell)$ is the link number of the parent of link $\ell$ in the tree, taking the base link as the root node, can be used to model a general robot mechanism with a tree structure. The numbers are selected to ensure that $\lambda(\ell) < \ell$. An unbranched kinematic chain is a specific instance when $\lambda(\ell) = \ell-1$ and the links and joints are numbered sequentially from the base to the tip.

If define $v_l$ as the speed of link $\ell$ and $v_\ell^z$ as the speed across joint $\ell$ respectively

$$v_\ell^z = v_\ell - v_{\lambda(\ell)} \tag{5}$$

The following form is another way to express the joint velocity:

$$v_\ell^z = m_\ell Q_\ell \tag{6}$$

where $\ell$, $Q_l$ is a $S_\ell \times 1$ vector of joint velocity variables, $S_\ell$ is the degree of freedom of joint $\ell$ and $m_\ell$ is a $4 \times S_\ell$ matrix spanning the motion freedom subspace of joint

$\ell$. $m_\ell$ is a vector that describes the joint's axis of motion in the exceptional case of a 1-DoF joint.

Eqs. 1 and 2 together yield

$$v_\ell = v_{\lambda(\ell)} + m\ell_Q \ell \tag{7}$$

This is the common recursive equation to calculate connection velocities. The time-derivative of equation 3 is the analogous formula for accelerations.

$$A_\ell = A_{\lambda(\ell)} + H_\ell Q_\ell + m_\ell q_\ell \tag{8}$$

where $q_\ell$ is a vector of joint acceleration variables and $A_\ell$ is the link $\ell$ acceleration. For prismatic and revolute joints, among many other unique instances,

$$M_\ell = v_\ell \times h_\ell$$

These formulas calculate the velocity and acceleration of each link in turn, proceeding outward from the base to the terminal links, given the velocity and acceleration of the base, $v_0$ and $A_0$. A typical algorithm appears:

for $\ell = 1$ to $n$ do

$v_\ell = v_{\lambda(\ell)} + m_\ell Q_\ell$

$A_\ell = A_{\lambda(\ell)} + M_\ell q_\ell + m_\ell q_i$

The characteristic $\lambda(\ell) < \ell$ makes sure that $v_{\lambda(\ell)}$ is determined before $v_\ell$.

### 5.1.3 Image Recognition

To make understanding easier, divided the detailed formulation of T into three formulations before providing it. First, simply suppose that each chosen component is preceded by image recognition. Therefore, the following is how the time spent taking the images would be expressed:

$$T_m = \sum_{p=1}^{P} \sum_{n=1}^{N} \sum_{l=1}^{L} M_b * T_a \quad \forall w \tag{9}$$

The technological time required for a single acquisition is denoted by $T_{a.}$

The formulation would be as follows if assumed that portion of this process might be completed as a background work, or in "hidden time," while the robot is moving or dropping a component on the conveyor:

$$T_m = \sum_{p=1}^{P}\sum_{n=1}^{N}\sum_{l=1}^{L}(1-k)*M_b*T_a \quad \forall w \tag{10}$$

In this case, k denotes the portion of technological time that is executed in hidden time. If k = 100%, or if the acquisition is fully completed in concealed time, $T_m$ may be zero in this case.

Combine the failure rate to create a more precise formulation, which results in the following:

$$T_m = \sum_{p=1}^{P}\sum_{n=1}^{N}\sum_{\ell=1}^{L}\left[\tau_q + \left(1-\tau_q\right)*\left(1-k\right)\right]*M_b*T_a \quad \forall w \tag{11}$$

When the robot is unable to select an item from component q($\tau q_{>}0$), it continues to wait until the camera takes another picture of the contents of the bin, which takes more time. This is indicated by the phrase:

$$\tau_q*M_b*T_a \tag{12}$$

However, the time acquisition component for successful picks ($I-\tau q_{)}$ may be carried out in concealed time:

$$(1-\tau_q)*(1-k)*M_b*T_a$$

## 5.1.4 Contact Detection

A frame's location and orientation pose in relation to another frame can be mathematically described using the generic homogeneous transformation sometimes called base axes. It is represented by the 4x4 matrix K, where m is the fourth column vector and L is a 3x3 orientation submatrix.

$$K = \begin{matrix} L & m \\ 0 & 1 \end{matrix}$$

The first three of the four vectors that describe a second frame are represented by the columns of K; they are the direction cosines of the referred or secondary frame in relation to the base frame when they are normalised. The final vector shows the secondary origin's location.

The first three columns show how the secondary frame axes are oriented in relation to the base axes in the more general scenario of both rotation and displacement, and the fourth column shows where the secondary origin is in relation to the base origin.

L = I = the identity 3x3 matrix if there is no rotation. The origin of a second frame with respect to a reference frame is represented by the fourth column m (ma, mb, mc). M = 0 if rotation is all that exists.

It is common practise to rotate the base axes first, then translate the origin or, alternatively, to translate the frame first then rotate around the translated frame rather than the reference frame when locating points in translated and rotated reference frames relative to base axes. This order will be represented by K(L,m).

Multiple rotations are common in real rigid body rotations.

### 5.1.5 Bins

The following assumptions are used to evaluate this term:

1. During the reference period, presum that the robot only eats bins that are completely filled.
2. Believe that there are no picking failures while retrieving empty bins, in contrast to picking parts.
3. Regardless of the type of bin, the time taken into account for a single tool change is distinct.
4. The time taken into account for a single empty bin pick has a different value for each bin.

Bin removal timing estimates are provided by:

$$T_P = (T_s + T_r + T_x + 2*T_g)*N_y \tag{13}$$

The time required for a single tool change is denoted by $T_s$. Before deleting a bin, presume that a tool change is performed consistently.

Pick one container and place it on the evacuation ramp at time $T_r$.

$T_g$ is increased to account for the stops at the evacuation ramp and the first halt where the robot grabs the trash can. Keep in mind that the time spent pausing when returning to pick up a part is already factored into the pick and place time.

Also take into account the time it takes to move a box from one place to the evacuation ramp. This comes from:

$$T_x = \frac{\text{Span}_{\text{rack}}}{\text{R}} \tag{14}$$

The typical number of bins taken out in order to prepare a set of boxes is $N_y$.

$$N_y = \frac{N_b}{N_r} \quad \forall w \tag{15}$$

During the reference period, component $\ell$ was consumed from a total of $N_b$ bins. It is said in the following way:

$$N_b = \sum_{l=1}^{E_{kit}} \frac{R_s}{N_q} \tag{16}$$

Where $N_q$ is the number of parts contained in a complete bin of component $\ell$, Rs is the number of parts of component $\ell$ consumed during the reference period, and E kit is the total number of components in the robotic kitting system.

The number of sets that were prepared during the reference period is $N_n$. The formula is: $N_n = \frac{N_P}{V}$. $P_n$ generated during the reference period is represented by the number $N_P$.

## 5.1.6 Time to Change

The following presumptions are taken into account for this term:

1. For every item on the choosing list, the robot uses a different tool.
2. No matter the tool, the $T_d$ for a single tool change is the same.
3. A tool change doesn't require any travel time.

The following provides the average tool-changing time:

$T_c = T_d{*}N_c$, ∀w

where $N_c$ represents the overall number of tool changes required to prepare the set s of boxes. The value of F specifies whether or not the component is in the set kit box and on the picking list. This number is obtained by adding the entire number of components to be chosen, and it is written as follows:

$$N_c = \sum_{p=1}^{P} \sum_{n=1}^{N} \sum_{\ell=1}^{L} \alpha_F$$

## 6. RESULT AND DISCUSSION

The occy robot chose for robotic systems for older people, which suggests that it might make older people feel more comfortable around one another. Occy make an effort to include a variety of contexts related to ageing, including healthcare, leisure, and employment. The best robot for each older person's specific task is this one. The advantage of being able to handle many jobs is that it can carry out multiple activities. This robot is employed during this period to avoid pressuring people who are still debating their options.

The information received from the respondents research-related activities. The actions brought about by the meeting's modified transcript or the chalkboard came from two different places. The entire group meeting was captured on tape and depicted on the chalkboard through the assigned work. Because the primary purpose of this study is to shed light on a variety of activities, only routine activities will be considered.

- Willingness to Use

Robots are more accurate and rarely make mistakes than humans. Occy can generate more in a shorter period of time. It is capable of operating continuously with no breaks, days off, or holidays. Compared to people, it is more repeatable in its application performance.

- Basic Activities

This would be the bare minimum of everyday duties that someone living independently should be capable of performing.

- Telephone calls and E-mail

Numerous characteristics of contact information make it exploitable. These characteristics can be used by occy to correct contact information.

- A database is used to store contact information.
- Elder person can send messages for work or in an emergency using occy.
- Email offers the ability to block the spam, delete, and reply to an important email. This work he do very quickly.

- Leisure and Sports Activities

Occy new and cheerful entertainment is one of its nicest features. Older folks have more time for leisure activities and hobbies, yet contemporary digital technologies succumb to human loneliness. Robots called occy will join, support, and help humans. It marks the beginning of the occy in hobbies, leisure, and amusement.

The sport that is utilised to demonstrate that AI has attained human-level intellect is increasingly thought of as a measure of that intelligence. Embodied intelligence is necessary for human-level intelligence, making the creation of sports by humanoid robots a better sport to test human-level intellect. The activity is picked because team sports are simple to identify, uniquely human, and predicted to emerge under straightforward, manageable conditions.

*Figure 4. Numerous activities*

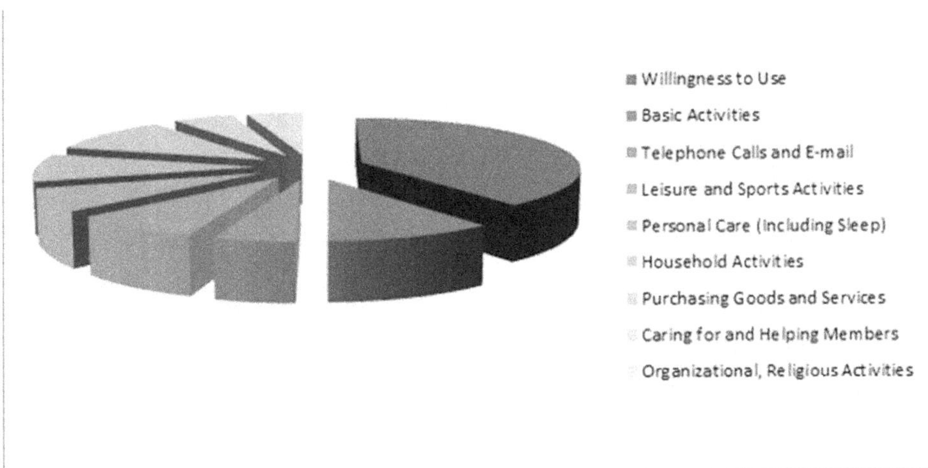

- Personal Care (Including Sleep)

Robotic assistants may frequently be present in the homes of elderly people, helping with heavy lifting, acting as a communication tool to help them set reminders and monitor and take care of themselves, bridging care gaps and offering emotional support, and enabling remote access for medical professionals to maintain independence and dignity.

Occy can make alarming predictions about how the amount and quality of sleep a person gets affects their health because it lowers their risk of acquiring mental and chronic illnesses. Only a few of the most cutting-edge methods for tracking sleep quality offer solutions from academic research; instead, most of them concentrate on comparative analyses and evaluations of currently available commercial products.

- Household Activities

Occy that assist people by handling goods or communicating information. Occy actively helps human do physical tasks by enhancing or supplanting the user's natural abilities.

- Purchasing Goods and Services

Robots can significantly reduce income loss and the old irritation caused by out-of-stock or misplaced items since he can check thousands of SKUs faster and more correctly than humans.

- Caring for and Helping Members

In residential institutions for people with dementia, the use of occy has demonstrated beneficial psychosocial advantages, such as enhancing mood and social interaction and lowering agitation.

- Organizational, Religious Activities

Individuals are able to plan and scan a path on their own, which enhances workflow. Increasingly, businesses will rely on robots to increase productivity by automating routine tasks, eliminating human error, and enhancing operational skills.

Some people are beginning to use technology in their prayers. Occy can carry messages, read prayers, and even offer consolation to people going through a spiritual crisis.

## 6.1 Accuracy Response of Occy

The name and appropriate value of the activity are used to analyse the ageing activity in the robot model. The robot model is used to evaluate various activity types. The graph model and performance states of occy. The accuracy response is obtained based on the service type, and the occy robot activities provide the highest accuracy response with a percentage of 96%.

True Positive: Actually correctly classified true positive
True Negative: Actually correctly classified negative positive
False Positive: Actually negative misclassified
False Negative: Actually true misclassified

### Metrics

Here, there are no False Positives and there are 117 True Negatives (TN), 126 True Positives (TP), 10 False Negatives (FN), and 253 total test samples (FP).

Accuracy

$$= \frac{TP + TN}{TP + FP + TN + FN} = \frac{\text{Number of Activites}}{\text{Total Activities Undergoing Examination}}$$

*Figure 5. Accuracy*

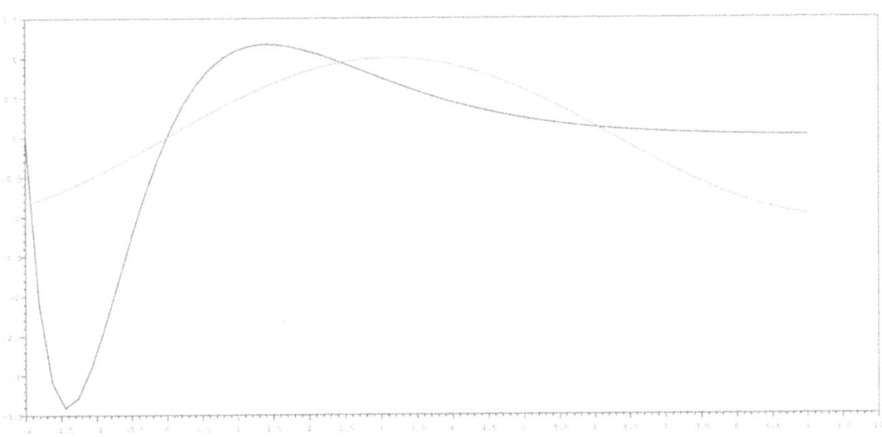

The ratio of instances correctly identified to the total number of cases being evaluated can also be used to measure accuracy. The greatest value of accuracy is 1 and the worst value is 0.

It produces a result of 96%.

## 7. CONCLUSION

The context and prevalent attitudes and conceptions regarding occy robot are given, the introduction of care robots in elder care services appears more positive than the introduction of welfare technology in general. Significant challenges must be addressed for the elderly, and at all levels of society, occy robot can become an integral part of daily care and assist the elderly.

## REFERENCES

Aceto, G., Persico, V., & Pescapé, A. (2019). A survey on information and communication technologies for industry 4.0: State-of-the-art, taxonomies, perspectives, and challenges. *IEEE Communications Surveys and Tutorials*, *21*(4), 3467–3501. doi:10.1109/COMST.2019.2938259

Barreiro, C. A., & Treglown, L. (2020). What makes an engaged employee? A facet-level approach to trait emotional intelligence as a predictor of employee engagement. *Personality and Individual Differences*, *159*, 109892. doi:10.1016/j.paid.2020.109892

Bossone, E., Ranieri, B., Coscioni, E., & Baliga, R. R. (2019). Community health and prevention: It takes a village to reduce cardiovascular risk! Let us do it together! *European Journal of Preventive Cardiology*, *26*(17), 1840–1842. doi:10.1177/2047487319867505 PMID:31382807

Carra, G., Argiolas, A., Bellissima, A., Niccolini, M., & Ragaglia, M. (2018). *Robotics in the construction industry: State of the art and future opportunities*. In ISARC. Proceedings of the International Symposium on Automation and Robotics in Construction (Vol. 35, pp. 1-8). IAARC Publications.

Chang, V., Baudier, P., Zhang, H., Xu, Q., Zhang, J., & Arami, M. (2020). How Blockchain can impact financial services–The overview, challenges and recommendations from expert interviewees. *Technological Forecasting and Social Change*, *158*, 120166. doi:10.1016/j.techfore.2020.120166 PMID:32834134

Chiacchio, F., Petropoulos, G., & Pichler, D. (2018). *The impact of industrial robots on EU employment and wages: A local labour market approach (No. 2018/02).* Bruegel working paper.

de Kervenoael, R., Hasan, R., Schwob, A., & Goh, E. (2020). Leveraging human-robot interaction in hospitality services: Incorporating the role of perceived value, empathy, and information sharing into visitors' intentions to use social robots. *Tourism Management, 78*, 104042. doi:10.1016/j.tourman.2019.104042

Demir, K. A., Döven, G., & Sezen, B. (2019). Industry 5.0 and human-robot co-working. *Procedia Computer Science, 158*, 688–695. doi:10.1016/j.procs.2019.09.104

Dunjko, V., & Briegel, H. J. (2018). Machine learning & artificial intelligence in the quantum domain: A review of recent progress. *Reports on Progress in Physics, 81*(7), 074001. doi:10.1088/1361-6633/aab406 PMID:29504942

Ellitan, L. (2020). Competing in the era of industrial revolution 4.0 and society 5.0. *Jurnal Maksipreneur: Manajemen, Koperasi, dan Entrepreneurship, 10*(1), 1-12.

Getson, C., & Nejat, G. (2021). Socially Assistive Robots Helping Older Adults through the Pandemic and Life after COVID-19. *Robotics, 10*(3), 106. doi:10.3390/robotics10030106

Holroyd, C. (2022). Technological innovation and building a 'super smart'society: Japan's vision of society 5.0. *Journal of Asian Public Policy, 15*(1), 18–31. doi:10.1080/17516234.2020.1749340

Jannat, U. K., Kumar, M. M., & Islam, S. A. (2022). *A Case Study On The Cybersecurity Of Home Appliance Robots*. AGPE.

Jin, B. E., & Shin, D. C. (2020). Changing the game to compete: Innovations in the fashion retail industry from the disruptive business model. *Business Horizons, 63*(3), 301–311. doi:10.1016/j.bushor.2020.01.004

Kang, H. S., Makimoto, K., Konno, R., & Koh, I. S. (2020). Review of outcome measures in PARO robot intervention studies for dementia care. *Geriatric Nursing, 41*(3), 207–214. doi:10.1016/j.gerinurse.2019.09.003 PMID:31668459

Patil, R. V., Thote, A. M., & Chavan, S. T. (2022). Pre-Detection and Classification of Coronavirus Disease by Artificial Intelligence and Computer Vision. In Pandemic Detection and Analysis Through Smart Computing Technologies (pp. 169-194). Apple Academic Press. doi:10.1201/9781003281610-8

Song, I. Y., Song, M., Timakum, T., Ryu, S. R., & Lee, H. (2018). The landscape of smart aging: Topics, applications, and agenda. *Data & Knowledge Engineering*, *115*, 68–79. doi:10.1016/j.datak.2018.02.003

## ADDITIONAL READING

Boubaker, O. (Ed.). (2020). *Control Theory in Biomedical Engineering: Applications in Physiology and Medical Robotics*. Academic Press.

Fosch-Villaronga, E., & Mahler, T. (2021). Cybersecurity, safety and robots: Strengthening the link between cybersecurity and safety in the context of care robots. *Computer Law & Security Review*, *41*, 105528. doi:10.1016/j.clsr.2021.105528

Grischke, J., Johannsmeier, L., Eich, L., Griga, L., & Haddadin, S. (2020). Dentronics: Towards robotics and artificial intelligence in dentistry. *Dental Materials*, *36*(6), 765–778. doi:10.1016/j.dental.2020.03.021 PMID:32349877

Kaur, R. & Kautish, S. (2022). Multimodal sentiment analysis: A survey and comparison. *Research Anthology on Implementing Sentiment Analysis Across Multiple Disciplines*, (pp. 1846-1870).

Kautish, S., Peng, S. L., & Obaid, A. J. (Eds.). (2021). *Computational intelligence techniques for combating COVID-19*. Springer International Publishing. doi:10.1007/978-3-030-68936-0

Kautish, S., Reyana, A., & Vidyarthi, A. (2022). SDMTA: Attack detection and mitigation mechanism for DDoS vulnerabilities in hybrid cloud environment. *IEEE Transactions on Industrial Informatics*, *18*(9), 6455–6463. doi:10.1109/TII.2022.3146290

Madhu, G., Govardhan, A., Ravi, V., Kautish, S., Srinivas, B. S., Chaudhary, T., & Kumar, M. (2022). DSCN-net: A deep Siamese capsule neural network model for automatic diagnosis of malaria parasites detection. *Multimedia Tools and Applications*, *81*(23), 34105–34127. doi:10.100711042-022-13008-6

Mayoral-Vilches, V. (2022). Robot cybersecurity, a review. *International Journal of Cyber Forensics and Advanced Threat Investigations*.

Özşeker, D. B., Aktaş, E., & Kurgun, O. A. (2022). Applications and Implications of Service Robots in Hospitality Sector: A Case Study. *İşletme Araştırmaları Dergisi*, *14*(4), 3081-3104.

Pradhan, B., Bharti, D., Chakravarty, S., Ray, S. S., Voinova, V. V., Bonartsev, A. P., & Pal, K. (2021). Internet of things and robotics in transforming current-day healthcare services. *Journal ofHealthcare Engineering*, *2021*, 1–15. doi:10.1155/2021/9999504 PMID:34104368

Sampathkumar, A., Rastogi, R., Arukonda, S., Shankar, A., Kautish, S., & Sivaram, M. (2020). An efficient hybrid methodology for detection of cancer-causing gene using CSC for micro array data. *Journal of Ambient Intelligence and Humanized Computing*, *11*(11), 4743–4751. doi:10.100712652-020-01731-7

Tasioulas, J. (2019). First steps towards an ethics of robots and artificial intelligence. *Journal of Practical Ethics*, *7*(1).

Teresa, Z. (2014). History of service robots. In Robotics: Concepts, Methodologies, Tools, and Applications (pp. 1-14). IGI Global. doi:10.4018/978-1-4666-4607-0.ch001

Tuomi, A., Tussyadiah, I. P., & Stienmetz, J. (2021). Applications and implications of service robots in hospitality. *Cornell Hospitality Quarterly*, *62*(2), 232–247. doi:10.1177/1938965520923961

Wentao, C. H. U., Kuok-Tiung, L. E. E., Wei, L. U. O., Bhambri, P., & Kautish, S. (2021). Predicting the security threats of internet rumors and spread of false information based on sociological principle. *Computer Standards & Interfaces*, *73*, 103454. doi:10.1016/j.csi.2020.103454

Zhu, Q., Rass, S., Dieber, B., & Vilches, V. M. (2021). Cybersecurity in robotics: Challenges, quantitative modeling, and practice. *Foundations and Trends® in Robotics, 9*(1), 1-129.

## KEY TERMS AND DEFINITIONS

**AI:** AI stands for "Artificial Intelligence." It is the field of computer science and engineering focused on creating machines that can perform tasks that would normally require human intelligence, such as recognizing speech, understanding natural language, making decisions, and learning from experience.

**Bins:** Bins typically consist of a combination of hardware and software components, such as robotic arms, sensors, and control systems.

**Contact Detection:** In robotics, contact detection is essential for ensuring safe and accurate operation of robots.

**Elder Care:** Elder care refers to the specialized care and support provided to older adults who may require assistance with activities of daily living or who have health conditions that require ongoing management.

**Facial Recognition:** Facial recognition is a type of biometric technology that uses artificial intelligence and machine learning algorithms to analyze and identify human faces. The technology works by capturing an image or video of a person's face and then comparing it to a database of stored images to find a match.

**Image Recognition:** Image recognition is the process of identifying and detecting objects or features within an image or a video sequence. It is a subfield of computer vision and artificial intelligence that involves training algorithms to recognize patterns and features in visual data.

**Occy Robot:** Occy robots are often designed to be user-friendly and able to interact with humans in a natural way.

**Real-Time Analysis:** Real-time analysis typically involves the use of sophisticated data processing and analysis tools, such as data streaming platforms, machine learning algorithms, and artificial intelligence models.

# Chapter 3
# Learning Framework for Real-World Facial Emotion Recognition

**Rohan Appasaheb Borgalli**
*Fr. Conceicao Rodrigues College of Engineering, India*

**Sunil Surve**
*Fr. Conceicao Rodrigues College of Engineering, India*

## ABSTRACT

*Facial expression recognition (FER) is an important research area in the fields of computer vision and artificial intelligence due to its application in academics as well as in industry. Research shows that using facial images/videos for recognition of facial expression is better because visual expressions carry major information through which emotions can be conveyed. Past research on FER has focused on the study of seven basic emotions; however, many more facial expressions are exhibited by humans that are considered compound emotions. State of art results shows machine learning and deep learning-based approaches are powerful over conventional FER approaches. This chapter focuses on surveying past work done in the field of real-world compound facial emotion recognition and implementing various learning frameworks such as machine learning and deep learning for real-world facial emotion recognition systems for detecting compound emotion using the facial expression image dataset RAF-DB for a real wild scenario.*

DOI: 10.4018/978-1-6684-8171-4.ch003

## 1 INTRODUCTION

Facial expression recognition (FER) is a vital research area as it is helpful in many applications. A FER system model mentioned in this paper detects the human facial expression and identifies the corresponding induced emotion for a static image or sequence of images. Though much research has been done, recognizing basic and compound facial expressions with a high accuracy remains to be difficult due to the complexity and varieties of facial expressions.

Generally, human beings can convey emotions and intentions through nonverbal ways such as facial expressions, gestures, and involuntary language. According to Darwin & Prodger (1998), A person's facial expressions reflect their emotional states and motives. Because of its importance in systems such as machine vision and machine learning, many attempts have been made in the past to use FER system effectively. Various FER systems have been developed based on facial expressions to integrate facial information. Because of its accessibility and efficiency, facial expression recognition has become one of the most important methods to detect emotions. It has also been widely used in lie detection, medical assessment, driver safety, human-machine interaction, and solving other complex real-world problems with FER.

In 1978 automatic facial expression recognition was published first by Suwa (1978). Since then, many researchers have been working on developing FER methods that are robust and accurate. There are two categories of FER based on the environment of a captured image. The first one is in the lab-controlled with facial images usually frontal posed and with fixed illumination. The other one is in the real wild. Researchers focused on facial expression recognition in the lab-control environment in the first few years.

For lab-controlled FER, some databases have been proposed, such as FER2013 (Wolfram Research, 2018), CK+ (Lucey et al., 2010), MMI (Pantic et al., 2005), and JAFFE (Lyons et al., 1998) . Recently the paradigm of researchers shifted from a lab-controlled database to a real-world database. Facial expression recognition in the wild is more challenging because the facial images have arbitrary illuminations, non-frontal faces, and partially occluded faces. From a research point of view, generalizing FER for the real world makes more sense to study it in the wild. As a result, researchers focus is on developing facial expression recognition systems in the wild. To fill this gap of the FER system in lab-controlled and real wild scenarios, some databases in the wild, such as RAF-DB (Li et al., 2017), RAF-ML (Li & Deng, 2019) and AffectNet (Fabian Benitez-Quiroz et al., 2016) are, have been collected and made available for research. Even though the usage of data from the wild for facial expression recognition was increased, it was still challenging due to non-frontal and partially occluded faces.

At the beginning of the twentieth century, Ekman & Friesen (1971) deðned six basic emotions based on a study of different cultures: anger, disgust, fear, happiness, sadness, and surprise,which showed that people deduce certain basic emotions in the same way regardless of culture. Most of the past works mentioned in a survey by Li & Deng (2020) focused on these six basic emotions and Neutral.

However, some facial expressions induced by humans are more complex and cannot be simply considered a basic emotion, leading to compound emotion (Lindquist et al., 2016).

Recently, advanced research on FER focused on compound emotion by mentioning a variety set of compound emotions along with basic emotion given by Gunes & Schuller (2013), Du et al. (2014), Guo et al. (2018) and Li & Deng (2019).

To solve the FER problem, several image processing techniques were employed (Revina et al., 2021), followed by machine learning techniques such as the Multi-layer Perceptron Model, k-Nearest Neighbors, and Support Vector Machines mentioned in a survey of Li & Deng (2020). These traditional methods works on the principle of extracting various features like Local Binary Patterns (LBP) discussed by Zhao & Pietikainen (2007), Histogram of Gradient (HOG) features, Eigenfaces, Gradient and Laplacian (Pandey et al., 2019) Face-landmark features, edge, skin color, face muscle motion and Texture features (Lekdioui et al., 2017) . Such features are effective for facial images. However, due to the increase in the amount and complexity of facial images in the wild, handcrafted features only cannot achieve good performance. With the evolvement of research in deep learning techniques, this problem is solved. Deep learning techniques automatically extract features specific to the task, especially the Convolutional Neural Network (CNN) is very popular among them.

Literature survey by Li & Deng (2020) showed the process of FER begins with the pre-processing of images to detect face and extract features before training models either through the handcraft method or with CNN along with transfer learning can also be tried by Lim et al. (2018) using standard Models.

To implement the FER system effectively, it should work accurately not only on lab-controlled environment images but also on wild environments and detect basic as well as compound emotions.

This paper implemented different learning frameworks for facial expression recognition systems on the RAF-DB dataset (Li et al., 2017) to detect basic and compound emotion and perform well in a real-world scenario.

The rest of the paper is arrenged as follows. Section 2 presents a literature review of related works.

In Section 3, the details of the RAF-DB database used for the experiment is mentioned, Section 4 proposes methodology used to detect compound emotion on real-world image datasets, Section 5 discusses the result achieved based on experiments carried out, Section 6 concludes the paper and future work.

## 2 RELATED WORK

To move the FER system from a lab-controlled to a wild scenario, researchers developed datasets such as Li et al. (2017) introduced the RAF-DB (wild) database, which combines 7 basic emotions and 12 compound emotions. Li & Deng (2019) introduced a new real-world facial expression database, RAF-ML, to explore and address the challenges of multi-label visual recognition in the field. Emotion Recognition in the Wild (EmotiW), presented by Fabian Benitez-Quiroz et al. (2016), collected sufficient training data from challenging real-world situations, which clearly promoted the transition of FER from laboratory-controlled in in-field settings.

Also, advanced research in FER made the transition from basic to compound emotion recognition given by Gunes & Schuller (2013) indicates that the face displays compound emotion, a combination of the six basic emotions results to 22 emotions, including seven basic emotions and twelve compound emotions along with the three additional emotions (appall, hate, and awe) and gave Compound Facial Expressions of Emotion (CFEE) database. They used the Gabor filter to extract Gabor features of each image, and Compound emotion detection is made by applying the Multi-Class Support Vector Machine (mSVM) classifier with the 10-fold cross-validation and leave-one-sample-out to get an accuracy of 73.61%, 70.03%, and 76.91% when using only geometric features, only appearance features, and when both features are combined in a single feature space respectively.

Du & Martinez (2022) usually examined people who exhibited mixed facial expressions which is combination of more than one basic emotion (for example, a composite facial expression of happily surprised people who have both a sense of joy and surprise at the same time, as happens on a birthday. Group). In their previous study of Du et al. (2014), they found 15 compound emotions. However, their study (Du & Martinez., 2022) added two more emotions and came up with 17 compound emotions based on different categories of people from different cultures who had been tested. An extensive database of auto-response data was collected on Internet channels, websites, and newspaper websites. Therefore, these information sites focus on non-verbal communication between people through their facial expressions, and this they apply to psychological research.

In Li et al. (2017), researchers proposed a deep locality-preserving learning method for emotion recognition using the RBM-DB dataset. To address the problem of a non-lab-controlled database, they proposed a new DLP-CNN (Deep Locality-Preserving CNN) approach to extract deeper features, aimed at maximizing the power of deep features to discriminate different emotions by maintaining spatial similarity while enhancing class diversity. The benchmark test on 11-class compound expressions Happily Surprised, Happily Disgusted, Sadly Fearful, Sadly Angry, Sadly Surprised, Sadly Disgusted, Fearfully Angry, Fearfully Surprised, Angrily

Surprised, Angrily Disgusted, and Disgustedly Surprised. They also reported that DLP-CNN transcends the state-of-the-art handcrafted features and deep learning-based methods for expression recognition in the wild. Experiments showed that the average accuracy of compound emotion recognition was around 44.55%, which indicated that compound (wild) emotion recognition was the most challenging task.

Li & Deng (2019) proposed DeepBi-Manifold (DBM)-CNN features extracted from RAF-ML achieve similar performance on cross databases which indicating that the RAF-ML can be used as a 'standard' database containing a wide diversity of data for training for analysis of facial expression. DBM-Domain Adaptation CNN can improve the performance on cross-dataset validation.

In 2016, EmotionNet Challenge (Benitez-Quiroz et al., 2017) introduced an accurate, real-time algorithm to automatically annotate one million facial expressions in the wild. This challenge has two tracks. The first track was related to the automatic categories of 11 Facial Action Units (AUs), and the second one idetifies compound emotions.

As per our work, only the compound emotions track is briefly reviewed. The EmotionNet challenge employed the dataset defined in Fabian Benitez-Quiroz et al., (2016). The training, validation, and test sets were already described with 950K, 2K, and 40K facial images, respectively. The validation and test sets were annotated manually. However, the training set was annotated automatically using proposed algorithm in Benitez-Quiroz et al. (2017). Finally, 16 basic and compound emotion categories now available for recognition.

Also, Guo et al. (2018) addressed the problems of limited compound emotions to datasets with a limited number of categories and unbalanced data distribution, whose labels are automatically annoted by machine-based algorithms, which can lead to error. They released the iCV-MEFED database, which includes 50 classes of compound emotions and labels examined by psychologists. Researchers have used the Convolutional Neural Network (CNN) method to automatically extract features from a given datasset. Next, they explored three strategies as follows: (1) the first method is based on multi-modality with texture and geometric features used for CNN trained end to end. (2) The second method consisted of unsupervised learning via shallow CNN to extract the features classified with different SVMs. And (3) the third method is based on the structure of CNN Inception-v3 which has a discriminatory loss function (central loss). All methods reflect the misclassified dominant and complementary emotions present in the compound emotions (happy or happily surprised) instead of the seven basic emotions.

Some of the researchers developed advanced architectures like Conditional Random Field (CRF) (Liliana et al., 2018) for multi-modality that is a mixture of the Active Appearance Model (AAM) and CNNs based on visual and geometrical facial information. The proposed architecture includes two sets of features that are visual and geometry features from 68 facial landmarks. After that, both visual and geometry features are concatenated into a long vector of 392 features. Finally, the concatenated vector of facial extracted features is fed to the Hidden Conditional Random Field (HCRF) as a classifier with a hinge loss layer to detect compound emotions. This experient is carried out using the iCV-MEFED dataset (Guo et al., 2018) . They got an accuracy of 93,19% on CK+ and 85.71% on their dataset.

Caramihale et al. (2018) proposed a generative adversarial network (GAN) classifier for emotion detection to improve the classification accuracy combining five facial datasets and overall accuracy for identification of true/generated images are 75.2% and 82.9% respectively.

In Radford et al. (2015), they discussed Deep convolutional generative adversarial networks (DCGANs) is a strong candidate for unsupervised learning with certain architectural constraints and demonstrated same on various image datasets. The proposed DC-GANs pair that learns a hierarchy of representations from object parts to scenes in the generator and discriminator. To use these extracted features can be used for general image representations.

To tackle the problem of source retaining, Zhu et al. (2019) propose a new method, untraceable GAN (UGAN), which introduces untraceable constraints and prototype injection. It is used to synthesize the source domain image that is untraceable.

Most FER solutions are focused on lab-controlled databases and are used to detect basic emotion. Hence, there is a need for an efficient and robust real-life FER system that detects particular basic and compound facial expressions on lab-controlled images and wild scenario images.

To correctly recognize basic and compound human emotion in real-world images on the RAF-DB dataset (Li et al., 2017), the proposed learning framework based on machine learning and deep learning architecture for FER systems is performing better than the state of the art results.

## 3 DATASET OVERVIEW

### Real-world Affective Faces Database (RAF-DB) (Li et al., 2017)

RAF-DB is a large-scale well diverse facial images facial expression database consisting 29672 number of real-world images downloaded from the Internet. Around 40 annotators independently labels each image based on the crowdsourcing annotation. Images in this database are full of varity in subjects, age, gender and ethnicity, head poses, lighting conditions, occlusions (e.g., facial hair, glasses, or self-occlusion), post-processing operations (e.g., various filters and special effects), etc. RAF-DB has enormous diversities rich annotated images with a 7-dimensional expression distribution vector for each image, has two different subsets: single-label subset, including 7 classes of basic emotions; multi-label subset, including 12 classes of compound emotions,

To objectively measure performance, the database has been split into a training set and a test set where the size of the training set is five times larger than the test set, and expressions in both sets are distributed equally. A few sample images of the dataset is shown in below Fig. 1

*Figure 1. RAF-DB dataset sample images*

RAF-DB is the first database that contains compound expressions in the wild. They reported that a cross-database study shows that the action units of basic emotions in RAF-DB are much more diverse than, or even deviate from, lab-controlled ones.

Also, handcrafted Features are available along with the Image Database. These Handcrafted features are as follows

1. HOG features: HOG features containing 4000-dimensional HOG features of each image for training and testing set, respectively.
2. Gabor features: Gabor features containing 4000-dimensional Gabor features of each image for training and testing set, respectively.
3. BaseDCNN features: baseDCNN features containing 2000-dimensional baseDCNN feature of each image for training and testing set, respectively.
4. DLP-CNN features: DLP-CNN features containing 2000-dimensional DLP-CNN feature of each image for training and testing set, respectively.

# 4 PROPOSED METHODOLOGY

## 4.1 The Framework for Compound Facial Expression Recognition

For Compound Facial expression recognition, we are using various frameworks starting from traditional Machine Learning, based on Feature Engineering as shown in Fig.2 a]. Feature engineering is a process of using domain knowledge for extracting features to reduce the data complexity and learning algorithms can easily work on visible patterns. This process is expensive in terms of time and difficult as expertise required.

In Machine learning, an expert needs to identify the most applied features and then as per the required and data type handcrafted. For example, features can be pixel values, shape, textures, position, and orientation. The performance of most Machine Learning algorithms depends on how accurately the features are identified and extracted. For our application, we are using Handcrafted features such as HOG, GABOR, baseDCNN, and DLP-CNN, which are available in the RAF-DB database for each image. After Feature extraction, we use a Traditional Machine Learning algorithm to classify Compound Facial Expression in 11 Classes.

*Figure 2. Traditional machine learning flow and deep learning flow for compound facial expression recognition*

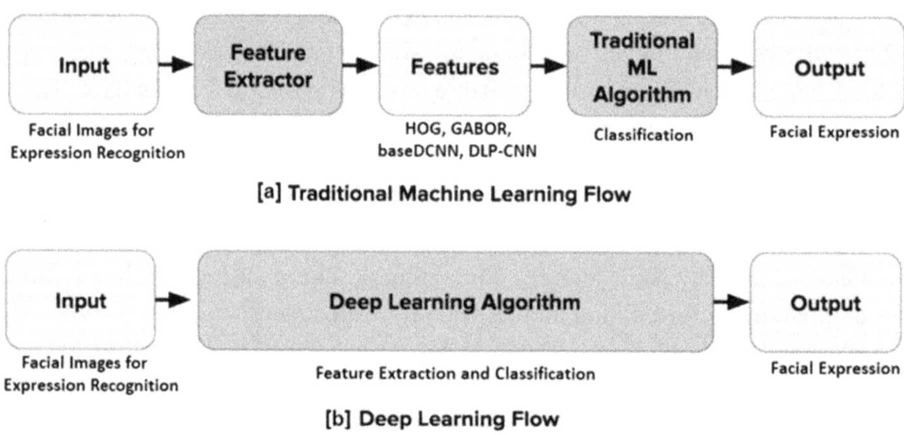

On the other hand, we are also using Deep Learning algorithms for Feature extraction and classification, as shown in Fig. 2 b]. Deep learning algorithms attempts to extract high-level features from data. This is major difference in Deep Learning and traditional Machine Learning. Therefore, deep learning reduces the task of developing a new feature extractor for every problem. CNN will try to learn basic features such as edges and lines in early layers, parts of human face, and then a high-level representation of a face.

Deep learning perform better than traditional machine learning as scale of data increases. Deep learning algorithms don't perform that well when the data is small. This is because deep learning algorithms need a huge amount of data to learn perfectly. On the other hand, traditional machine learning algorithms with handcrafted features work well.

## 4.2 The Framework for Compound Facial Expression Recognition With Handcrafted Feature

For our application of Facial Expression Recognition, we are using RAF-DB Dataset, which has Facial images with Compound Facial Emotion as labels. Figure 3 gives a compound facial expression recognition framework with a handcrafted feature. Pre-processing techniques can be applied before extracting handcrafted features. HOG, GABOR, DLP-CNN, and baseDCNN handcrafted features which are available in RAF-DB dataset itself for all Training and Test Images These features are used with traditional Machine Learning Algorithms such as K-Nearest Neighbor (KNN), Linear SVM and Non-Linear SVM, and Custom shallow CNN to the Predicted output shown in below Fig. 3.

For Example, Consider the image shown under the pre-processed image as a Test image for that handcrafted features are calculated, and using one of the classifiers, we can predict output for a given image it's angrily disgusting.

*Figure 3. The framework for compound facial expression recognition with handcrafted feature*

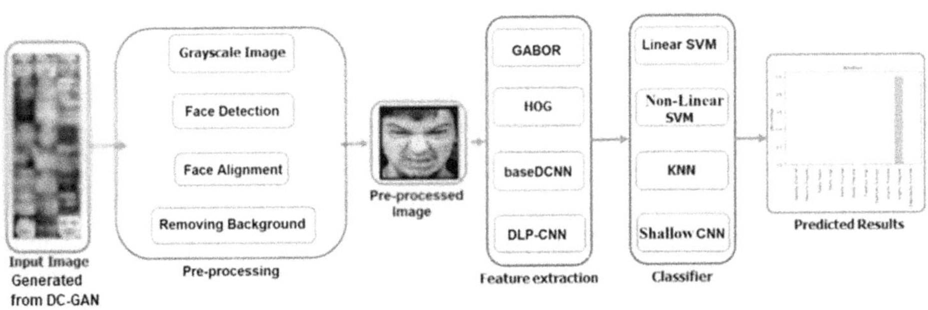

## 4.3 CNN Based Deep Learning Framework for Compound Facial Expression Recognition

The proposed methodology is based on the Convolution Neural Network-based deep learning framework given in Fig. 4. In this, we are using standard CNN Architectures such VGG19 (Simonyan & Zisserman, 2014), inception V3 (Szegedy et al., 2016), Inception ResNet V2 (Szegedy et al., 2017), Xception (Chollet, 2017), and MobileNet (Howard et al., 2017) architecture which is well known for Image classification applications. These Standard Networks have starting layers that are used to learn Low-level features such as edges, blobs, and colors, and the last layers learn task-specific features. In our case, it's Compound Facial Expressions.

By considering this fact, we modified the standard network by replacing the last layers so that specific features to our dataset are learned. In the RAF-DB database, we have a variety of images, but to learn features faster, we are pre-processing images to capture only face part using the Haar Cascade algorithm for detection of frontal face.

The training set consists of different facial Compound Expressions; However, the training data distribution is usually uneven. The imbalance of training data distribution will harm network performance. To eliminate this adverse effect, we performed data augmentation and Data Generation operations to balance the data distribution and to generate more training data.

The proposed methodology is based on the Convolution Neural Network-based deep learning framework given in Fig. 4. In this, we are using standard CNN Architectures such VGG19 (Simonyan & Zisserman, 2014), inception V3 (Szegedy et al., 2016), Inception ResNet V2 (Szegedy et al., 2017), Xception (Chollet, 2017), and MobileNet (Howard et al., 2017) architecture which is well known for Image classification applications. These Standard Networks have starting layers that are used to learn Low-level features such as edges, blobs, and colors, and the last layers learn task-specific features. In our case, it's Compound Facial Expressions.

By considering this fact, we are modifying the standard network by replacing the last layers so that features are specific to our dataset. In the RAF-DB database, we have a variety of images, but to learn features faster, we are pre-processing Images so only the required part of the face is captured using the Haar Cascade algorithm for detection of frontal face.

The training set consists of different facial Compound Expressions; However, the training data distribution is usually uneven. The imbalance of training data distribution will harm network performance. To eliminate this adverse effect, we performed data augmentation and Data Generation operations to balance the data distribution and to generate more training data.

To train, validate and test, network images are generated using image DataGenerator, which Generates batches of tensor image data with real-time data augmentation.

While training network with 2118 samples of training and 1044 samples of validation, accuracy, and loss is calculated so gets an idea how very network is training and checking network finally around 792 samples test images are supplied and Accuracy and Binary cross-entropy loss calculated.

*Figure 4. Proposed methodology for compound facial emotion recognition using standard CNN architecture*

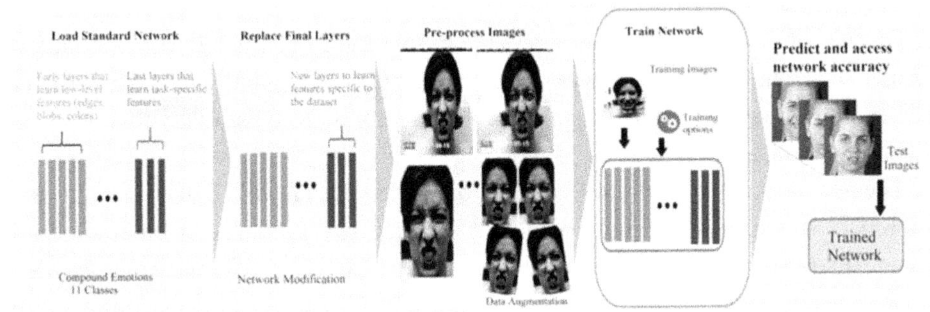


## 4.4 Deep Convolutional Generative Adversary Network (DC-GAN) Based Deep Learning Framework for Compound Facial Expression Recognition

GAN-based methods have been used successfully in synthesizing images to produce real faces images that are useful in training to augment data and help recognition functions. Many works have been proposed for novel models based on GAN (Goodfellow, 2016) of pose and identity-invariant FER.

DCGAN (RadfordGuo et al., 2015) is one of the most popular and effective GAN network designs, especially built on convolution layers without max pooling or fully connected layers. It uses the stride and transposed convolution to down sample and upsample. Fig.5 below is the design of a generator network.

1. DCGAN consists of two neural networks corresponding to each GAN, that is, one generator and one discriminator.
2. The generator captures random noise as input and outputs a generated fake image.
3. The discriminator takes real and false images as input with output values b/w 0 and 1, i.e., a the image confidence level is real or false.
4. DCGAN uses loss function as binary cross-entropy.

5. The generator aims to fool the discriminator by generating realistic fake images. In contrast, the goal of the discriminator is to identify real and fake images correctly. The DC-GAN working model indicates from the RAF-DB facial dataset that real images are shown with real labels, and on the other hand, the noise random distribution generator generates fake images G (Z), which has fake labels. Discriminators try to discriminate between real and fake images with labels, and by calculating Discriminator loss and Generator loss, training is carried out to achieve a good result.

For the generator's notation let z be a latent space vector sampled from a standard normal distribution. G(z), generator function maps the latent space vector z to data-space. The goal of G is to generate fake samples from estimated distribution of training data.

The generator is made up of Convolutional-transpose layers, batch norm layers, and ReLU activations. The input is a latent vector, z, that is choosen from a standard normal distribution of size 100, and the output is a 64x64x3 RGB image. The strided Conv-transpose layers allow the latent vector to be transformed into a volume with the same shape that of image.

*Figure 5. DC-GAN working model for Compound Facial expression recognition*

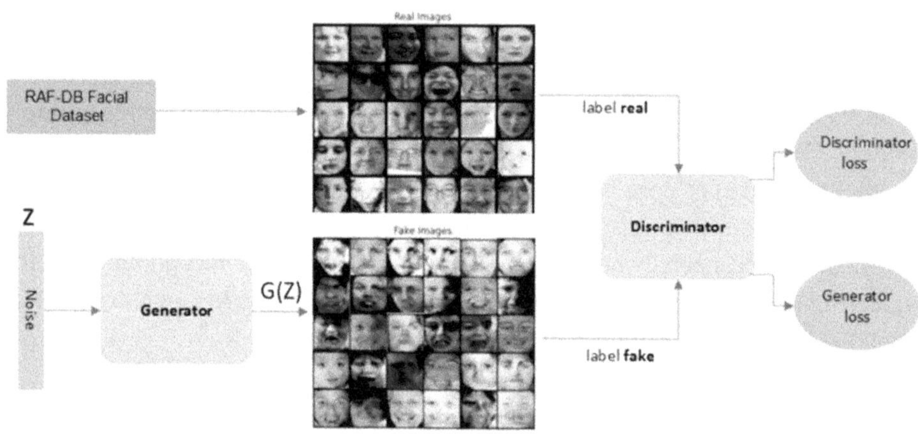


DC-GAN Generator took noise as latent space vector sampled from a standard normal distribution of size 100 and passed through series of Convolution Layers and activation Layers to generate output which is fed through a tanh function to return it to the input data range of $[-1,1]$ to generate fake images G(Z) of size (64,64,3).

Batch norm functions after the conv-transpose layers help to flow of gradients during training. Fig. 6 shows details of the generator for DCGAN.

*Figure 6. DC-GAN generator*

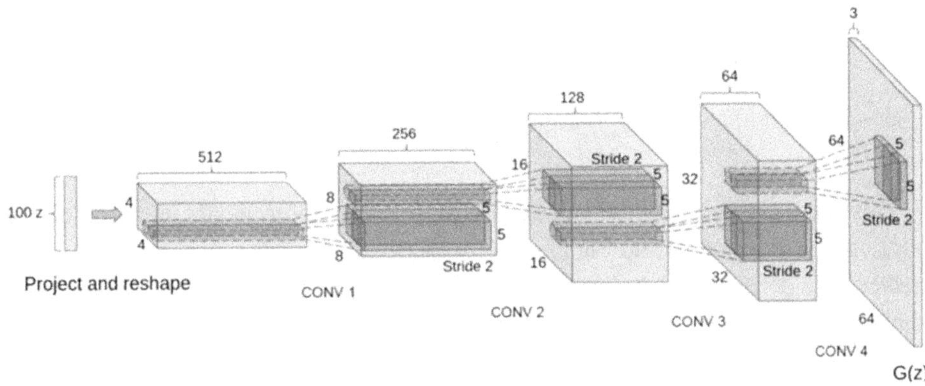


Let $x$ be data representing an image. $D(x)$ is the discriminator network that outputs the probability (scalar value) indicates that $x$ belongs to training data rather than the generator. Here, input to $D(x)$ is an image size 64x64x3. Intuitively, $D(x)$ should be HIGH when $x$ comes from training data and LOW when $x$ comes from the generator. $D(x)$ works as binary classification network that takes an image as input and outputs a scalar probability that the input image is real or fake. Here, D takes a 64x64x3 input image, processes it through a series of Conv2d, BatchNorm2d, and LeakyReLU layers, and outputs the final probability through a sigmoid activation function. Fig. 7 gives details of the DC-GAN discriminator

*Figure 7. DC-GAN discriminator*

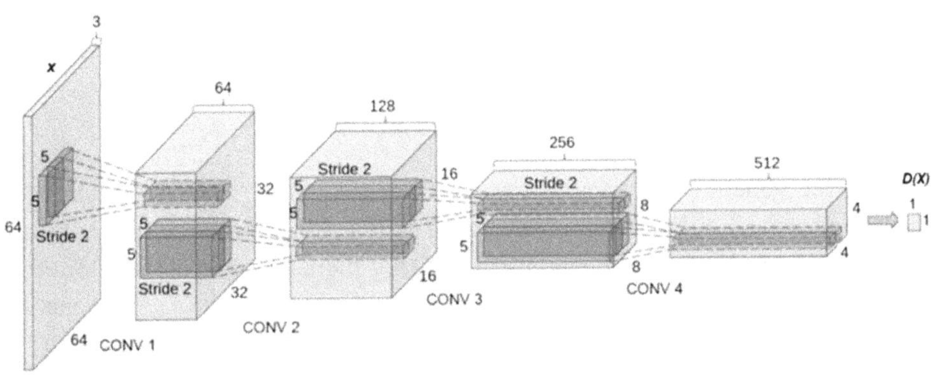

## Loss Functions and Optimizers

Discriminator and Generator learn through the loss functions and optimizers. We are using the Binary Cross Entropy loss (BCELoss) function, which is defined in Eq. (1) & (2)

$$l(x,y) = L = \{l_1, l_2, \ldots, l_N\}^T \tag{1}$$

$$l_n = -[y_n \cdot \log(x_n) + (1 - y_n) \cdot \log(1 - x_n)] \tag{2}$$

This function provides the calculation of both log components in the objective function (i.e. log (D(x)) and log (1−D (G (z))).

Next, we define real label as 1 and the fake label as 0. These labels will be used when calculating the losses of D and G. Finally, two separate optimizers is uesd, one for D and one for G. Both are Adam optimizers with learning rate 0.0002 and β= 0.5. To keep track of the generator's learning progress, In the training loop fixed batch of latent vectors drawn from a Gaussian distribution (i.e., fixed noise), periodically input this fixed noise into G, and over the iterations, images generated from the noise.

## Train the Discriminator

Objective of training the discriminator is to maximize the probability of correctly classifying a given image as real or fake. To update the discriminator stochastic gradient is used to maximize (D(x)) and log (1−D (G (z))). Due to the separate mini-batch, its calculated in two steps. First, batch of real samples created from the training set, forward pass through D, loss log D(x) calculated, then in a backward pass gradients is calculated. Secondly, batch of fake samples created with the current generator, then forward pass this through D, loss log (1−D(G(z)) calculated, and with backword pass gradients is *accumulate*d. Now, finally discriminator's optimizer step with the gradients accumulated from both the all-real and all-fake batches.

## Train the Generator

To train the generator by minimizing *log* (1−*D*(*G*(*z*))) to generate better fake images and provide sufficient gradients in the early learning process, To maximize *log*(*D*(*G*(*z*))) can be accomplished by classifying the Generator output from the discriminator, computing G's loss *using real labels*, computing G's gradients in a backward pass, and finally updating G's parameters with an optimizer step. It may seem counter-intuitive to use the real labels as Generator labels for the loss function, but this allows us to use the *log*(*x*) part of the BCELoss (rather than the *log*(1−*x*) part). The complete algorithm of Generator and Discriminator training and optimization as shown in Fig. 8, taken from Goodfellow (2016) .

*Figure 8. DC-GAN generator (Goodfellow, 2016)*

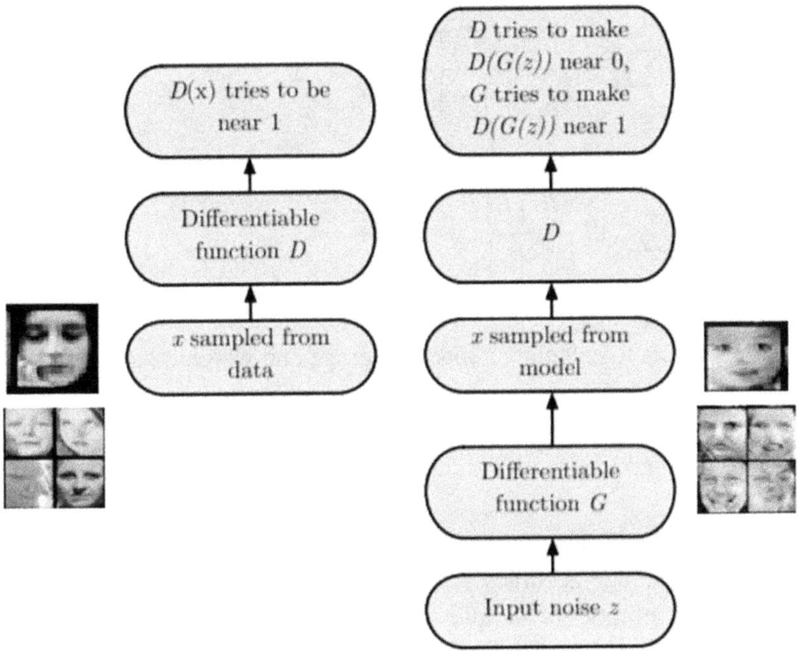


# 5 RESULTS

## 5.1 Available Metrics

Here we define set D of N images of a database as given in Equation (3)

$$D = \{d_0, d_1, \ldots, d_N\} \ldots. \tag{3}$$

We define $L_0, L_1, \ldots, L_{N-1}$ to be a family of compound emotion label (L) sets and $P_0, P_1, \ldots, P_{N-1}$ to be a family of prediction (P) sets where Li and Pi are the label set and prediction set, respectively, that correspond to Image di.

The set of all unique labels is given by equation (4)

$$L = \bigcup_{k=0}^{N-1} Lk \;.... \tag{4}$$

Mathematical representation of accuracy and f1-score is given in equations (5) & (6), respectively.

$$Accuracy = \frac{1}{N}\sum_{i=0}^{N-1} \frac{\left|L_i \cap P_i\right|}{\left|L_i\right| + \left|P_i\right| - \left|L_i \cap P_i\right|} \;.... \tag{5}$$

$$F1\,Score = \frac{1}{N}\sum_{i=0}^{N-1} 2\frac{\left|P_i \cap L_i\right|}{\left|P_i\right| \cdot \left|L_i\right|} \;.... \tag{6}$$

The next step is to calculate a macro average and weighted average

1. Macro average accuracy
   Macro average accuracy is an arithmetic mean of the accuracy.

2. Weighted average accuracy
   Weighted average accuracy is calculated by weighing the accuracy of each class by the number of samples from that class.

## 5.2 Confusion Matrix of RAF-DB Database

The confusion matrix is a error table that records indicates true/actual classification, and the predicted classification. For consistency throughout the paper, the columns indicates model prediction, whereas the rows indicates true classification. The classes are arranged in the same order in the rows and columns. Therefore, main diagonal indicates correctly classified elements. Confusion matrix is used to evaluate the accuracy between actual and predicted labels.

RAF-DB Compound Emotion Database has 11 Facial emotions {0:Happily Surprised, 1:Happily Disgusted, 2:Sadly Fearful, 3:Sadly Angry, 4:Sadly Surprised, 5:Sadly Disgusted, 6:Fearfully Angry, 7:Fearfully Surprised, 8:Angrily Surprised, 9:Angrily Disgusted, and 10:Disgustedly Surprised}. Confusion Matrix shows that most classes are classified with good accuracy except Sadly Surprised and Disgustedly Surprised. Sadly Surprised is misclassified as Sadly Disgusted and

Disgustedly Surprised is misclassified as Angrily Disgusted for most cases. Fig. 9 shows the normalized Confusion Matrix for a few machine learning methods and deep learning on the RAF-DB Database.

*Figure 9. Normalized confusion matrix for a few machine learning methods and deep learning on the RAF-DB Database*

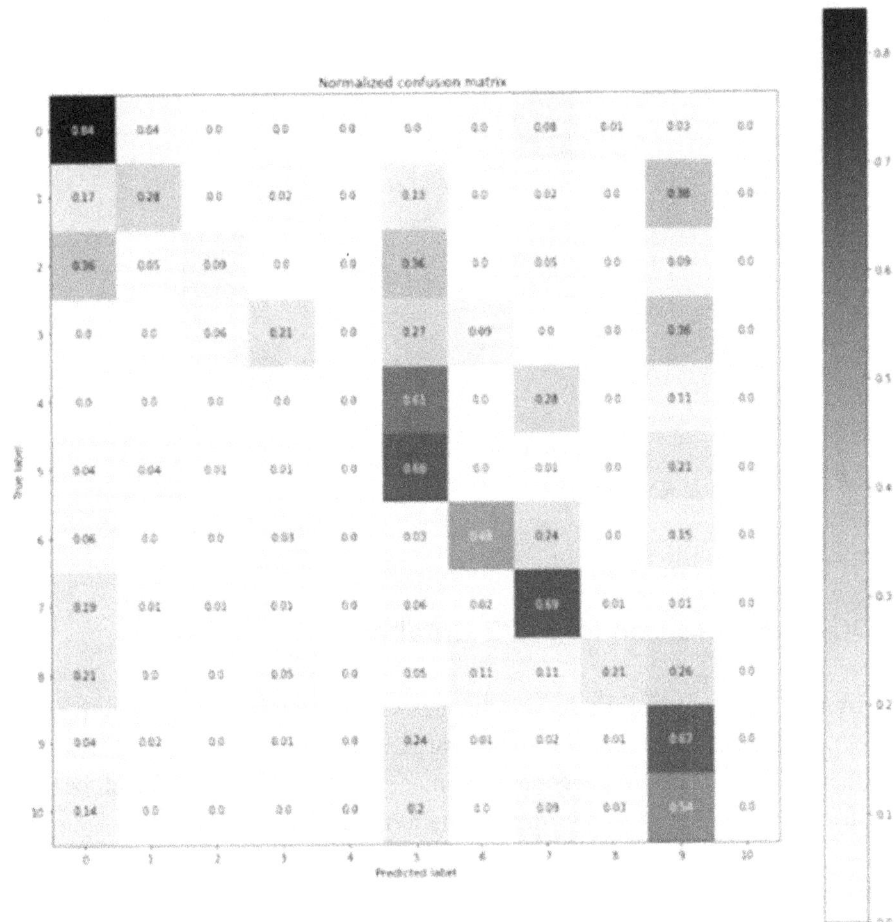

*Figure 10. DLP-CNN feature with Shallow CNN*

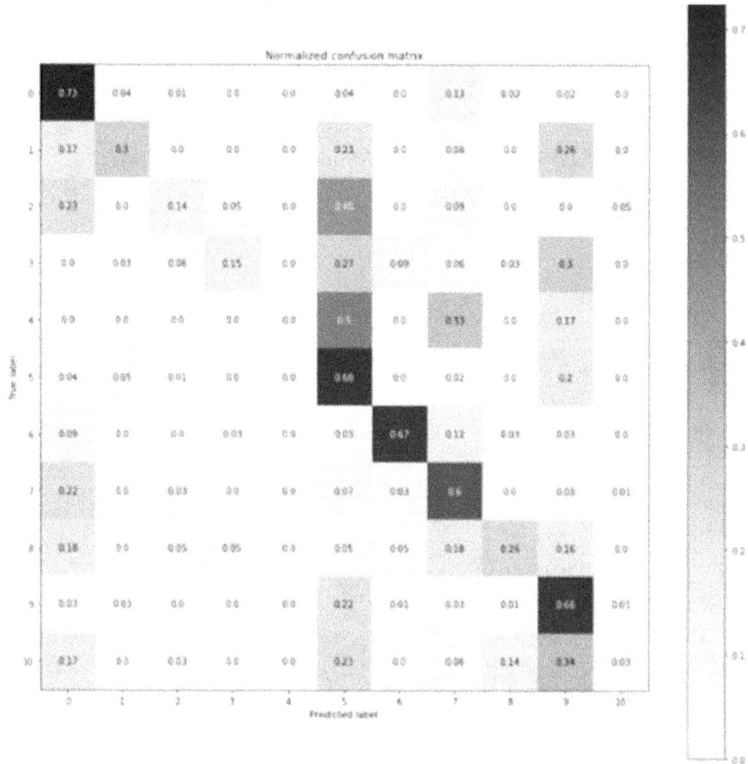


## 5.3 DC-GAN result on RAF-DB Database

Discriminator and generator's losses changes during training are shown in Figure 11. Second, Figure 12 visualizes a batch of real data next to a batch of fake data from the generator. These fake generated images are used to train model for better generalization and that results in improvement in testing accuracy.

## 5.4 Emotion Analysis on RAF-DB Sample Test Image

Emotion Analysis of RAF-DB sample test Images is done at the end to check the model's working. It shows for sample test image softmax output indicating a probability of each emotion with a plotted graph the maximum probability class will be considered as predicted emotion by the proposed model.

*Figure 11. Inception ResNetV2 architecture*

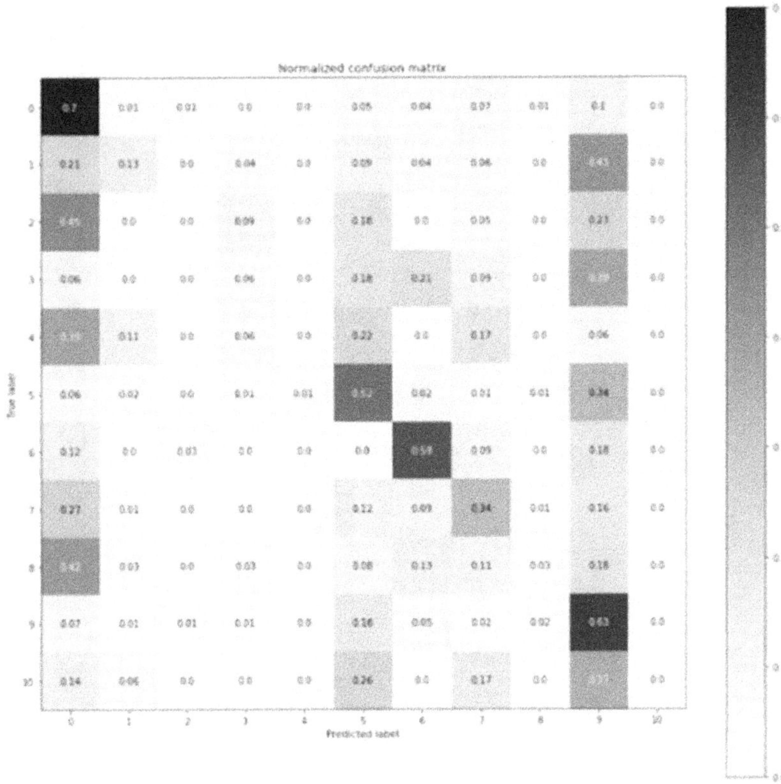

## 5.5 Accuracy Score

The score is calculated to evaluate model accuracy, which has overall accuracy, macro average, and weighted average. To summarize the results of the implemented methodology, refer to Table 1 which indicates the f1-score and accuracy on RAF-DB compound emotion dataset using handcrafted features HOG, GABOR, DLP-CNN, and baseDCNN that are available in the dataset itself for all training and test images. Bold numbers indicate the best f1-score on particular compound emotion of the RAF-DB dataset. These features are used with traditional Machine Learning Algorithms such as K-Nearest Neighbor (KNN), Linear SVM and Non-Linear SVM, and Custom shallow CNN to Predicted output.

*Figure 12. Xception architecture*

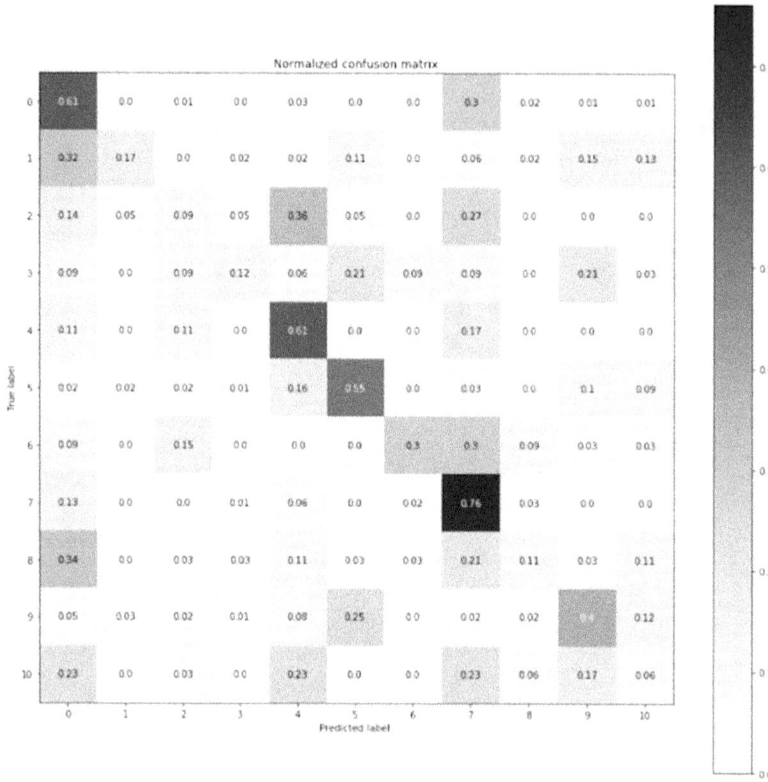


*Figure 13. Discriminator and generator's losses changes during training*

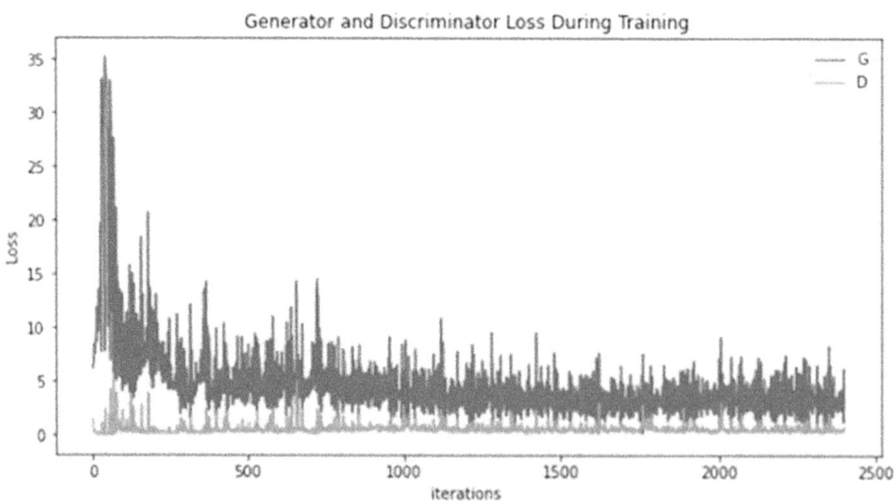

*Figure 14. A batch of real data next to a batch of fake data from the generator*

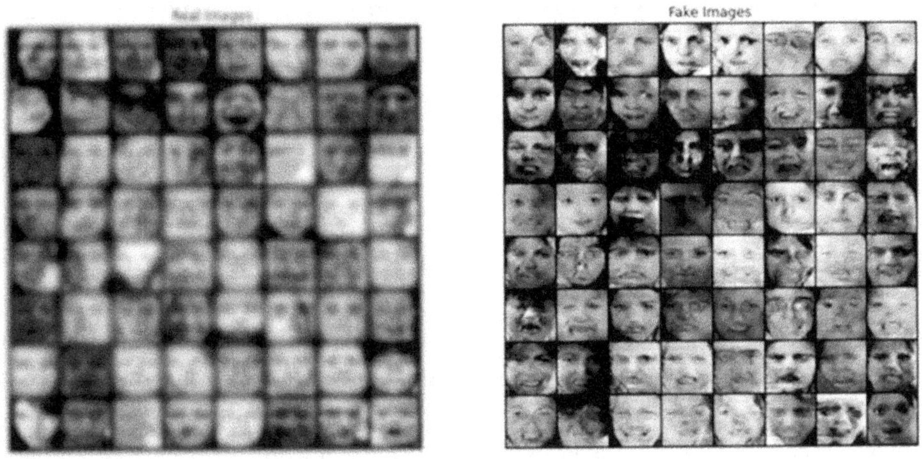


*Figure 15. Emotion Analysis on RAF-DB Sample Test Image*

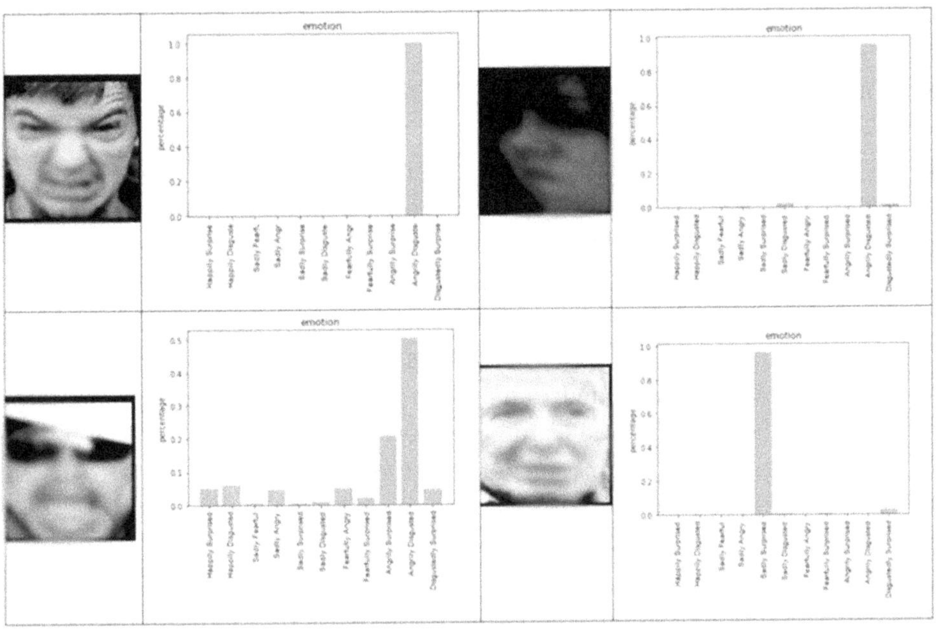


Standard CNN model architecture based on VGG19, inception V3, Inception ResNet V2, Xception, and MobileNet architecture is modified to compare results with deep learning methodology. Models are trained from scratch on a compound emotion of 11 classes. The model is trained on the Training and Validation dataset on each epoch to measure performance, and accuracy and loss are calculated for each epoch. Once training is done, indicating no more improvement of accuracy on the validation dataset. A trained model is finally used to calculate accuracy Performance (%) on the test dataset. Table 2 summarizes results of f1-score and accuracy with the implemented methodology of the Deep learning framework with standard CNN. Bold numbers indicate the best f1-score on particular compound emotion of the RAF-DB dataset.

## 5.6 Comparison with State-of-the-Art Methods

We compare our methods against state-of-the-art methods. For compound emotion detection on the RAF-DB dataset, we compare our model with the recent related work taken from Li et al. (2017), Li & Deng, (2019), and Li et al., (2019). Table 3 shows a comparison with state-of-the-art results in terms of accuracy. Table 4 shows comparison with state of art results in terms of the mean diagonal value of confusion matrix, and it shows the proposed method outperforms compared to state-of-the-art methods.

# 6 CONCLUSION AND FUTURE WORK

In this paper, various learning frameworks for Facial Emotion recognition based on traditional machine learning and deep learning frameworks were proposed to improve performance on the RAF-DB database for real-world images.

Experiment shows, for RAF-DB dataset handcrafted features with a machine learning approach, gives comparatively better result over deep learning-based standard CNN approach. DLP-CNN feature with m-SVM gives an accuracy of 57% maximum among all, and in deep learning based on Inception ResNetV2 architecture got an accuracy of 49%. The proposed learning framework for the RAF-DB dataset with DLP-CNN features outperforms state-of-the-art results. An advanced GAN-based DC-GAN model is trained to generate fake images in a deep learning framework. We also observe that even though the real-world dataset has non-uniform illumination conditions and occlusions, our method consistently generates high-quality images. This demonstrates that our method is very robust.

*Table 1. Summary results of accuracy on compound facial emotion recognition on RAF-DB Datasets with various handcrafted features classified with a traditional machine learning approach*

| Compound Emotion with #ID | HOG+ KNN | HOG+ m-SVM | HOG+ CNN | GABOR+ m-SVM | GABOR+ CNN | baseDCNN+ CNN | baseDCNN+ m-SVM | DLP-CNN+ CNN | DLP-CNN+ m-SVM | Support |
|---|---|---|---|---|---|---|---|---|---|---|
| #0: Happily Surprised | 0.58 | 0.6 | 0.53 | 0.64 | 0.57 | 0.37 | 0.62 | 0.66 | 0.72 | 135 |
| #1: Happily Disgusted | 0.26 | 0.23 | 0.15 | 0.31 | 0.19 | 0.07 | 0.27 | 0.35 | 0.34 | 47 |
| #2: Sadly Fearful | 0.19 | 0.08 | 0 | 0.06 | 0 | 0 | 0 | 0.16 | 0.14 | 22 |
| #3: Sadly Angry | 0.22 | 0.15 | 0.17 | 0.13 | 0.09 | 0.24 | 0.15 | 0.24 | 0.3 | 33 |
| #4: Sadly Surprised | 0 | 0 | 0.09 | 0.09 | 0 | 0 | 0 | 0 | 0 | 18 |
| #5: Sadly Disgusted | 0.53 | 0.61 | 0.52 | 0.59 | 0.49 | 0.46 | 0.6 | 0.57 | 0.58 | 141 |
| #6: Fearfully Angry | 0.43 | 0.61 | 0.44 | 0.61 | 0.4 | 0.28 | 0.36 | 0.68 | 0.54 | 33 |
| #7: Fearfully Surprised | 0.49 | 0.6 | 0.51 | 0.59 | 0.41 | 0.39 | 0.6 | 0.59 | 0.69 | 116 |
| #8: Angrily Surprised | 0.26 | 0.04 | 0.18 | 0.28 | 0.04 | 0.04 | 0 | 0.33 | 0.32 | 38 |
| #9: Angrily Disgusted | 0.49 | 0.6 | 0.48 | 0.6 | 0.51 | 0.4 | 0.62 | 0.62 | 0.59 | 174 |
| #10: Disgustedly Surprised | 0.05 | 0.05 | 0 | 0.04 | 0 | 0 | 0 | 0.05 | 0 | 35 |
| Accuracy | 0.45 | 0.53 | 0.44 | 0.53 | 0.44 | 0.35 | 0.53 | 0.55 | 0.57 | 792 |
| macro average accuracy | 0.32 | 0.32 | 0.28 | 0.36 | 0.25 | 0.2 | 0.29 | 0.39 | 0.38 | 792 |
| weighted average accuracy | 0.43 | 0.48 | 0.41 | 0.5 | 0.39 | 0.32 | 0.47 | 0.52 | 0.53 | 792 |

*Table 2. Summary of accuracy on compound facial emotion recognition on RAF-DB Datasets with various CNN based Standard architecture*

| Compound Emotion with #ID | VGG19 | Inception ResNetV2 | InceptionV3 | MobileNet | XceptionNet | Support |
|---|---|---|---|---|---|---|
| #0: Happily Surprised | 0.35 | 0.61 | 0.41 | 0.58 | 0.57 | 135 |
| #1: Happily Disgusted | 0.04 | 0.22 | 0.18 | 0.22 | 0.25 | 47 |
| #2: Sadly Fearful | 0 | 0.06 | 0.09 | 0.17 | 0.09 | 22 |
| #3: Sadly Angry | 0.09 | 0.27 | 0.05 | 0.21 | 0.18 | 33 |
| #4: Sadly Surprised | 0 | 0 | 0 | 0 | 0.22 | 18 |
| #5: Sadly Disgusted | 0.29 | 0.59 | 0.61 | 0.47 | 0.56 | 141 |
| #6: Fearfully Angry | 0.47 | 0.5 | 0.45 | 0.52 | 0.41 | 33 |
| #7: Fearfully Surprised | 0.41 | 0.62 | 0.55 | 0.64 | 0.6 | 116 |
| #8: Angrily Surprised | 0 | 0.17 | 0.21 | 0.05 | 0.14 | 38 |
| #9: Angrily Disgusted | 0.46 | 0.5 | 0.58 | 0.49 | 0.5 | 174 |
| #10: Disgustedly Surprised | 0 | 0 | 0 | 0 | 0.05 | 35 |
| Accuracy | 0.35 | 0.49 | 0.47 | 0.43 | 0.45 | 792 |
| macro average accuracy | 0.19 | 0.32 | 0.28 | 0.3 | 0.32 | 792 |
| weighted average accuracy | 0.3 | 0.47 | 0.43 | 0.43 | 0.45 | 792 |

*Table 3. Performance comparison with the state of the art results in terms of accuracy*

| Paper | Method | Accuracy |
|---|---|---|
| Li et al., (2017) | DLP-CNN | 57.95% |
| Li et al., (2019) | ResNet18(softmax loss) | 56.94% |
| | ResNet18 (softmax and separate loss) | 58.84% |
| Ours | DLP-CNN+m-SVM | 57.00% |

*Table 4. Performance comparison with the state of the art results in terms of Mean Diagonal value of Confusion Matrix*

| Paper | Method | Feature | Mean Diagonal value of Confusion Matrix |
|---|---|---|---|
| Li et al., (2017) | m-SVM | LBP | 28.84 |
| | | HOG | 33.65 |
| | | GABOR | 35.76 |
| | LDA | LBP | 22.89 |
| | | HOG | 24.01 |
| | | GABOR | 23.81 |
| Ours | m-SVM | HOG | 32.90 |
| | | GABOR | 35.36 |
| | | baseDCNN | 30.36 |
| | | DLP-CNN | 37.72 |
| | Shallow CNN | DLP-CNN | 38.36 |
| | MobileNet | ----- | 31.90 |
| | InceptionV3 | ----- | 32.82 |
| | Inception ResNetV2 | ----- | 34.18 |
| | XceptionNet | ----- | 34.36 |

In the future, to improve the accuracy of a non-lab controlled dataset for compound emotion detection, we can consider using an unsupervised pre-training strategy from transfer learning (Bengio, 2012), which may further decrease the recognition error rate. To improve in facial expression analysis deep belief network-based technique can be used Liu et al. (2014). Transfer learning can also be tried on RAF-DB Dataset (Riaz et al., 2020). The use of RNN and LSTM (Jarraya et al., 2020) can also be a good solution. FER based on Facial Action Units (AUs) showed promising results in state-of-the-art that can also be tried (Shao et al., 2019). Also, a separate loss function for compound emotion can be used, as mentioned in Li et al. (2019).

To extend this research of FER in the wild scenario, the RAF-AU dataset given by Yan et al. (2020) is very useful for detecting appropriate basic and compound facial emotion through its facial action units.

## 7 DATA AVAILABILITY STATEMENT

RAF-DB Dataset is subject to third-party restrictions. The RAF-DB Dataset that supports the findings of this study is available from Shan Li, and Deng W. Restrictions apply to the availability of these data, which were used under license for this study. This dataset is suitable for those who want to study real-world datasets of facial expressions to detect basic and compound facial expressions. The database is available for distribution for research purposes from Shan Li and Deng W. at URL link http://www.whdeng.cn/raf/model1.html with the permission of Shan Li and Deng W.

## ACKNOWLEDGMENT

We thank all the authors whose work contributed to the meta-analysis, especially Shan Li and Deng W., who took the time to respond to our requests and were given access to RAF-DB Dataset.

## REFERENCES

Bengio, Y. (2012, June). *Deep learning of representations for unsupervised and transfer learning*. In *Proceedings of ICML workshop on unsupervised and transfer learning* (pp. 17-36). JMLR.

Benitez-Quiroz, C. F., Srinivasan, R., Feng, Q., Wang, Y., & Martinez, A. M. (2017). Emotionet challenge: Recognition of facial expressions of emotion in the wild. arXiv preprint arXiv:1703.01210

Caramihale, T., Popescu, D., & Ichim, L. (2018). Emotion classification using a tensorflow generative adversarial network implementation. *Symmetry*, *10*(9), 414. doi:10.3390ym10090414

Chollet, F. (2017). Xception: Deep learning with depthwise separable convolutions. In Proceedings of the IEEE conference on computer vision and pattern recognition (pp. 1251-1258). IEEE. arXiv:1610.02357v3 10.1109/CVPR.2017.195

Darwin, C., & Prodger, P. (1998). *The expression of the emotions in man and animals*. Oxford University Press.

Du, S., & Martinez, A. M. (2022). Compound facial expressions of emotion: From basic research to clinical applications. *Dialogues in Clinical Neuroscience*. PMID:26869845

Du, S., Tao, Y., & Martinez, A. M. (2014). Compound facial expressions of emotion. *Proceedings of the National Academy of Sciences of the United States of America, 111*(15), E1454–E1462. doi:10.1073/pnas.1322355111 PMID:24706770

Ekman, P., & Friesen, W. V. (1971). Constants across cultures in the face and emotion. *Journal of Personality and Social Psychology, 17*(2), 124–129. doi:10.1037/h0030377 PMID:5542557

Fabian Benitez-Quiroz, C., Srinivasan, R., & Martinez, A. M. (2016). Emotionet: An accurate, real-time algorithm for the automatic annotation of a million facial expressions in the wild. In *Proceedings of the IEEE conference on computer vision and pattern recognition* (pp. 5562-5570). IEEE. 10.1109/CVPR.2016.600

Goodfellow, I. (2016). Nips 2016 tutorial: Generative adversarial networks. arXiv preprint arXiv:1701.00160

Gunes, H., & Schuller, B. (2013). Categorical and dimensional affect analysis in continuous input: Current trends and future directions. *Image and Vision Computing, 31*(2), 120–136. doi:10.1016/j.imavis.2012.06.016

Guo, J., Lei, Z., Wan, J., Avots, E., Hajarolasvadi, N., Knyazev, B., Kuharenko, A., Jacques Junior, J. C. S., Baro, X., Demirel, H., Escalera, S., Allik, J., & Anbarjafari, G. (2018). Dominant and complementary emotion recognition from still images of faces. *IEEE Access : Practical Innovations, Open Solutions, 6*, 26391–26403. doi:10.1109/ACCESS.2018.2831927

Howard, A. G., Zhu, M., Chen, B., Kalenichenko, D., Wang, W., Weyand, T., Andreetto, M., & Adam, H. (2017). Mobilenets: Efficient convolutional neural networks for mobile vision applications. arXiv preprint arXiv:1704.04861

Jarraya, S. K., Masmoudi, M., & Hammami, M. (2020). Compound Emotion Recognition of Autistic Children During Meltdown Crisis Based on Deep Spatio-Temporal Analysis of Facial Geometric Features. *IEEE Access : Practical Innovations, Open Solutions, 8*, 69311–69326. doi:10.1109/ACCESS.2020.2986654

Lekdioui, K., Messoussi, R., Ruichek, Y., Chaabi, Y., & Touahni, R. (2017). Facial decomposition for expression recognition using texture/shape descriptors and SVM classifier. *Signal Processing Image Communication, 58*, 300–312. doi:10.1016/j.image.2017.08.001

Li, S., & Deng, W. (2019). Blended emotion in-the-wild: Multi-label facial expression recognition using crowdsourced annotations and deep locality feature learning. *International Journal of Computer Vision, 127*(6), 884–906. doi:10.100711263-018-1131-1

Li, S., & Deng, W. (2020). Deep facial expression recognition: A survey. *IEEE Transactions on Affective Computing*, *13*(3), 1195–1215. doi:10.1109/TAFFC.2020.2981446

Li, S., & Deng, W. (2020). Deep facial expression recognition: A survey. *IEEE Transactions on Affective Computing*.

Li, S., Deng, W., & Du, J. (2017). Reliable crowdsourcing and deep locality-preserving learning for expression recognition in the wild. In *Proceedings of the IEEE conference on computer vision and pattern recognition* (pp. 2852-2861). 10.1109/CVPR.2017.277

Li, S., Deng, W., & Du, J. (2017). Reliable crowdsourcing and deep locality-preserving learning for expression recognition in the wild. In *Proceedings of the IEEE conference on computer vision and pattern recognition* (pp. 2852-2861). IEEE. 10.1109/CVPR.2017.277

Li, Y., Lu, Y., Li, J., & Lu, G. (2019, October). Separate loss for basic and compound facial expression recognition in the wild. In *Asian Conference on Machine Learning* (pp. 897-911). PMLR.

Liliana, D., Basaruddin, T., & Widyanto, M. (2018). Mixed facial emotion recognition using active appearance model and hidden conditional random fields. *International Journal of Pure and Applied Mathematics, 118*, 3159-3167.

Lim, Y. K., Liao, Z., Petridis, S., & Pantic, M. (2018). Transfer learning for action unit recognition.

Lindquist, K. A., Wager, T. D., Kober, H., Bliss-Moreau, E., & Barrett, L. F. (2012). The brain basis of emotion: A meta-analytic review. *Behavioral and Brain Sciences*, *35*(3), 121–143. doi:10.1017/S0140525X11000446 PMID:22617651

Liu, P., Han, S., Meng, Z., & Tong, Y. (2014). Facial expression recognition via a boosted deep belief network. In *Proceedings of the IEEE conference on computer vision and pattern recognition* (pp. 1805-1812). IEEE. 10.1109/CVPR.2014.233

Lucey, P., Cohn, J. F., Kanade, T., Saragih, J., Ambadar, Z., & Matthews, I. (2010, June). The extended cohn-kanade dataset (ck+): A complete dataset for action unit and emotion-specified expression. In *2010 IEEE computer society conference on computer vision and pattern recognition-workshops* (pp. 94-101). IEEE.

Lyons, M., Kamachi, M., & Gyoba, J. (1998). *The Japanese Female Facial Expression (JAFFE)*. Dataset.

Pandey, R. K., Karmakar, S., Ramakrishnan, A. G., & Saha, N. (2019). Improving facial emotion recognition systems using gradient and laplacian images. arXiv preprint arXiv:1902.05411

Pantic, M., Valstar, M., Rademaker, R., & Maat, L. (2005, July). Web-based database for facial expression analysis. In *2005 IEEE international conference on multimedia and Expo* (pp. 5-pp). IEEE.

Radford, A., Metz, L., & Chintala, S. (2015). Unsupervised representation learning with deep convolutional generative adversarial networks. arXiv preprint arXiv:1511.06434.

Revina, I. M., & Emmanuel, W. S. (2021). A survey on human face expression recognition techniques. *Journal of King Saud University-Computer and Information Sciences*, *33*(6), 619–628. doi:10.1016/j.jksuci.2018.09.002

Riaz, M. N., Shen, Y., Sohail, M., & Guo, M. (2020). Exnet: An efficient approach for emotion recognition in the wild. *Sensors (Basel)*, *20*(4), 1087. doi:10.339020041087 PMID:32079319

Shao, Z., Liu, Z., Cai, J., Wu, Y., & Ma, L. (2019). Facial action unit detection using attention and relation learning. *IEEE Transactions on Affective Computing*, *13*(3), 1274–1289. doi:10.1109/TAFFC.2019.2948635

Simonyan, K., & Zisserman, A. (2014). Very deep convolutional networks for large-scale image recognition. arXiv preprint arXiv:1409.1556

Suwa, M. (1978). A preliminary note on pattern recognition of human emotional expression. *Proc. IJPR*, (pp. 408-410). IEEE.

Szegedy, C., Ioffe, S., Vanhoucke, V., & Alemi, A. A. (2017, February). Inception-v4, inception-resnet and the impact of residual connections on learning. In *Thirty-first AAAI conference on artificial intelligence.* 10.1609/aaai.v31i1.11231

Szegedy, C., Vanhoucke, V., Ioffe, S., Shlens, J., & Wojna, Z. (2016). Rethinking the inception architecture for computer vision. In *Proceedings of the IEEE conference on computer vision and pattern recognition* (pp. 2818-2826). IEEE. 10.1109/CVPR.2016.308

Wolfram Research. (2018). *FER-2013*. Wolfram Data Repository.

Yan, W. J., Li, S., Que, C., Pei, J., & Deng, W. (2020). RAF-AU Database: In-the-Wild Facial Expressions with Subjective Emotion Judgement and Objective AU Annotations. In *Proceedings of the Asian Conference on Computer Vision*. Springer.

Zhao, G., & Pietikainen, M. (2007). Dynamic texture recognition using local binary patterns with an application to facial expressions. *IEEE Transactions on Pattern Analysis and Machine Intelligence*, *29*(6), 915–928. doi:10.1109/TPAMI.2007.1110 PMID:17431293

Zhu, D., Liu, S., Jiang, W., Gao, C., Wu, T., Wang, Q., & Guo, G. (2019). *Ugan: Untraceable gan for multi-domain face translation.*

## ADDITIONAL READING

Bartlett, M. S., Littlewort, G., Fasel, I., & Movellan, J. R. (2003). Real Time Face Detection and Facial Expression Recognition: Development and Applications to Human Computer Interaction. *Conference on Computer Vision and Pattern Recognition Workshop, CVPRW'03*. IEEE. 10.1109/CVPRW.2003.10057

Borgalli, M. R. A., & Surve, S. (2022, March). Deep learning for facial emotion recognition using custom CNN architecture. [). IOP Publishing.]. *Journal of Physics: Conference Series*, *2236*(1), 012004. doi:10.1088/1742-6596/2236/1/012004

Borgalli, R. A., & Surve, S. (2022). Compound Facial Expression Recognition and Pain Intensity Measurement Using Optimized Deep Neuro Fuzzy Network. [IJSIR]. *International Journal of Swarm Intelligence Research*, *13*(1), 1–27. doi:10.4018/IJSIR.304721

Borgalli, R. A., & Surve, S. (2022). Facial Emotion Recognition Through Detection of Facial Action Units and Their Intensity. *Science Vision*, *14*(1). http://sv-journal.org/2022-1/06/. doi:10.26583v.14.1.06

Borgalli, R. A., & Surve, S. (2023). Deep Learning Framework for Compound Facial Emotion Recognition. In A. Shukla, B. K. Murthy, N. Hasteer, & J. P. Van Belle (Eds.), *Computational Intelligence. Lecture Notes in Electrical Engineering* (Vol. 968). Springer. doi:10.1007/978-981-19-7346-8_65

Gao, S. (2022). A two-channel attention mechanism-based mobileNetV2 and bidirectional long short memory network for multi-modal dimension dance emotion recognition. *Journal of Applied Science and Engineering*, *26*(4), 455–464.

Kamińska, D., Aktas, K., Rizhinashvili, D., Kuklyanov, D., Sham, A. H., Escalera, S., Nasrollahi, K., Moeslund, T. B., & Anbarjafari, G. (2021). Two-Stage Recognition and beyond for Compound Facial Emotion Recognition. *Electronics (Basel)*, *10*(22), 2847. doi:10.3390/electronics10222847

Leite, D., Frigeri, V. Jr, & Medeiros, R. (2021). Adaptive Gaussian fuzzy classifier for real-time emotion recognition in computer games. arXiv preprint arXiv:2103.03488.

Egede, J., Valstar, M., & Martinez, B. (2017). Fusing deep learned and handcrafted features of appearance, shape, and dynamics for automatic pain estimation. *12th IEEE International Conference on Automatic Face & Gesture Recognition (FG 2017)*. IEEE.

Shen, W. (2022). A Novel Conditional Generative Adversarial Network Based On Graph Attention Network For Moving Image Denoising. *Journal of Applied Science and Engineering*, *26*(6), 831–841.

Teng, L., & Qiao, Y. (2022). BiSeNet-oriented context attention model for image semantic segmentation. *Computer Science and Information Systems*, (00), 40–40. doi:10.2298/CSIS220321040T

Yang, M.-T., Cheng, Y.-J., & Shih, Y.-C. (2001). Facial expression recognition for learning status analysis. *International Conference on Human-Computer Interaction*. IEEE.

## KEY TERMS AND DEFINITIONS

**Compound Emotion:** Compound emotions, also known as blended emotions, are emotional experiences that result from the simultaneous or sequential activation of two or more primary emotions. They can be thought of as complex emotional states that are composed of multiple emotions.

**Confusion Matrix:** The confusion matrix is a cross table that counts the instances of the true/actual classification and the anticipated classification within two classes. Columns are used to show model predictions, while rows are used to show the actual classifications, in order to maintain consistency. The amount of times one rater agreed with the other can be found on a diagonal from top left to bottom right, which is where correctly categorised things can be discovered. Using a confusion matrix, the accuracy of the actual labels in comparison to those anticipated is evaluated.

**Convolution Neural Network (CNN):** A convolutional neural network (CNN) is a type of deep learning neural network that is specifically designed for image processing and computer vision tasks. CNNs are based on the idea of applying convolutional filters (also known as kernels or feature detectors) to the input image to extract meaningful features.

**Deep Convolutional Generative Adversary Network (DC-GAN):** The deep convolutional generative adversary network is one of the most popular and successful GAN designs, consisting mainly of convolutional layers without any max pooling or fully connected layers. For downsampling and upsampling, it uses a convolutional stride and transposed convolutions.

**Deep Learning:** Deep learning is a subset of machine learning that involves the use of artificial neural networks with multiple layers to process and analyze data. Deep learning algorithms are designed to automatically learn representations of data through the use of multiple layers of artificial neurons, each layer building upon the previous one to create more complex and abstract representations.

**Facial Emotion Recognition:** Facial emotion recognition (FER) is the technology that analyses facial expressions from sensors, static images, speech, videos and other kind of available data to reveal information on emotional state of humans.

**Feature Extraction:** Feature extraction is the process of selecting and transforming raw data into a set of features, or measurable attributes, that are relevant and useful for a specific task or application. It is a common technique used in machine learning, computer vision, and signal processing to reduce the dimensionality and complexity of data while retaining important information.

**Generative Adversary Network (GAN):** Generative adversary network trains models through a minimax two-player game between a generator G(z) that generates synthesised input data by mapping latents to a data space with z as p(z) and a discriminator D(x) that assigns a probability y = Dis(x) € [0,1] that is a real training sample to distinguish real from fake input data.

**Learning Framework:** A learning framework is a structured approach or methodology that provides a systematic process for designing, developing, delivering, and evaluating learning programs or activities. It outlines the key principles, strategies, and methods used to facilitate learning and ensure that learning objectives are achieved.

**Machine Learning:** Machine learning is a subfield of artificial intelligence that focuses on building algorithms and models that can learn from and make predictions or decisions based on data, without being explicitly programmed. It involves the use of statistical and computational techniques to enable machines to learn from experience and improve their performance on a given task.

# Chapter 4
# Power of Robotics:
## Applications From Healthcare to Spatial Research - An Exploration

**S. Uma**
*Hindusthan College of Engineering and Technology, India*

**Saranya V.**
*Hindusthan College of Engineering and Technology, India*

## ABSTRACT

*Robotics is developing at a rate that exceeds the understanding of science at this time. As their integration into society develops, stronger intellectual underpinnings will be needed to reach greater, more dependable, and safer capabilities. Robots are utilized in a variety of fields today, including industry, health, science, space exploration, construction, food packaging, and even surgery. Robotics is the design, creation, and application of devices (robots) to carry out tasks that have historically been completed by humans. Robots are frequently utilised in fields where repetitive, uncomplicated activities must be completed, and in fields where work must be done under hazardous conditions for humans. Robots make it easier to respond to disasters, improve physical capabilities, work in settings where human interaction is required, and permit exploration beyond Earth's orbit. Robots, unlike humans, don't become bored, therefore in many scenarios they can boost production, efficiency, quality, and consistency of products.*

DOI: 10.4018/978-1-6684-8171-4.ch004

## 1.0 INTRODUCTION

Early in the 1950s, George C. Devol, an inventor from Louisville, Kentucky, created the first robots that we are familiar with today. A programmable manipulator known as "Unimate" from "Universal Automation" was designed by him, and it was granted a patent (Stanford University, 2019). He attempted to advertise his products in the industry for the next ten years. Thanks to enhanced sensor technologies and even more amazing advancements in machine learning and artificial intelligence (AI), robots will keep evolving from being rote machines to collaborators with cognitive capacities. These advancements, which have a positive future outlook for these and allied industries, will be extremely beneficial to robotics.

Science fiction and research and development are giving way to reality with the fourth industrial revolution (4IR) of robotic and automation technologies. In every sphere of the economy and every aspect of daily life, machines are becoming more powerful, thanks to enormous advances in computing power, the proliferation of data obtained by strong algorithms incorporated into digital platforms, the development of new materials, and the integration of the urban environment (Uma & Suganthi, 2015). The technical challenges happening in our culture will have a great impact in our lives in the decades to come. The booming artificial intelligence and robotics industries are one important factor that has an impact on and alters many facets of daily life. The brisk growth of the robotics and artificial intelligence industries is one important factor that has an impact on and alters different facets of daily life.

Robots can operate more accurately, which decreases the amount of time and materials lost, and they can also work quicker than people. Robotics aims to create devices that can aid and support people. Mechatronics, electronics, bioengineering, computer engineering, control engineering, software engineering, mathematics, and other disciplines are all integrated into robotics. Robotics will boost productivity and economic growth while giving many people throughout the world new work opportunities. However, there are still cautions about significant job losses, such as predictions of 20 million manufacturing job losses by 2030 or the possibility of 30% of all employment being automated by that year (Taylor, 2019).

Technology improvements within and outside of the business can considerably increase the usage of robotics and automation in manufacturing. Scientists, businesspeople, and normal people all have diverse perspectives on the potential impacts of the continued development of AI (Uma & Chitra, 2012) and robots. As robots go from our dreams into our homes, offices, and production facilities, they will become our companions, enabling us to accomplish far more than we could on our own. Additionally, while some see the power and benefits of smart technology as limitless, others are concerned (Uma, 2018). The expanding presence of robotics and automation systems will specifically (re)shape the logics, materialities, practises, processes, and impacts of the urban context by building on the digital turn and mediation of the city through corporate data platforms.

The world is rapidly changing for modern warehouses, e-commerce companies, retailers, hospitals, and other third-party logistics companies considering autonomous mobile robots (AMRs) as a critical technology to control excessive labour costs and improve throughput to reduce fulfilment time. Factory owners and managers want quick, easy-to-deploy AMRs that can be modified as required. In contrast to its predecessor, the automated guided vehicle (AGV), an AMR can grasp commands, dynamically identify and avoid obstacles, and go through a variety of work environments without adhering to a predetermined path or requiring operator direction.

We might expect to see more, more intelligent robots coexisting with us in more settings. Despite the warnings of grim doomsday prophets, these enhanced robots won't replace workers. Some industries experience ups and downs and even become obsolete as a result of evolving technology, opening up new employment and educational opportunities (Terra, 2022).

## 2.0 NEED FOR ROBOTS

Humans no longer need to perform hazardous jobs because robots can work in those conditions. They are capable of handling repetitious jobs, dangerous materials, and heavy lifting. Robots are frequently utilised in fields like manufacturing automobiles to carry out simple, repetitive activities and in fields where work must be done in hazardous conditions for people. Here are some instances where robots outperform people: automate laborious or routine tasks. In a variety of situations, robots can increase output, effectiveness, product homogeneity, and quality. Robots, unlike people, never get bored. They can repeatedly perform the same action and they never get exhausted.

They can achieve accuracy down to fractions of an inch, which is necessary for manufacturing microelectronics, for instance (Mitchell, n.d.). Without a doubt, robots are much more productive than people, especially when it comes to tasks like manufacturing items. Robots not only operate more accurately, which cuts down on time and material waste, but they can also work quicker (and for longer) than people (Han, 2020). Robots might start travelling to strange planets and operate remotely in the year 2030 (Poverly, 2022). Robotics is an area of technology that is rapidly developing and also influences how people will travel, work, and explore in the future.

Internet of Things (IoT), AI, and other related peripheral developments are assisting in further elevating the situation. The world will eventually successfully transition to using entirely renewable energy, thanks to robotics, and the fifth industrial revolution (5IR)—a carbon-free environment—will start. The polar ice caps will freeze, temperatures will drop, and the oceans will retreat (Pollock, 2019).

Robots are general-purpose, programmable machines with the ability to work with humanistic traits and quality assurance in hazardous working conditions

- With repetitive tasks
- With accuracy and consistency
- With challenging human handling tasks
- Flexible and reprogrammable features
- Interfacing with other computer systems
- In multi-shift operations

The following characteristics make industrial robots crucial.

- Routine jobs that robots can complete around-the-clock.
- Robots don't get ill or require vacations.
- Robots can perform duties deemed too risky for people to perform.
- Compared to humans, robots are far more precise while operating machinery.
- May be able to complete activities that are hard for people to complete
- May be more affordable in the long run

Robots are employed to help the disabled, perform dirty tasks, repetitive tasks, dangerous tasks, and tasks which are impossible and beyond human ability. Robots are dependable and diligent workers. They may perform labour that is dangerous, monotonous, or extremely taxing on people. They are capable of working nonstop without grumbling or requiring breaks for food or rest. Robots can also travel to places that people cannot, like as the surface of Mars, deep inside the ocean, or within a nuclear power plant's radioactive components.

## 3.0 HISTORY OF ROBOTICS

Around 3000 B.C., Egyptian water clocks used human figurines to strike the hour bells. Archytus of Taremtum, the creator of the pulley and screw, also constructed a flying wooden pigeon about 400 B.C. Hydraulic sculptures that could speak, gesture, and prophecy were common in Hellenistic Egypt during the second century B.C (Robotics, 2020).

Petronius Arbiter created a doll that could move like a human being in the first century A.D. In 1557, Giovanni Torriani built a wooden robot that could fetch the Emperor's daily bread from the store (History of Robots Timeline I Preceden, n.d.). Robotic inventions peaked (before the twentieth century) in the 1700s, when innumerable brilliant but unworkable automata (i.e. robots) were constructed. The nineteenth century was also rich with new robotic inventions, such as Edison's talking doll and Canadians' steam-powered robot (*Robot Technologies from a to Z at a Glance - Mobilefest*, 2021). Although these historical inventions may have planted the first seeds of inspiration for the modern robot, scientific progress in the field of robotics in the twentieth century outnumbers previous advances by a thousand fold.

The first robots as we know them were created in the early 1950s by Louisville, Kentucky-born inventor George C. Devol (*The Invention of the Industrial Robot | National Inventors Hall of Fame®*, n.d.). He designed and patented "Unimate," a reprogrammable manipulator from "Universal Automation." In the late 1960s, businessman/engineer Joseph Engleberger purchased Devol's robot patent and modified it into an industrial robot, forming Unimation to manufacture and market the robots. Engleberger is renowned in the industry as "the Father of Robotics" for his efforts and results (*About Joseph Engelberger - Father of Robotics*, n.d.).

Academia has also made significant advances in the development of new robots. Charles Rosen headed a research team at Stanford Research Institute in 1958 in constructing a robot called "Shakey," which was significantly more advanced than the original Unimate, which was built for specific, industrial uses (Reynoso, 2021). Shakey could roam around the room, view the scene with his television "eyes," navigate unfamiliar terrain, and, to some extent, respond to his surroundings.

A brief history of the robots (Dinh Thai Bao, 2019) is given in table 1.

*Table 1. History of robots*

| | |
|---|---|
| 1956 | George Devol sought a patent for the first programmable robot, eventually dubbed 'Unimate'. |
| 1957 | Sputnik 1, the first artificial satellite, is launched. |
| 1960`s | Industrial robots have been developed. According to the Robotic Industries Association, a "industrial robot is a re-programmable, multipurpose manipulator designed to move materials, parts, tools, or specialised devices via varied programmed motions to accomplish a variety of tasks". |
| 1961 | The first Unimate robot is deployed in a General Motors plant in Trenton, NJ to tend a die casting process. The ability of the machine to be reprogrammed and retooled to do multiple jobs was critical. The Unimate robot was a revolutionary mechanical design based on a cantilever beam with several degrees of freedom. The beam's flexibility posed control issues. Hydraulic actuation was eventually employed to address precision issues (Norman, n.d.). |
| 1962–1963 | The addition of sensors is viewed as a technique to improve the operation of robots. This incorporates force sensing for stacking blocks (Dinh Thai Bao, 2019), a visual system for binary decision making about the presence of impediments in the surroundings (Dinh Thai Bao, 2019), and pressure sensors for grabbing (Dinh Thai Bao, 2019). At MIT's AI lab, a robot interacts with an unstructured environment (Dinh Thai Bao, 2019). |
| 1966–1968 | SRI (Stanford Research Institute) created 'Shakey,' a mobile robot. 'Shakey' is capable of planning, navigation, and object movement. It was placed in a designated room with things of various colors. A vision system would distinguish and push items in accordance with a plan. |

*continues on following page*

*Table 1. Continued*

| | |
|---|---|
| 1969 | The Apollo 11 mission places man on the moon for the first time. The landing took place within the Lunar Module 'Eagle' (*Apollo 11:Destination Sea of Tranquility - 1st Lunar Landing*, n.d.). |
| 1970 | Luna 17 lands on the moon, bringing Lunokhod 1, a roaming remote-controlled robot (*Lunokhod 1*, 2021). |
| 1971-1973 | The Stanford Arm is created, as well as the first programming language for robots, WAVE (Lozano-Perez, 1983). |
| 1972 | Hirose, Tokyo Institute of Technology, created the first snake-like robot, ACM III (Rezapour, 2015). |
| 1970's | Jet Propulsion Laboratory (JPL) creates its first planetary exploration Rover with the help of a TV camera, a laser<br>range finder, and touch sensors (https://jpl.nasa.gov, n.d.). |
| 1975 | Viking 1 and 2 were launched with an articulated robot arm each. |
| 1976 | The film Star Wars debuts R2-D2 and C-3PO characters. |
| 1977 | Hilaire, a mobile robot, was developed at Toulouse's Laboratoire d'Automatique et d'Analyse des Systemes (LAAS). This three-wheeled mobile robot is still in service today. |
| 1978 | Unimation PUMA (Programmable Universal Machine for Assembly). |
| 1979 | SCARA (Selective Compliant Articulated Robot for Assembly) is a robot that has been introduced in Japan and the United States (by Adept Technologies). |
| 1980's | Feedback control to enhance accuracy, programme compliance, the introduction of personal computers as controllers, and the commercialization of robots by a significant number of companies: KUKA (Germany), IBM 7535, Adept Robot (USA), Hitachi, Seiko (Japan). |
| Early 1980's | Steve Jacobsen built the Utah-MIT arm (16 Degrees of Freedom (DOF)), and Salisbury's hand (9 DOF). |
| 1977-1983 | Hans Moravec's Stanford cart/CMU Rover evolved into the Nomad mobile robot. |
| 1980's | Raibert 1986 and legged and jumping robots (BIPER - Shimoyama). |
| 1984-1991 | V. Braitenberg reintroduced W. Grey Walter's tortoise mobile robots, creating autonomous robots with behaviors. Hogg, Martin, and Resnick of Massachusetts Institute of Technology (MIT)build mobile robots out of LEGO bricks (precursor to LEGO Mindstorms). Rodney Brooks of MIT develops the first insect robots at the MIT AI Lab, heralding the start of behavioral robotics. |
| 1986 | Honda begins development of its first humanoid robot, dubbed 'E0' (later to become Advanced Step in Innovative Mobility (ASIMO)). |
| 1988 | **Stanford Climbing and Aerial Maneuvering Platform** (SCAMP) was created as the first emotional robot pet. |
| 1989 | BEAM (from **biology, electronics, aesthetics and mechanics)** robots is introduced by Mark Tilden, with beam standing for Biology, Electronics, Aesthetics, and Mechanics. |
| 1990 | Modifiable assembly robots, Autonomous mobile robots, Robots with vision control, Robots that walk are introduced. |
| 1991 | The HelpMate mobile autonomous robot was the first to be utilized in a hospital. |
| 1990's | Cog, Kismet (developed at Massachusetts Institute of Technology (MIT)), Wasubot, WHL-I - Japan, Honda P2 (1.82m, 210kg) and P3 (1.6m, 130kg), ASIMO are examples of humanoid robots developed during this period. |
| | SARCOS ("Jurassic Park"), Sony AIBO, LEGO Mindstorms, Khepera, and Parallax are examples of entertainment and education robots. |
| 1997 | ROBOCUP, a competition that simulates the game of soccer between two teams of robots, has been organized all over the world since 1997. |
| 1997 | As part of the Mars Pathfinder mission, Sojourner becomes the first rover to land on Mars. |
| 1998 | Lego Mindstorms, the company's first robotics product, enters the market. |
| 1999 | Sony debuts **Artificial Intelligence Robot** (AIBO), an autonomous robotic dog that can see, walk, and interact with its surroundings. A year later, the SDR-3X humanoid robot, subsequently known as QRIO, was released (both discontinued in 2006). |

*continues on following page*

*Table 1. Continued*

| | |
|---|---|
| 1990's | Introduction of space robots (both manipulators and rovers - the MARS rover in 1996), parallel manipulators (Stewart-Gough Platforms), multiple manipulators, precision robots ("Robotworld"), surgical robots ("RoboDoc"), and first service robots (as couriers in hospitals, etc) |
| 2000 | **Advanced Step in Innovative Mobility** (ASIMO), Honda's first non-prototype humanoid robot, is unveiled. |
| 2001 | The United States Air Force tested the MQ-1 Predator, the first armed unmanned aerial vehicle (UAV) outfitted with two Hellfire missiles. |
| 2000's. | IRobot unveils "Roomba," the world's first self-driving vacuum. |
| | Pister @ Berkeley, University of Texas at Arlington (UTA), **Ecole polytechnique federale de Lausanne** (EPFL) a research institute and university in Lausanne, Switzerland (EPFL/Lausanne), microfactories, mini and micro robots, "Smart Dust" |
| | Military applications include robotic helpers for risky settings and reconnaissance, as well as AUVs and Unmanned Underwater Vehicles (UUVs). |
| | The Da Vinci surgical robot is introduced by Intuitive Surgical. |
| | Sensor Network Robotic Deployment |
| 2002 | Roomba, a personal robotic vacuum cleaner, is introduced by iRobot. |
| 2003 | Osaka University has unveiled its first 'Actroid,' a humanoid robot with strong visual human features. |
| 2004 | The inaugural DARPA Grand Challenge is to assist. The challenge, sponsored by the US Department of Defense, aims to develop autonomous vehicles for battle. |
| | Spirit and Opportunity, two Mars rovers, touch down on the planet. The rover Spirit had completed 2150 days of its 92-day (90-sol) mission as of November 25th, 2009. |
| 2010 | NASA and General Motors are collaborating to create Robonaut-2, a new version of NASA's humanoid robot astronaut. |

## 4.0 TYPES OF ROBOTS

While there are a wide range of robotics applications in medicine, space, industries, manufacturing, agriculture, stocking shelves, welding metal in hazardous locations, and much more, robots today's may typically be divided into a number of categories. Robotics solutions are being used in an increasing variety of industries and applications as robotics manufacturers continue to bring breakthroughs in terms of capabilities, pricing, and form factor. Thanks to improvements in processing speed and AI capabilities, we can now employ robots in a variety of ways to carry out essential tasks. In general, the different classifications for robots are as follows.

i. Fixed or mobile.
ii. Serial or parallel.
iii. According to degree of freedom (DOF).
iv. Rigid or flexible.
v. Control – point-to-point, autonomy and "intelligence"

The classification of robots based on their application, or by their locomotion / kinematics is as given below.

- Mobile Robots
- Industrial Robots
- Autonomous Robots
- Autonomous Mobile Robots (AMRs)
- Remote Controlled Robots
- Pre-programmed Robots
- Tele operated Robots
- Augmenting Robots
- Humanoid robots
- Cobots
- Hybrids
- Automated Guided Vehicles (AGVs)
- Articulated Robots

## 4.1 Mobile Robots

A mobile robot is a machine that can move around in its environment under the control of software and uses sensors and other technology to recognise its surroundings. It is an automated machine that can move across an uncontrolled environment without the need of mechanical or electrical guiding aids (*Types of Robot - Javatpoint*, n.d.). Mobile robots can be categorised in two different ways (*Mobile Robots*, 2020):

i. Working Environment Based
ii. Based on Device Used to Work

### 4.1.1 Working Environment Based

Mobile Robots can be divided into the following categories based on the environments in which they operate.

- Aerial Robots

Unmanned aerial vehicles (UAVs) are another name for aerial robots. These machines have the ability to fly. E.g., helicopters, drones, etc.

- Land or Home Robots

Unmanned Ground Robots (UGVs) are another name for Land Robots. These robots are capable of navigating on dry land and within homes, offices, and other structures. These are employed to provide equipment. E.g., dirtdog

- Underwater Robots

Autonomous underwater vehicles (AUVs) are another name for underwater robots. These machines can steer themselves and move swiftly across water. E.g., submarines

### 4.1.2 Based on Device Used to Work

On the basis of the tools they employ for their jobs, mobile robots can be categorised as follows.

- Wheeled Robots

Robots on wheels are frequently referred to as autonomous intelligent vehicles (AIVs). These robots are the ones that move by rolling on wheels. Wheeled robots come in a variety of varieties, including one-wheeled and two-wheeled robots.

- Humanoid Robots

Robots with a body shape that resembles a human being are said to be humanoid. They rely heavily on sensors to operate. E.g., androids, etc.

- Legged Robots

Robots that have legs or other articulated limbs that move in accordance with a mechanism are said to have legs. E.g., haxapod robot.

## 4.2 Industrial Robots

Industrial robots (Figure 1) continuously complete the same jobs without moving. These robots are employed in fields where repetitive, boring operations are necessary and are well suited for robots (*Types of Robot - Javatpoint*, n.d.). Industrial robots were the earliest robots, and they took the place of humans in simple repetitive activities. Without human intervention, factory assembly lines can run in a controlled environment where the robot must carry out duties in a specific order and react to things that are precisely positioned in front of it. These could be considered automata rather than robots, according to some. But modern automata frequently rely on sensors to the point that they can be regarded as robots.

However, because they operate in a special environment that people are not permitted to enter while the robot is at work, their design is more straightforward. Today's robots, however, require more flexibility, such as the capacity to manipulate objects in various orientations or to identify various objects that must be packaged in the proper order. The robot might be needed to move items between warehouses. Although this increases autonomy, the fundamental characteristic—that the environment is largely limited and may be tailored to the robot—remains.

When industrial robots interact with people, more flexibility is needed, and this imposes strict safety standards on both mobile robots and robotic arms. In particular, the robot's pace needs to be slowed down, and the mechanical design needs to make sure that there are no moving elements that could endanger the user. The benefit of humans collaborating with robots is that each may focus on what they do best: humans do more sophisticated processes and define the overall goals of the robot since they are fast to see mistakes and possibilities for improvement, while robots handle repetitive or dangerous duties (Ben-Ari & Mondada, 2017).

*Figure 1. Industrial robots*

## 4.3 Autonomous Robots

Robots with autonomy are self-sufficient. They employ a programme that gives them the option to choose the course of action based on their environment. With the aid of artificial intelligence, these robots frequently pick up new skills. They begin with a brief routine and modify it to improve their performance of a task. As a result, the most effective routine will be used again (*Types of Robot - Javatpoint*, n.d.). In contexts that are less strictly regulated than those in which pre-programmed robots work, autonomous robots function without the assistance of human operators (Javaid et al., 2021). They are said to have "autonomy" because it is ultimately the machine's duty to detect environmental changes and adjust to them. When a robot is functioning properly, humans do not need to constantly watch over and change the environment to account for the robot's limits. In addition to being employed in industry and as labor-saving appliances in homes, autonomous robots are also gaining popularity as toys(Gottlieb & Anderson, n.d.).

Robots with autonomy work without human supervision. These robots are typically made to complete duties in public spaces without human supervision. They are highly distinctive in that they employ decision-making systems (often a computer) to choose the best course of action based on their data and mission and then use sensors to perceive the world around them. The Roomba vacuum cleaner, which uses sensors to move freely throughout a home, is one example of an autonomous robot (Builtin, 2019). A lot of mobile robots can be operated remotely and are used to carry out operations like aerial photography, bomb disposal, and pipe inspection. These robots are not autonomous; instead, they allow their operator remote access to hazardous, remote, or inaccessible locations by using their sensors.

Some of them may be semi-autonomous, carrying out supporting functions on their own. While a person chooses the flight path for a drone, the autopilot stabilises the aircraft. While a person inspects the pipe for problems that need to be fixed, a robot inside of it can manoeuvre. Fully autonomous mobile robots can accomplish activities such as carrying stuff while navigating through unpredictable terrain (such as walls and doors inside buildings and crossroads on streets) and in a constantly changing environment without the need for an operator (people walking around, cars moving on the streets).

The first mobile robots were created for straightforward surroundings, such as robotic lawn mowers or swimming pool cleaners. Because it has been demonstrated that it is possible to construct fairly cost robots that can navigate an indoor environment crowded with obstacles, robotic vacuum cleaners are currently commercially accessible. Many autonomous mobile robots are built to assist workers in regulated environments, including warehouses. An intriguing illustration is a robot that weed-eats fields. Although this environment is largely structured, in order to locate and get rid of weeds, enhanced sensing is needed.

Robots interact with humans in their surroundings, even in highly organised factories, therefore their sensory systems must be exceedingly accurate. The self-driving automobile is arguably the autonomous mobile robot receiving the most attention right now. Due to the very complex, uncertain environment of motorised traffic and the stringent safety requirements, these are incredibly challenging to build.

Space is an even more challenging and hazardous environment. Mobile robots with a degree of autonomy include the Mars rovers Curiosity and Sojourner. The Sojourner was operational in 1997 for three months (Desai, 2022). Since the Curiosity's 2012 Mars landing, it has been in operation! The missions (the driving routes and the scientific investigations to be carried out) are managed by humans on Earth, but the rovers are capable of autonomous hazard avoidance.

Today, a large portion of robotics research and development is devoted to enhancing sensors and enabling more intelligent control of the robot in order to make it more autonomous. Better sensors can detect the specifics of increasingly complicated circumstances, but handling these scenarios requires very flexible and adaptable robot behaviour control. Due to the low cost of cameras and the wealth of information they can capture, research into vision is particularly active. Systems are being worked on to increase their adaptability so they can learn from people or change to new circumstances. The interaction between people and robots is a topic of active research. This calls for both intellect and sensing, but it also needs to consider the psychology and sociology of the relationships(Ben-Ari & Mondada, 2017).

## 4.4 Autonomous Mobile Robots (AMRs)

AMRs travel the globe while making judgments almost immediately. They are able to absorb information about their surroundings with the aid of technologies like sensors and cameras. Whether they are moving to escape an approaching employee, picking the ideal parcel, or choosing the best surface to disinfect, on-board processing equipment enables them to assess it and make an intelligent decision. They are mobile solutions that work with little assistance from people (*Types of Robots and Industry Applications of Robotics Technology* -..., n.d.).

## 4.5 Remote Controlled Robots

Robots that are controlled remotely are utilised to complete difficult and ambiguous jobs that autonomous robots are unable to complete owing to operational uncertainty. The greatest people for complex tasks are those with true brainpower. As a result, a remote control can be used to control a robot. Without physically being present at the work location, a human can undertake dangerous tasks through remote control operation (*Types of Robot - Javatpoint*, n.d.).

## 4.6 Pre-Programmed Robots

Pre-programmed robots work in a straightforward, regulated environment, so they don't need a lot of intelligent control systems (also known as "artificial intelligence") to function well. Pre-programmed robots carry out routine, easy duties in a controlled setting. A mechanical arm in an auto assembly line would be an illustration of a pre-programmed robot. Its task is to complete that task longer, quicker, and more effectively than a person could (Builtin, 2019). The pre-programmed robots that are utilised to construct cars in the majority of auto factories throughout the world are probably the most well-known. Similar robots exist that work on the human body rather than automobiles. Learn more about industrial and medical applications of pre-programmed robots (Gottlieb & Anderson, n.d.).

## 4.7 Tele Operated Robots

Robots that can be operated remotely by a person are teleoperated. Teleoperated underwater robots were used to assist stop the Gulf of Mexico oil leak. Another illustration is the space shuttle's robotic arm. There are numerous examples of this, and every year, more and different kinds of robots are created. The application of these robots in the realm of medicine will receive special attention. Even if your doctor is travelling to China, she can now be tele-present in your hometown (Gottlieb & Anderson, n.d.). Semi-autonomous robots that can be controlled by humans from a safe distance are known as teleoperated robots. Typically, these robots operate in harsh environmental factors including weather and terrain. Teleoperated robots include, for instance, the human-controlled submarines that were used to repair undersea pipe breaks during the BP oil spill or drones that were used to find landmines in a war zone (Builtin, 2019).

## 4.8 Augmenting Robots

Some robots are directly attached to the human user's body, either by having it in contact with the user's body in some other way or by the user holding the controls in their hand. The user's physical movements—or, in some cases, just their mental activity—control some sort of robotic effector in the end. By extending the user's reach, for example, these robots can improve the user's inherent competence. Alternatively, they can teach the user a skill they lack or have lost (e.g., a robotic prosthetic arm). The lifting capacity of giants and the finesse of a master surgeon can be provided to people by these robots (Gottlieb & Anderson, n.d.).

Virtual reality (VR) robots, often known as augmentation robots, can supplement or replace lost human talents. Science fiction could soon become reality in the realm of robotics for human enhancement, where robots with the power to make people faster and stronger could completely rewrite what it means to be human. Robotic exoskeletons or prosthetic limbs are two examples of modern augmenting robots that can lift heavy objects (Builtin, 2019).

## 4.9 Humanoid Robots

Robots that resemble humans or act like them are what they are. The majority of the time, these robots mimic human actions (such as running, jumping, and carrying goods), and occasionally they are created to resemble us, even having human features and attitudes. Sophia from Hanson Robotics and Atlas from Boston Dynamics are two of the most well-known humanoid robots. Robots are frequently portrayed as humans in science fiction and the media. Even though the robotic figures R2-D2 and 3-CPO from the Star Wars films are well known, the idea has a long history (Ben-Ari & Mondada, 2017). Four Swiss watchmakers—Pierre, Henri-Louis, and Jean-Frédéric Jaquet-Droz—built humanoid automata in the eighteenth century to showcase their mechanical prowess and promote their watches (*Pierre Jaquet-Droz - Complete Biography, History and Inventions*, 2021). For similar purposes, a lot of businesses today produce humanoid robots.

Humanoid robots are a type of mobile, autonomous robot with incredibly intricate mechanical designs for the arms and legs. Researchers study the mechanics of walking and how humans and machines interact using humanoid robots. In a home or on a space station, humanoid robots have been proposed to do services and maintenance. They could be used to care for elderly people who would become uneasy around a machine that doesn't look human.

The uncanny valley is a term for the phenomena wherein robots that remarkably resemble humans can repel people. It can be exceedingly challenging to control and design humanoid robots. They are expensive to construct and have numerous joints that may move in a variety of ways. For the majority of applications, robots with wheels or tracks are favoured because they are more straightforward, affordable, and durable.

While many mobile humanoid robots could potentially be considered AMRs, the word is typically used to refer to robots that perform tasks that are human-centric and frequently have human-like forms. When performing duties like giving directions or providing concierge services, they employ many of the same technological elements as AMRs in order to sense, plan, and act(*Types of Robots and Industry Applications of Robotics Technology* -..., n.d.).

## 4.10 Cobots

Cobots are machines that work with or directly for humans. Cobots can share workspaces with workers to enable them to work more effectively, in contrast to the majority of other types of robots, which carry out their tasks alone or in completely isolated work locations. They are frequently utilised to remove laborious, risky, or taxing operations from regular workflows. Cobots occasionally have the ability to react to and pick up on human movement (*Types of Robots and Industry Applications of Robotics Technology* -..., n.d.).

## 4.11 Hybrids

Robots of different kinds are frequently joined to produce hybrid systems that can do more difficult jobs. For instance, a robot for handling packages inside of a warehouse might be made using an AMR and a robotic arm. Consolidation of compute capabilities goes hand in hand with the consolidation of functionality into single solutions (*Types of Robots and Industry Applications of Robotics Technology* -..., n.d.).

## 4.12 Automated Guided Vehicles (AGVs)

While AMRs can move around freely, AGVs frequently need operator supervision and rely on rails or pre-established pathways. In regulated locations like warehouses and factory floors, these are frequently utilised to distribute commodities and move objects (*Types of Robots and Industry Applications of Robotics Technology* -..., n.d.).

## 4.13 Articulated Robots

Robotic arms and articulated robots are designed to mimic the actions of a human arm. These typically have between two and ten rotary joints. These are perfect for arc welding, material handling, machine tending, and packaging since each extra joint or axis provides for a wider range of motion (*Types of Robots and Industry Applications of Robotics Technology* -..., n.d.).

## 5.0 APPLICATIONS OF ROBOTS

Robots are employed in many various ways around the world today, and as technology develops, they are expected to become more and more prevalent in our daily lives. Many employment in sectors like manufacturing, agriculture, entertainment, etc., demand tedious, repetitive work that also demands a great deal of precision. Robots are more appropriate in these circumstances than people because they are precise, clever, and don't grow tired easily! There are other tasks that are extremely hazardous and harmful for humans, such as space and undersea exploration. Robots are ideal in this situation as well because they never get weary and cannot be killed. These benefits have led to numerous robot applications in practically every industry imaginable.

Human beings have been interested in robotics for more than a century. On the other side, the media and the global film industry have an impact on how we view robots (Hollywood). Robots currently carry out a wide range of duties in a wide range of industries, and this number is steadily growing. An application-based division of robots into types is the most effective method.

Robots are used for a number of applications listed below

(i) Machine loading and unloading
(ii) Inspection
(iii) Processing operations like,
  a. Spot welding
  b. Spray coating
  c. Laser drilling and cutting etc.
  d. Drilling, routing, machining operations
  e. Continuous arc welding
  f. Grinding, polishing debarring wire brushing
(iv) Assembly tasks, assembly cell designs, parts mating
(v) Automation.
(vi) Material transfer applications

The profile of the future robot based on the research activities will include the following,

i. Systems and integration and networking
ii. Robots in space
iii. Security guards
iv. Medical care and hospital duties
v. Garbage collection and waste disposal operations
vi. Universal gripper

vii. Household robots
viii. Intelligence
ix. Sensor capabilities
x. Telepresence
xi. Mechanical design
xii. Mobility and navigation (walking machines)
xiii. FMS (Flexible Manufacturing Systems)
xiv. Hazardous and inaccessible non-manufacturing environments
xv. Underground coal mining
xvi. Firefighting operations etc.

A brief discussion of some of these applications are given below.

## 5.1 Industrial Robots

These robots are useful in an environment where production is industrialised (Kautish et. al 2022). These are typically articulated arms designed specifically for tasks like material handling, painting, welding, and others. If we judge solely by the application, other robots and some automatically guided cars can also fall under this category of robot (www.facebook.com/elprocus, 2016). Although there are many different industrial robot types in use today, they are typically programmable, articulated arms that are stationary but may move on several axes. They are employed in a variety of processes throughout the production of goods, including welding, painting, and assembling. Industrial robots are incredibly swift and accurate and provide firms with significant cost-saving advantages. The International Federation of Robotics (IFR) predicts that by 2019, there will be about 2.6 million industrial robots in use. Additionally, construction and other duties are performed by industrial robots in space (Types of Robots and Their Applications - Guardforce, 2019).

## 5.2 Domestic Robots

Robots used in the home, often known as domestic or household robots. Many different pieces of equipment fall under this category of robot, including robotic pool cleaners, robotic sweepers, robotic vacuum cleaners, robotic sewer cleaners, and other robots that can carry out various domestic chores. Additionally, a number of surveillance and telepresence robots can be regarded as domestic robots if used in that kind of setting (www.facebook.com/elprocus, 2016). It's likely that you have noticed several home robots on sale in shops. Robot vacuum cleaners, pool cleaning robots, and gutter cleaning robots are some of the robots used to complete household duties. Additionally, Guardforce Hong Kong offers cleaning robots appropriate for larger commercial facilities (Types of Robots and Their Applications - Guardforce, 2019). Lawn mowing, vacuuming, or window cleaning are just a few of the routine household duties that robots can handle. The time and effort they save astounds people: In a survey conducted by the German Federal Association for Information Technology, Bitkom, 42% of the more than 1,000 participants said they could see themselves living with a robot(30). More than 80% of people would prefer help vacuuming or washing floors, and 41% want a robot to work in the garden (AG, n.d.).

## 5.3 Medical Robots

Robots used in hospitals and other medical facilities. surgical therapy robots are first and foremost. Several robotically controlled vehicles as well as potential lifting assistance. A surgery robot is the most typical sort of medical robot you will encounter (www.facebook.com/elprocus, 2016). In robot-assisted surgery, the surgeon directs a robot to carry out extremely delicate medical procedures using a telemanipulator or computer, therefore lowering the risks. Future surgical procedures may potentially be carried out remotely from anywhere in the world by a surgeon using robot assistance (Types of Robots and Their Applications - Guardforce, 2019).

## 5.4 Service Robots

Service robots are robots that, in practise, fall under no other category. These could include various robots for data collection, robots ready to demonstrate innovations, robots used for research, etc. Robots employed for a variety of service-related tasks fall under the wide category of service robots (www.facebook.com/elprocus, 2016). Concierge robots and hospitality robots are the two most prevalent forms of service robots. For instance, Guardforce Hong Kong has introduced a concierge robot that is ideal for commercial buildings, events, and exhibitions since it automates visitor registration, manages entry points by facial recognition, and provides multi-media information. Additionally, Guardforce sells a mall hospitality robot that lets patrons download coupons and serves as a shopping assistant as well as a handshake and chat feature. It creates the ideal one-stop customer service solution.

## 5.5 Military Robots

Robots are now used by the armed forces and military. These robots include drones that conduct exploration and robots that dispose of bombs. Robots that are initially developed for military and armed services applications are frequently used in fields like law enforcement, exploration, and salvage, among others. Robots are perfect for use in the military or in emergency situations where it could be too risky for humans to respond. In addition to drones and bomb disposal robots, emergency response teams also deploy aerial, ground, and underwater search and rescue robots. These robots can withstand severe environments and navigate treacherous terrain, making them appropriate for exploring the ocean below or even other planets, like the Mars Curiosity Rover (Types of Robots and Their Applications - Guardforce, 2019). There are numerous uses for robots in the military as well. They can be utilised as medics to assist friendly forces, as armed systems to attack the opposing forces, or as drones to maintain watch on the enemy.

The MAARS (Modular Advanced Armed Robotic System), which resembles a tank and is equipped with tear gas, lasers to confuse adversaries, and even a grenade launcher for emergency scenarios, is one of the most well-known robots employed in the military. DOGO is a tactical fighting robot that also comes equipped with a camera for spying on the enemy's actions and a 9-millimeter handgun for emergencies! (28)

## 5.6 Robots for Entertainment

These robots are used for entertainment. The scope of this category is enormous. Model robots like Robosapien and running photo frames are used in the opening, while genuine heavy hitters like articulated robot arms are used as movement simulators in the closing (www.facebook.com/elprocus, 2016).

Another popular topic in the entertainment sector is robots. They can be utilised behind the scenes in movies and television shows to manage the camera, offer special effects, etc., but they can't technically become actors and actresses. They can be employed for dull, routine jobs that are unsuitable for humans because, after all, the film industry is a creative one. Robots can also perform stunts that are extremely unsafe for people to perform yet look really spectacular in an action movie. The wonderful experiences of visitors are being improved by the use of autonomous robots in theme parks like Disney World (*Top 10 Applications of Robotics in 2020*, 2020).

## 5.7 Space Robots

As a separate category, I would like to separate off robots used in space. The robots used on the International Space Station, the Canadarm used in space shuttles, the Mars explorers, and other robots used in space exploration and other operations would all fall under this category (www.facebook.com/elprocus, 2016). There are a lot of activities in space that astronauts should avoid at all costs. Humans are unable to spend all day on Mars collecting soil samples or working on spacecraft repairs from the outside while they are in the void of space. Robots are a fantastic option in these circumstances because there is no risk of human life being lost.

Therefore, NASA and other space organisations commonly deploy robots and autonomous vehicles to perform tasks that humans cannot. For instance, the Mars Rover is an autonomous robot that travels on Mars and captures images of fascinating or significant Martian rock formations before sending them back to Earth for NASA scientists to investigate (Types of Robots and Their Applications - Guardforce, 2019).

Robots made by students are the eighth category, along with hobby and competition robots. Robots designed solely for study, entertainment, and competition include sumo-bots, Line followers, and robots (www.facebook.com/elprocus, 2016).

## 5.8 Security Robots

Robotics is becoming more and more common, and the contact between humans and machines that results from this interaction presents both benefits and concerns for the safety and security of people and data. The need for safety is particularly clear-cut when it comes to the deployment of industrial and collaborative robots at work.

Security robots have several benefits since they can work around the clock, in any weather, day or night. Security robots, like the Security Patrol Robot from Guardforce Hong Kong, can monitor industrial parks, power plants, malls, campuses, residential developments, and chemical storage facilities because they patrol on predetermined routes, provide remote CCTV surveillance with facial recognition, provide information, and even detect abnormal temperatures and gas leaks among other things. They can be set up in a variety of ways to suit the requirements of the location and the application, and they allow for significant human cost savings (Types of Robots and Their Applications - Guardforce, 2019). What if every security officer was a robot? Even criminals would be terrified!

Because they can safeguard people and wouldn't put themselves in risk like human security guards would, robots are being considered as security agents. Robotics businesses are presently experimenting with teaming human security advisors with robot guards. American company Knightscope is quite well known in this industry and has autonomous security robots that can provide real-time, useful intelligence to human security officers. Robots like these can assist in solving crimes like armed robberies, burglaries, domestic abuse, fraud, hit-and-runs, etc (*Top 10 Applications of Robotics in 2020*, 2020).

49 percent of the participants in the sample Bitkom study were willing to consider giving a robot control over their home's security, despite the fact that issues with data protection and security would need to be taken into account (30). While the homeowners are on vacation, on a business trip, or at work, a security robot takes care of the home. These robots can be managed from an app connected to the Internet. The robot notifies a smartphone if it detects any impulses using motion detection. The built-in camera offers a two-way voice intercom feature and records in HD.

## 5.9 Agriculture Robots

Another industry where robots have a lot of potential is agriculture (AG). Currently, pilot studies utilising multispectral cameras and robot arms mounted to harvesters are being conducted to improve the cucumber harvesting process. Small planting robots that are operated from a tablet are used for seeding. These robots not only sow the seeds, but they also record all the crucial information. Drones are useful for keeping an eye on weed growth and the ripeness of vegetable items. If necessary, they may also be used to spray vulnerable regions.

Agriculture Human civilisation is based on the agricultural industry. Agriculture, however, is a seasonal industry that depends on favourable weather, good soil, etc. Additionally, there are several repetitive chores in agriculture that waste farmers' time and are better handled by robots. These include planting seeds, eliminating weeds, harvesting, etc. Crop harvesting is typically done by robots, which makes farmers more productive. The Ecorobotix is an illustration of a robot that is employed to clear weeds from farmland. It has a sophisticated camera system that can be used to target and spray weeds while being fueled by solar energy (*Top 10 Applications of Robotics in 2020*, 2020).

## 5.10 Healthcare

The healthcare industry has undergone significant upheaval thanks to robots. And everything has improved! They can assist surgeons in executing procedures with more precision, serve as prosthetic limbs, give patients rehabilitation, etc. There are countless options. There are countless options. The da Vinci robot is one example of this, which may assist surgeons in carrying out difficult procedures on the heart, head, neck, and other delicate places. Exoskeletons are another type of robotic device that has been developed and can be used to help persons recovering from spinal injuries, strokes, etc. with additional support (*Top 10 Applications of Robotics in 2020*, 2020).

## 5.11 Underwater Exploration

Robots are a fantastic alternative for investigating locations that are difficult for people to access, like the ocean's depths! Deep beneath the ocean, there is intense water pressure, which prevents humans from going there and limits the depth to which robots like submarines may travel. Now that specialised robots have been created, it is possible to finally explore the hidden depths of the ocean. These remote-controlled robots can explore the ocean's depths and gather information and pictures of the aquatic plant and animal life (*Top 10 Applications of Robotics in 2020*, 2020).

## 5.12 Food Preparation

Want to avoid cooking? Don't worry, there are robots that can even prepare full meals for you to eat. These robot chefs are capable of cooking utilising a vast array of recipes. All that is required of humans is for them to select the recipe they desire and provide the robot pre-packaged containers containing each component. The food can subsequently be cooked by the robot by itself. One such robotics business, Moley Robotics, has developed a robotic kitchen featuring a robot that can prepare meals just like a professional chef! So don't be concerned if you can't cook. since a robot can now! (*Top 10 Applications of Robotics in 2020*, 2020).

## 5.13 Manufacturing

Welding, assembling, packing, and other routine jobs in the manufacturing sector don't call for the use of the mind at all. Robots can efficiently complete these chores, leaving humans to handle mentally demanding and creative duties. Under the direction and supervision of a human, these robots (Figure 2) can be taught to carry out these boring and repetitive duties with accuracy. This choice is also the ideal for hazardous production procedures that might hurt people(*Top 10 Applications of Robotics in 2020*, 2020).

## 5.14 Client Services

There are robots that have been designed to closely resemble humans for aesthetic reasons. To promote robotics, these robots are primarily employed in high-profile customer service roles. Nadine, a humanoid robot in Singapore, is one such example. She can continue a conversation based on prior interactions, shake hands, make eye contact, and recognise people from previous visits. Junko Chihira, a humanoid robot employed by the tourist information desk at Aqua City Odaiba, a shopping mall on Tokyo's waterfront, is another such customer service robot. She is from Japan (*Top 10 Applications of Robotics in 2020*, 2020).

# 6.0 ROBOTICS IN HEALTHCARE

Robotics has made significant advances in providing healthcare services in a variety of medical fields (Figure 3). Robots are now replacing human assistants as they learn to be more friendly and trustworthy. Robots have enhanced productivity and resource efficiency in the industrial and retail sectors, and there is growing interest in achieving a similar change in healthcare (Cresswell, Cunningham-Burley &

*Figure 2. Axis welding robots*

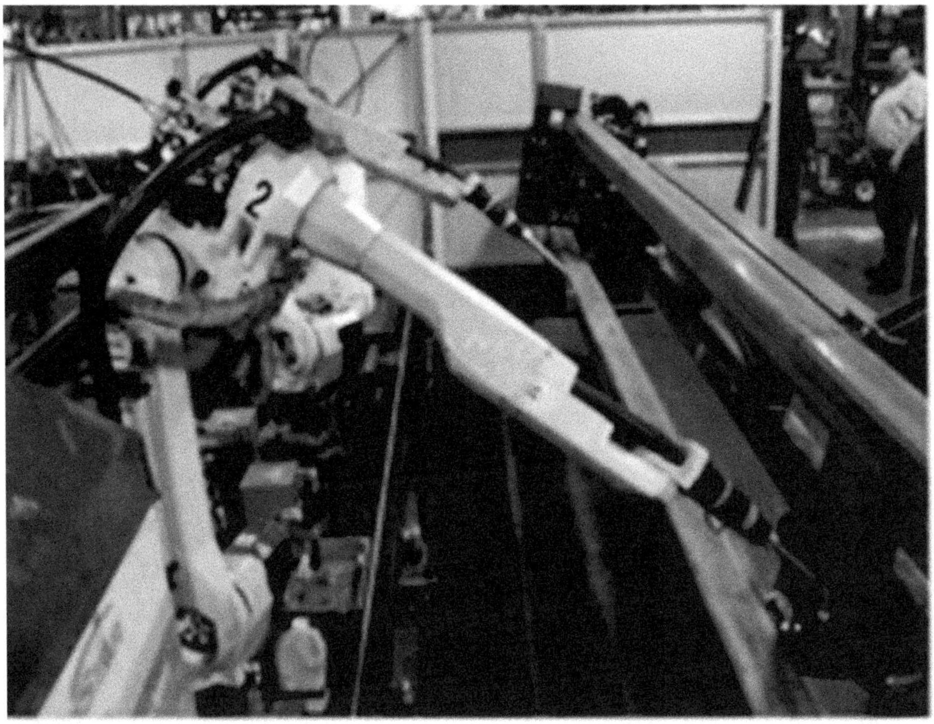

Sheikh, 2018). Given their success in the industrial sector, robotics and artificial intelligence (AI) are two of the most promising new technologies that are predicted to improve care quality and safety while reducing costs (Riek, 2017; Cresswell, Cunningham-Burley & Sheikh, 2018).

Healthcare robots (Figure 5) are expected to be deployed at an unprecedented rate in this regard (Simshaw et al., 2016) as a result of their lower cost and increased roles and capacities (COMEST, 2017) to perform medical interventions (Nouaille et al., 2017), support impaired patients (Tucker et al., 2015), provide therapy to children (Scassellati, Admoni & Matari, 2012), or keep the elderly company (Broekens, Heerink & Rosendal, 2009).

Products, technology, and applications make up the three segments of the medical robotics market. The market is divided into a number of product categories, including non-invasive radiosurgery robots, hospital and pharmacy robots, rehabilitation robots, and others (Partners, 2022). Applications such as laparoscopy, neurology, orthopaedics, gynaecology, urology, cardiology, and others make up the market (Partners, 2022). Depending on the end user, the market is segmented into hospitals, ambulatory surgery centres, and other entities (Partners, 2022).

*Figure 3. Robots in healthcare*

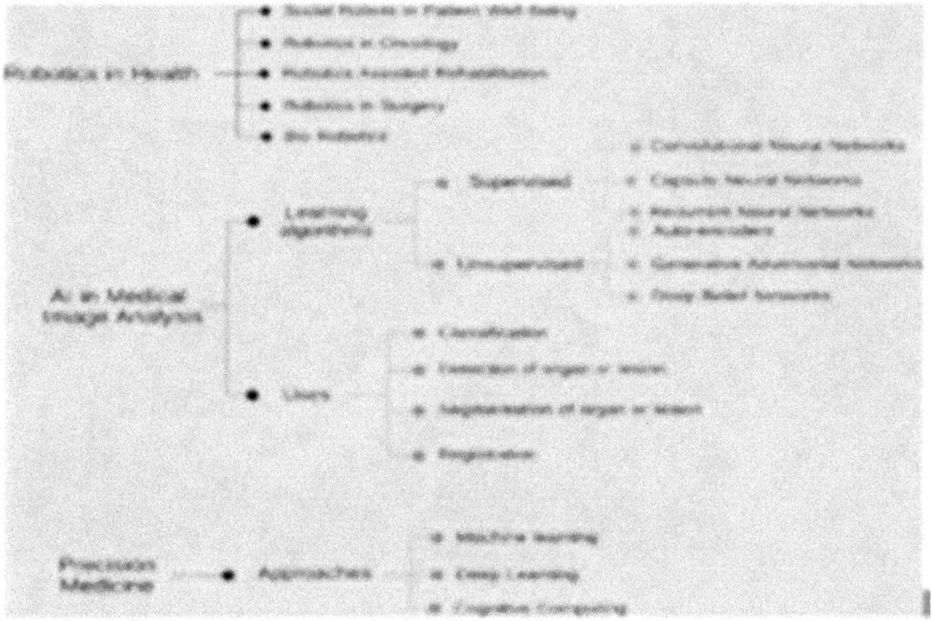

Robots are utilised to assist people maintain their independence for longer, as has been the case in hospitals and labs for more than 30 years. Aging and many diseases are associated. Robots are utilised in physical therapy, surgery, rehabilitation, and in the care of people with chronic diseases (Shah & Chircu, 2018). End-of-life care could be greatly improved by robots (Vitanza et al., 2019). Robots are now more socialised thanks to AI. Advanced robotic technology helps with minimally invasive surgical operations. Robotic surgery may seem like the stuff of the future, but the technology has been around for a while. Technology has advanced because curious surgeons have been investigating the potential uses and advantages of robotic aid (Vandemeulebroucke et al., 2018).

The movements of a surgeon can be stabilised and optimised by a surgical robot. Robots, for instance, can guarantee more precision and reduce hand tremors. This is important because procedures will get more complicated in the future. Oncological procedures, for instance, will concentrate on removing much smaller tumours, which demand more precision with better control and navigation.

Similar to this, interventional techniques like percutaneous coronary interventions (PCIs) demand greater precision and accuracy since they include negotiating intricate, highly twisted blood arteries, crossing lesions, and correctly sizing and installing stents on the first try (Mangusli, 2022). In a catheterization lab, robot-assisted devices can shield interventional radiologists from needless radiation.

*Figure 4. Laparoscopic surgery robot*

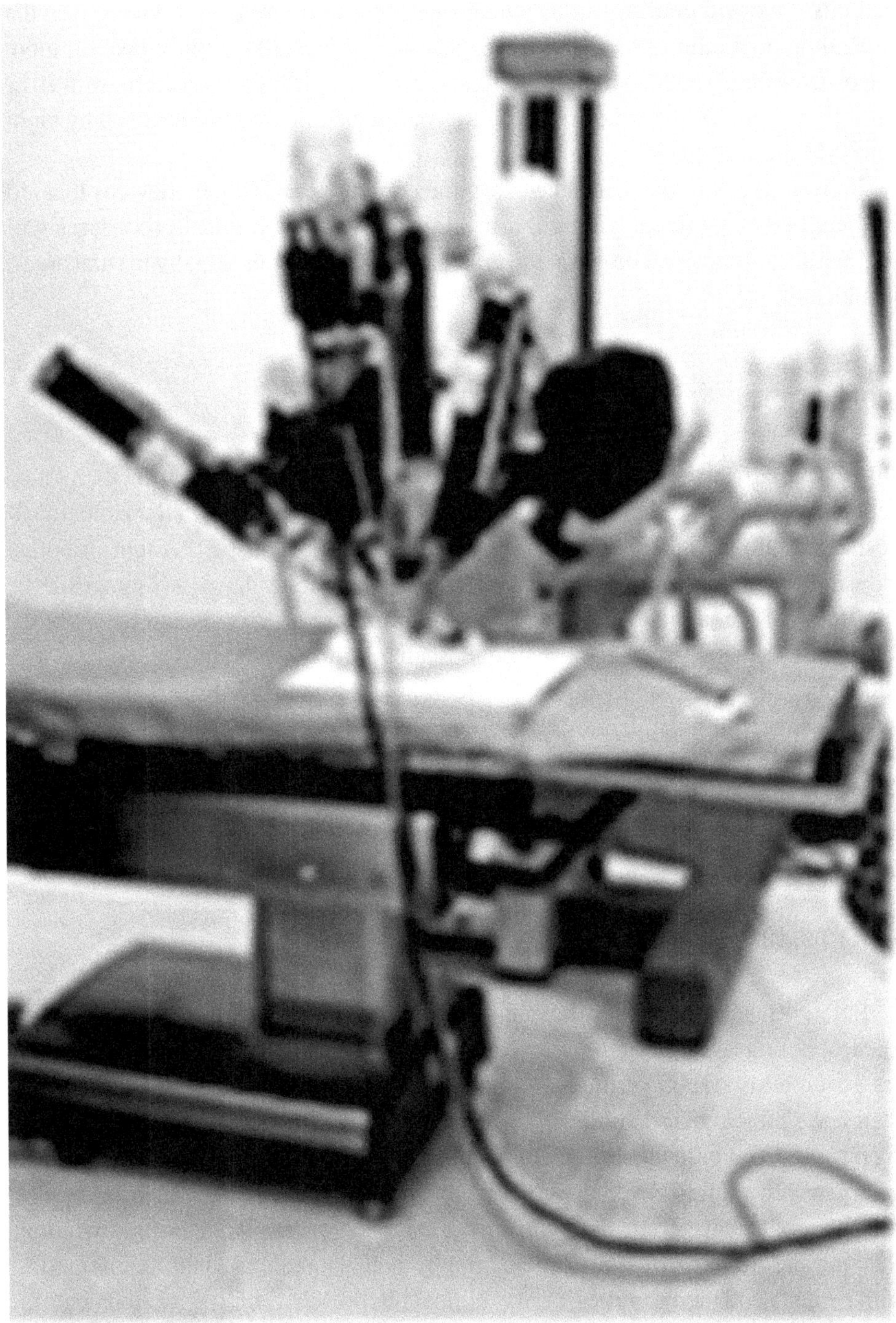

The healthcare sector is suffering from a skills gap, and current trends suggest that this issue will continue to be a major problem in the years to come. Given the expanding population and rising life expectancy, this challenge will take on more importance. And robots will help find a solution! According to research, from 2012 to 2018, the number of robotic procedures performed in the US increased by eight times (Mangusli, 2022).

Patient care is frequently delayed. In the coming years, robotic devices that aid medical personnel in carrying out crucial diagnostic and surgical procedures will proliferate. Patients will be given the greatest care possible, especially in rural areas, thanks to it.

## 7.0 ROBOTS AND COVID-19

Due to the hardships brought on by COVID-19 and the ensuing lockdowns, a lot of people and businesses have turned to robots to help them deal with the pandemic's obstacles. Although there are still numerous technological obstacles for the robotics industry to overcome, robotic and autonomous systems have advantages over conventional human labour, such as inherent immunity to viruses and the inability for disease-causing germs to pass from human to robot to human. The coronavirus pandemic's usage of robots offers a de facto work domain analysis that can direct R&D for upcoming pandemics and perhaps other disasters as well (Murphy et al., 2022). It may be possible to pinpoint cultural, economic, and policy variables that influence adoption and the state of the economy by looking at application variations by nation. The six socio-technical work domains that make up the general mission or goal of having robots are listed below.

- Clinical Care

Refers tasks associated with hospitals and patients that often take place in organised medical settings with trained employees. The usage of robots and cost reimbursement are governed by the government and medical insurance, who ultimately have the final say. Clinical care is utilised in the broadest sense possible and encompasses everything from patient assessment to intensive care units and the diagnostic and treatment procedures covered in robot surveys for medicine like (*Mobile Robots*, 2020). Robots have been used in clinical care for a variety of tasks, including cleaning the point of treatment, managing patient intake and visitors, interacting with patients and their families, interventional telepresence, delivery and inventory, and observational telepresence.

- Laboratory and Supply Chain Automation

This sociotechnical work domain represents the medical support industry, which carries out specific tasks like processing coronavirus test results and providing staff for public health agencies and hospitals. They are separate in terms of economics and functionality even though they are a component of the medical response. Automation of the supply chain and laboratories is a significant work area that cannot be avoided. Robots were used for a variety of functions during COVID-19, including the production and handling of infectious material, laboratory automation, and delivery (Vincent Wang & Wang, 2021).

- Non-Hospital Care

This includes healthcare establishments outside of hospitals, including nursing homes, private clinics, and incarceration facilities. These facilities use some of the same trained people and clinical care methods, but they are governed by distinct budgets and laws. During COVID-19, robots were employed for socialising, keeping tabs on the quarantined, and delivering both medical and non-medical supplies.

- Quality Of Life

This sociotechnical work domain describes the ways that robots are employed to support people, whether by commercial enterprises facilitating pandemic-appropriate social interactions or delivering goods and services. To deal with pandemic lockdowns, people may use personal robots in novel ways. For instance, they may walk their dogs using a small unmanned aircraft system. It was difficult to maintain a good, secure, and safe standard of living in the middle of the COVID-19 pandemic. Robots were used to deliver food and non-food, attend social gatherings in public, control inter-personal interactions, and do other personal tasks.

- Continuity of Work and Education

Activities carried out by commercial companies and educational institutions to keep their operations running are included in this sociotechnical work category. Despite having distinct funding sources, businesses and educational institutions have similar work envelopes and must deal with employees or students who may not be educated to utilise or communicate with robots. Robots were utilised in process automation, telepresence, construction, agriculture, private health surveillance, private security, and most crucially sanitation to provide continuity of work in various sectors and education.

- Public Safety

This relates to public health initiatives carried out by emergency management organisations or law enforcement. These tasks include cleaning public areas and having trained individuals enforce quarantine regulations. Robots were employed in the public safety work domain to enforce quarantines, clean public areas, identify infected people, broadcast public service messages, and monitor traffic.

The COVID-19 pandemic has put a tremendous strain on the world's healthcare systems, resulting in an explosive rise in positive cases and a sharp rise in the need for medical attention (WHO, 2021). The demand for COVID-19 case management in hospitals and other healthcare facilities has increased due to the rising need for coronavirus treatment and solutions. The delivery of medical care to every patient has become difficult due to the overstretched state of the health care systems. Additionally, this pandemic is having a severe effect on the medical device business. Medical equipment companies are having trouble running their businesses as the COVID-19 pandemic spreads. To support the human work force in the medical industry, it is crucial to have medical robots.

## 8.0 ROBOTICS IN SPACE RESEARCH

The universe has been a mystery to humans since the dawn of time. Humanity's space exploration is a reflection of one of our best qualities, curiosity (Agrawal, 2021). Over the years, space has occasionally aroused our interest, and we have only just started to vaguely fathom it. In order to answer some of science's most fundamental concerns about the universe's creation and ongoing evolution, it is thought necessary to explore space. Only the emerging field of space robotics has made such extensive investigation conceivable. In order to explore the harsh environment of space, robots are essential. The use of robotics in space places special demands on the technology.

The Moon is the only other celestial body that people have set foot on besides the Earth. Robotic missions, on the other hand, will soon be able to reach space more quickly and collect more data than human missions. Robots are superior to humans in many aspects because they never get tired, can work in an environment without air, and do not grow bored or distracted. These robots have served to fulfil peoples' increasing curiosity and have enabled notable advancements in the realm of space exploration. In order to do tough or repetitive jobs like repairs in hazardous locations (such as those with radiation or micrometeorite dangers), routine procedures (such as video collection), etc., NASA has created a variety of robotic equipment to assist, supplement, or replace astronauts.

There are several different categories into which robots can be categorised. The two main categories of space robots are remotely operated vehicles (ROV) and remote manipulator systems (RMS). An unmanned spaceship, a lander operated from a stationary location and in contact with an extra-terrestrial plain, or a rover travelling over the terrain after landing are all examples of conventional ROVs.

Some of the Innovative Robots used for space exploration and travel is given in this table.

Future uses of space can be categorised as either exploration, provision of global information services, or use.

Earth-orbiting satellites including space labs and massive antennae, solar and planetary orbiters, and surface probes for the moon and Mars make up exploration. Earth orbiting missions, solar and planetary orbital missions, and surface probes are the different types of space exploration missions.

Worldwide communications, global navigation, and monitoring of the land, the sea, and the atmosphere are all examples of global information services. Energy sources, raw materials, manufacturing, and human services are all included in space utilisation.

Construction of space, deployment of satellites. Space-based intelligent action and manipulation are required for retrieval, servicing, and space rescue missions. One idea to meet this need is teleoperation, which uses manipulator systems remotely operated by human operators in response to remote sensory input. This is because maintaining humans in space is expensive. Rigidly connected, tethered, surface, and free-flying systems are a few examples of tele operator types. Manipulator technology includes tele operator technology. Man-machine communication and sensors. However, tele operators have drawbacks including information flow restrictions and delayed transmission times. Limitations on operator performance and local labour expenses. Autonomous space robots are suggested as a replacement for tele operators because of these factors. Deep space probes, lunar or Mars rovers, earth orbiting robots for satellite maintenance or repair and space construction, and space rescue robots are examples of application scenarios.

Robots are superior to humans since they do not need to eat, drink, breathe properly, or have regular bowel movements, which are fundamental needs for humans. They won't need to be transported back to Earth for a very long time if they can survive in space. Second, we don't need to worry too much about the safety of the robots we launch into space. No human lives will be lost if the mission fails, even if the robots must last long enough to fulfil the task for which they were launched into space. In addition, robots are capable of a variety of activities that humans are not. Some robots can withstand high radiation levels or extreme temperatures. Humans offer operational flexibility, but they are also quite vulnerable and have a low tolerance for the harsh conditions of space.

*Table 2.*

| Sl. No. | Space Robot Name | Country of Origin | Date sent to space | Reference | Features |
|---|---|---|---|---|---|
| 1. | Sputnik1 | USSR | Oct 4, 1957 | NASA | *Sputnik 1(Figure 5) was the first man-made satellite to circle the planet and the first artificial object to do so. The infamous "space race" between the USSR and the US was officially started with the launch of Sputnik 1. Following that, engineers progressively started building robots to be dispatched to other planets for a number of goals, including taking close-up photos of planets and figuring out whether other planets support life(Chertok & Siddiqi, n.d.).* |
| 2. | Mariner 2 (Figure 6)<br>Mariner 4 | USA | December 14, 1962 | *Universe Today* | *The first robotic spacecraft, Mariner 2, successfully completed a flyby of Venus. On July 14, 1965, Mariner 4, the first orbiter launched into orbit, took the first close-up pictures of Mars (Robots in Space - Universe Today, 2009). The primary purpose of orbiters is to take pictures for scientists to view and analyse, whereas the primary purpose of landers is to look for indications of life on other planets. Because of the differences in these robots' functions, orbiters and landers have frequently been sent into space together(Bliley Technologies, 2019).* |
| 3 | Viking 1<br>Viking 2 | USA | August 20, 1975<br>September 9, 1975 | NASA<br>NASA | Both spacecrafts, which were launched to explore Mars, had a lander and an orbiter, the same robotic technology as Mariners 2 and 4. While the landers were responsible for landing on Mars and gathering data, the orbiters were tasked with taking pictures of the planet's surface. This allowed researchers to better understand whether there is life on Mars. Each spacecraft's orbiters flew together, but the landers eventually dispersed to study various areas of Mars' surface. |
| 4 | Voyager 1<br>(Figure 7)<br>Voyager 2 | | | | Launched to probe the solar system's furthest reaches. The two robots, according to NASA, were created to perform up-close observations of Jupiter, Saturn, Saturn's rings, and both planets' largest moons. The two robots have been travelling for more than 40 years and are now closer to Pluto than either the sun or Earth. Voyager 1(Figure 4) didn't enter interstellar space until August 2012; Voyager 2 eventually did so on November 5, 2018. Currently, information is still being exchanged between the two robots across the Deep Space Network (DSN) (Bliley Technologies, 2019). |
| 5 | Dextre | Canadian Space Association | 2008 | *Canadian Space Association* | *Dextre (Figure 8) is a robotic arm created to "install and replace small equipment, such as external cameras or the 100-kg batteries used on the Space Station, replace malfunctioning components in the Station's electrical systems, and test new tools and robotics techniques"(Bliley Technologies, 2019).*<br>*Currently, Dextre is managed by ground control crews from both NASA and the Canadian Space Association. Dextre's hands alone are equipped with a motorised wrench, a retractable power connector, a data and video link, a camera, and lights. Dextre can adapt to a variety of situations and projects because of the way that his hands are made to be able to hold both heavy and delicate equipment. Dextre was launched on Space Shuttle Endeavour in 2008 after being installed on the International Space Station(Bliley Technologies, 2019).* |

*Figure 5. Sputnik*

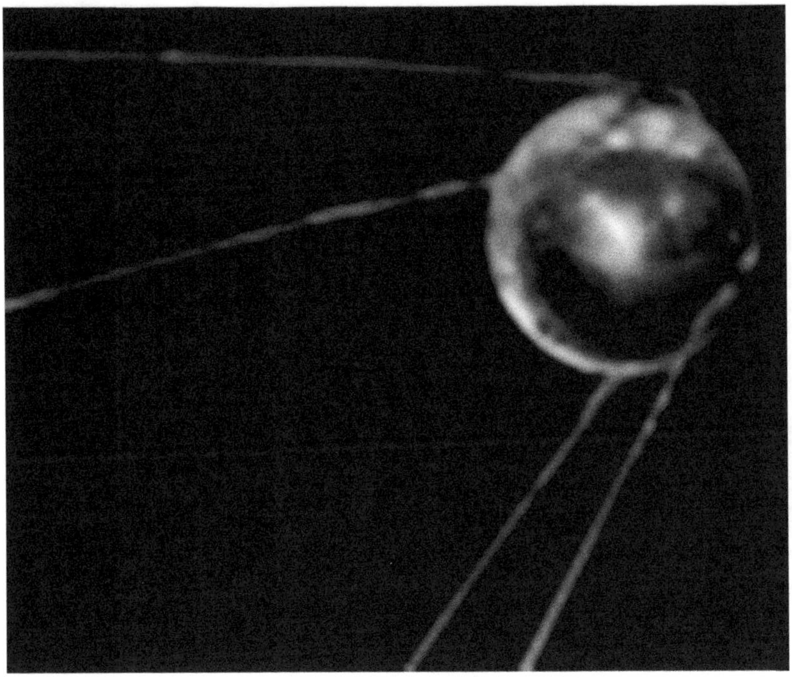

*Figure 6. Mariner 2*

*Figure 7. Voyager*

Every facet of human survival in ordinary life must be carefully considered while preparing for a manned space journey. Robots don't even endure the same health problems that astronauts do after spending months in space, like bone loss and muscle atrophy. Finally, robots can complete jobs that would be too risky or difficult for astronauts to complete. Some astronauts' space suits are unable to keep them safe from the extreme temperatures, preventing them from exploring the sun and other stars, which are extremely hot, blazing gaseous balls, or the cold, vast reaches of our outer solar system. For crewed spacecraft, Venus and Mercury are intolerably hot, whereas the asteroid belt and Jupiter are inconceivably cold. When humans reach an exploratory goal, their presence can be both a benefit and a curse.

It will become polluted in the new regions that humans are exploring since they are dispersing toxins and contaminants from the earth with each step and breath. Some scientists contend that although humans have greater operational flexibility in space, this comes at a cost. Because of their size, fragility, and vulnerability, humans are finicky about their surroundings and have a limited tolerance for open spaces (i.e., high energy radiation, extreme heat, and cold, etc.).

*Figure 8. Dextre*

Due to our vulnerability, our aversion to harming human life, and our need on consumables (food, water, and oxygen), it is necessary to invest a substantial number of money on additional engineering and several redundant systems in order to mitigate risk to the much larger support teams needed to take care of every aspect of daily life during a manned space flight, in addition to the astronauts. The amount of time an astronaut has to conduct experiments is limited by the need to monitor all of this equipment.

Space robots come in a variety of sizes and designs. The most typical robot used on space missions is the rover. This vehicle has the ability to transport research equipment over another planet's surface. Typically, the car and the instruments are controlled separately. Perseverance and Curiosity are two of NASA's most well-known Mars rovers, but there are many other fascinating ones as well. New robots are continually being created by NASA engineers. The name of the robot is A-PUFFER, which stands for Autonomous Pop-Up Flat Folding Explorer Robot.

One of the most recent robots is called A-PUFFER, which stands for Autonomous Pop-Up Flat Folding Explorer Robot (Agrawal, 2021). This traveller on two wheels is light and was inspired by origami. The robot can flatten itself down and squat down to investigate small spaces. Also fascinating is BRUIE, the Buoyant Rover for Under-Ice Exploration (Agrawal, 2021). This robot can float in the water and roll its wheels down the underside of an icy surface while taking pictures and gathering data. In the future, scientists plan to use a robot similar to this to search for signs of life on frozen bodies around the solar system. For instance, the surface waters beneath the moons Enceladus and Europa of Saturn (Agrawal, 2021).

## 9.0 FUTURE PROSPECTS

The market for medical robots is anticipated to increase at a CAGR of 18.5% from 2022 to 2028, from an estimated value of US$ 9,189.70 million in 2022 to US$ 25,443.36 million in 2028 (Partners, 2022). As its name suggests, autonomous mobile robots (AMRs) are capable of carrying out tasks without much assistance from humans. AMRs must accurately perceive and grasp their surroundings in order to achieve this, frequently using embedded vision, the technology that underpins this capability, to make decisions and judgements in real-time (Uma et al., 2013).

The integration of camera modules within embedded electronics and devices, such as a robot, is referred to as "embedded vision." These cameras are crucial in helping these robots navigate safely through their intended surroundings. To do this, it is necessary to use specialised camera types that have been painstakingly developed and integrated to provide the required photographs and films. The machine vision camera market is estimated to reach $2.2 billion by 2027 at a CAGR of 9.9% between FY 2022 and 2027, while the global AMR market is anticipated to grow to $8.70 billion by the year 2028 (Veerabahu, 2022).

AMRs are used nowadays in a wide range of industries and for a wide range of purposes, including:

- Warehouses used for storage, management of supplies, inventory control, cleaning, etc.
- Tasks related to agriculture, such as weeding, harvesting, and fertilising; deliveries; and patrols.
- In retail enterprises, concierge services, pickup services, and shelf management
- Businesses, hospitals, large construction projects, eateries, and more

In 2030, the service robots market is expected to develop at a compound annual growth rate (CAGR) of 21.5% and reach $216 billion, thanks to investments in a variety of consumer-facing sectors, such as travel and tourism. According to GlobalData, a top provider of data and analytics, the global robots market is expected to expand from $70 billion in 2022 to $568 billion in 2030. (Perez et al., 2017). Between 2020 and 2030, robotic intelligence and cloud robotics are expected to rise at a CAGR of 29% and 21%, respectively, with coordinated cleaning fleets and language translation check-in helpers becoming increasingly popular and helpful (Perez et al., 2017).

According to experts, powerful robots and **Robotic Process Automation** (RPA) may soon transform everything from international travel to environmental protection to health care. 16 members of Forbes Technology Council discuss recent and prospective advancements in robotics that have the potential to significantly alter how we live and work (Panel, 2022).

## 9.1 Getting Rid of Ocean Plastic Waste

Scientists have created a tiny robot-fish that swims around and absorbs microplastics on its soft, flexible, self-healing body in order to eliminate them from seas and oceans (Quaglia, 2022). "While humanoid robots seem to be in the headlines a lot lately, innovation in robotics focused on removing waste and microplastics from the water will have a very substantial impact," says Guillermo of NotiSphere. Globally, social activities to minimize plastic usage fall short, yet significant developments and initiatives can make up for the lack of coordination on a global scale.

## 9.2 Driving Commercial Autonomous Cars

According to Jean-Baptiste Coger, Plato, there is a significant opportunity in the deployment of autonomous vehicles for industrial reasons. He continues by saying that broader autonomous driving and fleet management could enhance the safety, performance, and reliability of farming long-haul land transportation (through trucks and trains).

## 9.3 Constructing Buildings

Robotics will have a significant impact on the construction of buildings, which serve as the backdrop to our daily activities. Despite being the largest business, construction has seen the smallest increases in productivity over the past few decades. According to Aleksandr Gampel of Cuby Technologies, Inc., the construction industry will need to use robotics to help with current procedures and eventually automate activities due to a manpower shortage and the need to double our building stock by 2060.

## 9.4 Providing Emergency Services

Globally, there is great potential for the use of robotics in emergency services, according to Peter Abualzolof of Mashvisor. First responders, whether in disaster zones or the medical industry, can easily be robots. They can travel more quickly, access places that are off limits, and operate in danger zones where humans cannot.

While preserving the lives of human first responders, robots can provide services more quickly and effectively.

## 9.5 Treating Illnesses

Innovative uses of nanorobots in healthcare, such as the treatment of diseases or interactions with DNA, are fresh and intriguing. According to Kim Gunnelius, Ficolo, nanorobots have already been utilized to cure mouse pneumonia, and "nanoswimming" applications for treating cancer and strokes are in the research phase.

## 9.6 Carrying Out Surgery

Robotic surgery has the potential to significantly alter the standard of care that patients throughout the world can access. This technology not only provides higher levels of precision, fewer problems, and shorter hospital stays, but it also increases access to first-rate care by enabling in-person or remote surgery, which expands the clinical reach of top surgeons. Stericycle Communication Solutions' Matt Dickson

## 9.7 Powering Intelligent Virtual Assistants

Intelligent virtual assistants could aid us as we become busier by keeping track of our duties, taking meeting notes, sending emails, creating actionable summaries, and assigning tasks based on action points. IVAs (as phone apps or live robot programs) could provide memory jogs by routinely prompting elders with important information, considering that 40% of seniors have memory deficits. Nickelytics, Judah Longgrear

## 9.8 Enhancing Human Capabilities

According to Markus Pflitsch of Terra Quantum AG, quantum sensors will power the following generation of human-machine interface and offer a genuine chance for human improvement. Control of robotics is possible when nerve signals can be reliably perceived. Applications could allow patients who are trapped inside to reclaim control over their communications, assist disabled persons in controlling their prosthetic limbs, or steer electronic gadgets with the power of thinking.

## 9.9 Providing Personal Assistance for the Elderly

Tools are required to assist an aging population by enabling people to live safe, fulfilling lives on their own. According to Dominique Bonte, ABI Research, the potential of robotic devices powered by artificial intelligence to give physical and social stimulation and interaction will make them an increasingly desirable tool, especially if they are used to deliver integrated healthcare.

## 9.10 Delivering Goods and Services

The future of delivery is precise robotic and drone delivery of products and services.

You'll be able to snap a photo of a certain spot in your garage or backyard, and a drone will transport your products to the precise table you showed in the photo.

According to Ayman Shoukry, Specright Inc., a mix of robots, image recognition, and machine learning will be able to achieve that degree of exact navigation and last-centimeter delivery.

## 9.11 Performing Back-Office Tasks

The back office is greatly impacted by robotic process automation, not just for data entry but also for data extraction from legacy systems.

Even better, RPA and AI systems work in tandem: RPA handles data entry and source extraction, while AI provides sophisticated decision-making.

According to Matt Berseth, NLP Logix 12, these technologies work best when combined to build the framework for digital transformation.

## 9.12 Enabling Scientific Research

Nano-sized controlled machines are opening a door to the future of zoology, ornithology, and surgical medicine by enabling us to view the world from a "nano" perspective that was previously unattainable. For instance, the 0.5-millimeter-wide "crab" robot from Northwestern University can bend, twist, walk, turn, and even jump, according to Aleks Farseev of SoMin.ai, and it can be (and already is) used to study insect life.

## 9.13 Monitoring Energy Systems in Commercial Buildings

Energy cost reduction for commercial buildings is one of the largest robotics advances. According to Buyan Thyagarajan, Eigen X, microrobots can access areas that people cannot; real-time monitoring of HVAC systems and sensors on building equipment can result in a considerable decrease in the amount of money commercial real estate operators spend on energy and maintenance.

## 9.14 Farming and Food Processing

Given the persistent labor scarcity in the industry, the use of robotics in agriculture is anticipated to increase further. Engineers are developing even more sophisticated technologies to use in the fields, where crops are already planted, tended to, and harvested by robots. According to John Giordani, robots are also being tried in the food processing sector, where they are being employed for jobs like slicing meat.

## 9.15 Manufacturing Small-Batch, Personalized Pharmaceuticals

Robotic manufacturing of small-batch, tailored medicines will revolutionize how humans age, live, and flourish. Pharmaceutical production nowadays is a mass-volume endeavor with little personalization. According to Mark Francis, Electronic Caregiver, robotic manufacturing will make it possible to create customized prescriptions in the age of personalized medicine that combine the appropriate medications in the appropriate quantity for usage at the appropriate time—all in a single tablet.

## 9.16 Assisting International Travelers

Robots can be utilized for both real-time information and translation, as has been demonstrated over a number of years in Seoul's Incheon International Airport. Despite being in a new setting, this can assist international tourists feel at ease and confident, according to Wendy Johansson of MiSalud Health.

# 10.0 CONCLUSION

More arduous, duplicate manual labor activities may be anticipated to be handled by robots in the future, which will increase transportation efficiency, improve healthcare, and free up human beings to pursue personal development. Time will ultimately determine how this all turns out, of course. Every technology has its ups and downs, but in the vastly large field of robots—which includes warehouse, home care, hotel reception, robotaxi, etc.—everything is certainly moving up. Nowadays robots are employed for a variety of jobs, including those of security guards, chefs, medical assistants, customer service representatives, and even one-man armies in battle! Due to their accuracy and programming, robots can also be used for a wide range of tasks that humans would find uninteresting, dangerous, or repetitive. Robots can perform all physically demanding tasks that are challenging or impossible for humans, and because to advancements in artificial intelligence, they are even getting smarter. All things considered, robots can assist people perfectly and offer numerous solutions across a wide range of industries (*Top 10 Applications of Robotics in 2020*, 2020). Surgical robots will become increasingly important in the years to come as a result of significant developments in the underserved areas of early diagnosis and treatment. The three components of a robot's environment-based autonomy are perception, planning, and execution (manipulating, navigating, and collaborating). The primary goal of combining robotics with AI is to maximise autonomy through learning. When planning a task or engaging (either by manipulating or navigating) with the world, this level of intelligence can be measured as the ability to foresee the future (Uma & T, 2019). Many attempts have been made to create intelligent robots. Robots that can perform specialised autonomous tasks like driving a car, flying in both natural and artificial environments, swimming, carrying boxes and material over various terrains, picking up objects and setting them down already exist. However, developing an intelligent system that exhibits advanced forecasting intelligence remains essential for financial trading markets to overcome the decision making challenges for nonlinear environments (Uma & Suganthi, 2012). A little more than ten years ago, AI was unable to break any records or complete many tasks. These developments taken together will continue to change how we think about robotic intelligence in a variety of different contexts. Robots will eventually perform a variety of tasks for you, including mowing your yard, delivering mail, providing room service, police, and giving directions. There will be countless examples of these kind, helpful creations. By keeping them updated and employed, they will establish a new sector of the labour market. Future investments in robotics from major market players will be fueled by robotic intelligence and cloud robotics. Robots will be able to work together and access vast amounts of data, resulting in intelligent robot fleets that do tasks together. Robotic intelligence will also be improved by AI developments, such as by enhancing facial recognition and translation abilities (Davenport et al., 2019).

All applications of robotics share the trait of making human jobs simpler and enhancing daily life through the usage of this technology. However, the widespread usage of robots has both advantages and disadvantages, just like every technical advancement. Because of this, it's crucial to learn how to utilize robots responsibly and consciously. This includes considering how to employ them and comprehending how they work.

## REFERENCES

*About Joseph Engelberger - Father of Robotics*. (n.d.). Automate. https://www.automate.org/a3-content/joseph-engelberger-about

Infineon. (n.d.). *Fundamentals of robotics*. Infineon Technologies. Www.infineon.com. https://www.infineon.com/cms/en/discoveries/fundamentals-robotics/

Agrawal, R. (2021). The future of space robots. *International Journal of Advance Research, Ideas and Innovations In Technology, 7*(3), 2429–2435. https://www.ijariit.com/manuscripts/v7i3/V7I3-2141.pdf

*Apollo 11:Destination Sea ofTranquility - 1st Lunar Landing*. (n.d.). DLRARTICLE DLR. https://www.dlr.de/content/en/articles/moon-special/the-apollo-program/apollo-11.html

AZoRobotics. (2013, June 6). *Robots in Space*. AZoRobotics.com. https://www.azorobotics.com/Article.aspx?ArticleID=132

Ben-Ari, M., & Mondada, F. (2017). Robots and Their Applications. *Elements of Robotics*, 1–20. doi:10.1007/978-3-319-62533-1_1

Bliley Technologies. (2019). *How are Robots Used in Space Exploration?* Bliley.com. https://blog.bliley.com/robots-used-in-space-exploration

Broekens, J., Heerink, M., & Rosendal, H. (2009). Assistive social robots in elderly care: A review. *Gerontechnology (Valkenswaard), 8*(2). doi:10.4017/gt.2009.08.02.002.00

Builtin. (2019). *What Is Robotics? Types Of Robots*. Built In. https://builtin.com/robotics

Chertok, B., & Siddiqi, A. (n.d.). *Rockets and People The Moon Race*. NASA. https://www.nasa.gov/pdf/621513main_RocketsPeopleVolume4-ebook.pdf

Cresswell, K., Cunningham-Burley, S., & Sheikh, A. (2018). Health Care Robotics: Qualitative Exploration of Key Challenges and Future Directions. *Journal of Medical Internet Research, 20*(7), e10410. doi:10.2196/10410 PMID:29973336

Davenport, T., Guha, A., Grewal, D., & Bressgott, T. (2019). How Artificial Intelligence Will Change the Future of Marketing. *Journal of the Academy of Marketing Science, 48*(1), 24–42. Springer. doi:10.1007/s11747-019-00696-0

Desai, R. (2022, August 29). *Robot*. Dr Rajiv Desai. https://drrajivdesaimd.com/2022/08/29/robot/

Dinh Thai Bao. (2019). *ROBOT*. Academia.edu. https://www.academia.edu/19892932/ROBOT

Gore, U. C. (n.d.). *Univerzitet Crne Gore - O Univerzitetu*. UCG - Univerzitet Crne Gore. https://www.ucg.ac.me/skladiste/blog_13268/objava_56689/fajlovi/1_1%20Introduction%20to%20Robotics.ppt

Gottlieb, J., & Anderson, D. L. (n.d.). *Kinds of Robots*. Mind.ilstu.edu. https://mind.ilstu.edu/curriculum/medical_robotics/kinds_of_robots.html

Guizzo, E. (2011). *Types of Robots - ROBOTS: Your Guide to the World of Robotics*. @Robotsapp; Robots. https://robots.ieee.org/learn/types-of-robots/

Han, J. (2020, January 23). *4 Ways Robots Will Improve Our Standard of Living*. MarTech Series. https://martechseries.com/mts-insights/guest-authors/4-ways-robots-will-improve-standard-living/ https://jpl.nasa.gov.

*History Of Robots Timeline*. (n.d.). Preceden. [REMOVED HYPERLINK FIELD] https://www.preceden.com/timelines/625227-history-of-robots

Javaid, M., Haleem, A., Singh, R. P., & Suman, R. (2021). *Substantial capabilities of robotics in enhancing industry 4.0 implementation. Cognitive Robotics, 1, 58–75*. Sciencedirect. doi:10.1016/j.cogr.2021.06.001

Kaur, R. & Kautish, S. (2022). Multimodal sentiment analysis: A survey and comparison. *Research Anthology on Implementing Sentiment Analysis Across Multiple Disciplines*, 1846-1870.

Kautish, S., Peng, S. L., & Obaid, A. J. (Eds.). (2021). *Computational intelligence techniques for combating COVID-19*. Springer International Publishing. doi:10.1007/978-3-030-68936-0

Kautish, S., Reyana, A., & Vidyarthi, A. (2022). SDMTA: Attack detection and mitigation mechanism for DDoS vulnerabilities in hybrid cloud environment. *IEEE Transactions on Industrial Informatics*, *18*(9), 6455–6463. doi:10.1109/TII.2022.3146290

Lozano-Perez, T. (1983). Robot programming. *Proceedings of the IEEE*, *71*(7), 821–841. doi:10.1109/PROC.1983.12681

*Lunokhod 1*. (2021, December 19). Wikipedia. https://en.wikipedia.org/wiki/Lunokhod_1

Madhu, G., Govardhan, A., Ravi, V., Kautish, S., Srinivas, B. S., Chaudhary, T., & Kumar, M. (2022). DSCN-net: A deep Siamese capsule neural network model for automatic diagnosis of malaria parasites detection. *Multimedia Tools and Applications*, *81*(23), 34105–34127. doi:10.100711042-022-13008-6

Mangusli, D. (2022, October 13). *The Emerging Role Of Robotics In Making Healthcare Accessible*. BW Defence. https://bwhealthcareworld.businessworld.in/article/The-Emerging-Role-Of-Robotics-In-Making-Healthcare-Accessible-/13-10-2022-450332/

Marquiss, R. (2018). *Five Types of Industrial Robots And How To Choose The Best Fit*. Valin.com. https://www.valin.com/resources/articles/five-types-of-industrial-robots-and-how-to-choose-the-best-fit

Mitchell, R. (n.d.). *Advantages and disadvantages of robots*. FutureLearn. https://www.futurelearn.com/info/courses/begin-robotics/0/steps/2845

*Mobile Robots*. (2020, December 24). GeeksforGeeks. https://www.geeksforgeeks.org/mobile-robots/

Murphy, R. R., Gandudi, V. B. M., Amin, T., Clendenin, A., & Moats, J. (2022). An analysis of international use of robots for COVID-19. *Robotics and Autonomous Systems, 148*, 103922. doi:10.1016/j.robot.2021.103922 PMID:34803220

Norman, J. (n.d.). *Geroge Devol Invents Unimate, the First Industrial Robot*. History of Information. Www.historyofinformation.com. https://www.historyofinformation.com/detail.php?entryid=4071

Nouaille, L., Laribi, M. A., Nelson, C. A., Zeghloul, S., & Poisson, G. (2017). Review of Kinematics for Minimally Invasive Surgery and Tele-Echography Robots. *Journal of Medical Devices, 11*(4), 040802. doi:10.1115/1.4037053

Panel, E. (2022, November 16). *Council Post: 16 Ways Robots And RPA May Soon Change How We Live And Work*. Forbes. https://www.forbes.com/sites/forbestechcouncil/2022/11/16/16-ways-robots-and-rpa-may-soon-change-how-we-live-and-work/?sh=10821e1b3a0d

Papadopoulos, I., Koulouglioti, C., & Ali, S. (2018). Views of nurses and other health and social care workers on the use of assistive humanoid and animal-like robots in health and social care: A scoping review. *Contemporary Nurse, 54*(4-5), 425–442. doi:10.1080/10376178.2018.1519374 PMID:30200824

Partners, T. I. (2022, September 22). *Medical Robots Market 2022 Recovering from Covid-19 Outbreak*.Open PR. Www.openpr.com. https://www.openpr.com/news/2744760/medical-robots-market-2022-recovering-from-covid-19-outbreak

Perez, J. A., Deligianni, F., Ravi, D., & Yang, G.-Z. (2017). *Artificial Intelligence and Robotics*. UKRAS White Papers., doi:10.31256/WP2017.1

*Pierre Jaquet-Droz - Complete Biography, History and Inventions*. (2021, January 4). History Computer. https://history-computer.com/pierre-jaquet-droz-complete-biography/

Pollock, B. (2019, August 5). *How Robots Will Save the Human Race*. Automation World. https://www.automationworld.com/factory/robotics/blog/13320046/how-robots-will-save-the-human-race

Poverly, E. (2022, April 26). *The Future of Robotics: How Will Robots Change The World?* Revolutionized. https://revolutionized.com/the-future-of-robotics/

Quaglia, S. (2022, June 22). *Scientists unveil bionic robo-fish to remove microplastics from seas*. The Guardian. https://www.theguardian.com/environment/2022/jun/22/scientists-unveil-bionic-robo-fish-to-remove-microplastics-from-seas

Reynoso, R. (2021, May 25). *A Complete History of Artificial Intelligence*. AI & Machine Learning Operationalization Category. https://www.g2.com/articles/history-of-artificial-intelligence

Rezapour, E. (2015, March). *Model-based Locomotion Control of Underactuated Snake Robots*. Research Gate. https://www.researchgate.net/publication/273770473_Model-based_Locomotion_Control_of_Underactuated_Snake_Robots

*Robot Technologies from A to Z at a Glance*. (2021, September 2). Mobilefest. https://www.mobilefest.net/robot-technologies-from-a-to-z-at-a-glance/

Robotics, P. (2020, December 25). *The Origin, History and Story of Robotics*. Medium. https://medium.com/@plusrobotics07/the-origin-history-and-story-of-robotics-ab2f060a4259#:~:text=One%20of%20the%20first%20instances

*Robotics Career*. (2021, June 7). Robotics Career. Www.roboticscareer.org. https://www.roboticscareer.org/news-and-events/news/23037

*Robots can be Classified by their Application or their Kinematics*. (n.d.). Design Tech Products. Www.designtechproducts.com. https://www.designtechproducts.com/articles/classification-of-robots

*Robots in Space - Universe Today*. (2009, November 2). Universe Today. https://www.universetoday.com/43750/robots-in-space/

*RobotWorx*. (2019). RobotWorx. https://www.robots.com/faq/what-are-the-main-types-of-robots

Sampathkumar, A., Rastogi, R., Arukonda, S., Shankar, A., Kautish, S., & Sivaram, M. (2020). An efficient hybrid methodology for detection of cancer-causing gene using CSC for micro array data. *Journal of Ambient Intelligence and Humanized Computing*, *11*(11), 4743–4751. doi:10.100712652-020-01731-7

Scassellati, B., Admoni, H., & Mataric, M. (2012). *Robots for Use in Autism Research. 14*, 275–294.

Shah, R., & Chircu, A. (2018). IOT AND AI IN HEALTHCARE: A SYSTEMATIC LITERATURE REVIEW. *Issues in Information Systems*, *19*(3), 33–41. https://iacis.org/iis/2018/3_iis_2018_33-41.pdf

Shen, Y., Guo, D., Long, F., Mateos, L. A., Ding, H., Xiu, Z., Hellman, R. B., King, A., Chen, S., Zhang, C., & Tan, H. (2021). Robots Under COVID-19 Pandemic: A Comprehensive Survey. *IEEE Access : Practical Innovations, Open Solutions*, *9*, 1590–1615. doi:10.1109/ACCESS.2020.3045792 PMID:34976569

Simshaw, D., Terry, N. P., Hauser, K., & Cummings, M. L. (2016). Regulating Healthcare Robots: Maximizing Opportunities While Minimizing Risks. *Publisher*. https://hdl.handle.net/1805/11587

Stanford University. (2019). *Robotics: A Brief History*. Stanford.edu. https://cs.stanford.edu/people/eroberts/courses/soco/projects/1998-99/robotics/history.html

Taylor, C. (2019, June 26). *Robots could take over 20 million jobs by 2030, study claims*. CNBC; CNBC. https://www.cnbc.com/2019/06/26/robots-could-take-over-20-million-jobs-by-2030-study-claims.html

Terra, J. (2022, April 18). *The Future of Robotics: Everything You Need to Know in 2022*. Simplilearn. https://www.simplilearn.com/future-of-robotics-article

*The Invention of the Industrial Robot*. (n.d.). Invent.org. Www.invent.org. https://www.invent.org/blog/inventors/George-Devol-Industrial-Robot

*Top 10 Applications of Robotics in 2020*. (2020, November 2). GeeksforGeeks. https://www.geeksforgeeks.org/top-10-applications-of-robotics-in-2020/

Tucker, M. R., Olivier, J., Pagel, A., Bleuler, H., Bouri, M., Lambercy, O., Millán, J. del R., Riener, R., Vallery, H., & Gassert, R. (2015). Control strategies for active lower extremity prosthetics and orthotics: A review. *Journal of Neuroengineering and Rehabilitation*, *12*(1), 1. doi:10.1186/1743-0003-12-1 PMID:25557982

*Types of Robot*. (n.d.). Javapoint. Www.javatpoint.com. https://www.javatpoint.com/types-of-robot

*Types of Robots and Industry Applications of Robotics Technology*. (n.d.). Intel. https://www.intel.com/content/www/us/en/robotics/types-and-applications.html

*Types of Robots and Their Applications - Guardforce*. (2019). Guardforce.com.hk. https://www.guardforce.com.hk/en/news/blog_115/Types-of-Robots-and-Their-Applications_1729

Uma, S. (2018). Latest Research Trends and Challenges of Computational Intelligence Using Artificial Intelligence and Augmented Reality. *Computational Intelligence and Sustainable Systems*, 43–59. doi:10.1007/978-3-030-02674-5_3

Uma, S., & Chitra, A. (2012). Pattern recognition using enhanced non-linear time-series models for predicting dynamic real-time decision making environments. *International Journal of Business Information Systems*, *11*(1), 69. doi:10.1504/IJBIS.2012.048342

Uma, S., Chitra, A., & Suganthi, J. (2013). Design of a non-linear time series prediction model for daily electricity demand forecasting. *International Journal of Business Innovation and Research*, *7*(3), 298. doi:10.1504/IJBIR.2013.053633

Uma, S., & Suganthi, J. (2012). An Intelligent and Dynamic Decision Support System for Nonlinear Environments. *International Journal of Intelligent Information Technologies*, *8*(4), 43–61. doi:10.4018/jiit.2012100104

Uma, S., & Suganthi, J. (2015). *A Smart and Dynamic Decision Support System for Nonlinear Environments*. Recent Advances in Intelligent Technologies and Information Systems. https://www.igi-global.com/chapter/a-smart-and-dynamic-decision-support-system-for-nonlinear-environments/125509

Uma, S., & T, S. (2019). *The Potential of Cognitive Social Mining Applications in Data Analytics for Protecting Social Media Users for National and International Security*. Cognitive Social Mining Applications in Data Analytics and Forensics. https://www.igi-global.com/chapter/the-potential-of-cognitive-social-mining-applications-in-data-analytics-for-protecting-social-media-users-for-national-and-international-security/218389

Vandemeulebroucke, T., Dierckx de Casterlé, B., & Gastmans, C. (2018). The use of care robots in aged care: A systematic review of argument-based ethics literature. *Archives of Gerontology and Geriatrics*, *74*, 15–25. doi:10.1016/j.archger.2017.08.014 PMID:28926749

Veerabahu, M. (2022, November 7). *The role of embedded vision in powering autonomous robotics*. Times of India Blog. https://timesofindia.indiatimes.com/blogs/voices/the-role-of-embedded-vision-in-powering-autonomous-robotics/

*Viking 1 and 2, NASA's first Mars landers.* (n.d.). The Planetary Society. https://www.planetary.org/space-missions/vikingReik doi:10.1145/3127874Vincent Wang, X., & Wang, L. (2021). A literature survey of the robotic technologies during the COVID-19 pandemic. *Journal of Manufacturing Systems, 60*, 823–836. Advance online publication. doi:10.1016/j.jmsy.2021.02.005 PMID:33612914

Vitanza, A., D'Onofrio, G., Ricciardi, F., Sancarlo, D., Greco, A., & Giuliani, F. (2019). Assistive Robots for the Elderly: Innovative Tools to Gather Health Relevant Data. *Data Science for Healthcare*, 195–215. doi:10.1007/978-3-030-05249-2_7

WHO. (2021, April 23). *COVID-19 continues to disrupt essential health services in 90% of countries.* Www.who.int. https://www.who.int/news/item/23-04-2021-covid-19-continues-to-disrupt-essential-health-services-in-90-of-countries

World Commission on the Ethics of Scientific Knowledge and Technology (COMEST). (2017). *Report of COMEST on Robotics Ethics.* SHS/YES/COMEST-10/17/2 REV. www.facebook.com/elprocus.

# Chapter 5
# Therapeutic Applications of Social Robots in Rehabilitation

**Radhika Mahajan**
*Vivekananda Institute of Professional Studies, India*

**Manoj Kumar**
*Maharaja Surajmal Institute, India*

**Ashima Bhatnagar Bhatia**
*Vivekananda Institute of Professional Studies, India*

**Pawan Whig**
*Vivekananda Institute of Professional Studies, India*

## ABSTRACT

*The Industrial Revolution 4.0 has produced a wide range of innovative possibilities for the employment of artificial intelligence and computer-guided devices in rehabilitation. In recent times, medical field has shown a keen interest in Human-Robot Interventions (HRI) due to their application in treatment and rehabilitation of various diseases. Social robots are designed to interact with people in a manner that is consistent with human social psychology because of their capacity to elicit social and emotional responses from users. They work in a human-centric environment by speaking, moving, making gestures, or utilizing facial expressions to communicate with their users while adhering to a set of social norms. We outline some intriguing possibilities for social robots in healthcare-related applications in this chapter. Some of these include treatment of social anxiety, dementia, autism spectrum disorder, depression, alzheimer's disease, anorexia nervosa, neurodevelopmental issues, cancer, diabetes and upper limb rehabilitation after stroke.*

DOI: 10.4018/978-1-6684-8171-4.ch005

## ROBOTICS IN HEALTHCARE AND ROBOTICS TAXONOMY

### Rehabilitation Technologies

According to Zhang et al. (2013), who made the initial claim, it is impossible to determine the efficacy of a robot-assisted rehabilitation intervention since there are no common assessment criteria and metrics for the numerous robotics devices and control schemes. However, the scientific health care development fundamental these belongings is still unwell unspoken. Robots are also believed to increase the amount of rehabilitation exercises. Robotic intervention and conventional therapy have no statistically significant differences (Babaiasl et al., 2016; Bustamante Valles et al., 2016; Piau,, 2019; Schröder et al., 2019; Breazeal, 2011; Frennert & Östlund, 2014; González-González et al., 2021; Rasouli et al., 2022).

Perhaps tritely, a study can assert that robot assisted rehabilitation can encourage CVA patients' return to their families and society by improving their lower limb recovery (Li et al., 2014; Heerink et al., 2016). We contend, however, that previous research has had difficulty demonstrating the societal impact of mechanical, digital, and assistive robots. Shamsuddin and others

Without offering any supporting data or critical societal consequences, (Shamsuddin, 2017; Fosch-Villaronga et al., 2019) asserts that "a robotic animal is the solution to provide constant mental support and create warm and compassionate feelings from the sufferers." Studies from the past and present have gaps since they don't use interdisciplinary methods to assess the true societal impact of rehabilitation robots. These assistive technologies may aid in rehabilitating a CVA patient's physical function, but until the robots significantly enhance the social lives of the patients, they remain a cold and uninteresting technology and cannot be considered true social robots (Vincent & Law, 2015; Góngora Alonso et al., 2018). For instance, rehabilitation robotics is a fascinating area of research that is moving away from the promotion of mobility and toward the socialising of infants and toddlers (Agrawal, 2018; Scassellati et al., 2018). Given the silo-disciplinary character of earlier studies, it is unclear how far this transition goes beyond the surface level.

### Can Robots Give You Back Your Life? Recommendations and Controversial Items

Empirical case studies from around the world showed some highly effective rehabilitation robots. Example Zora for the elderly (Tuisku et al., 2018), SAR for pediatric therpy of children (Carrillo, 2018), SLT for cerebral palsy and developing communication skills (Niyetkaliyev et al., 2017), Makro for disorders realted to motion.

These reasoning Automaton Enduring connections for speech and cognitive therapy appear to be restricted to geriatric or paediatric requests. Only a small number of research have compared across countries or addressed a wider demography. The current state of research makes it difficult to determine if the attention being paid to these two groups is due to the fact that they have the greatest medical needs, or whether it is purely a coincidence of researchers' incidental interests.

Future studies, according to Belda-Lois et al. (2011), should evaluate and maximise the effects of rehabilitation using different signals. In the field of rehabilitative robotics, brain-machine interfaces (BMI) are currently a popular topic (Khan & Hong, 2017; Tang et al., 2017). According to Miao et al. (2018), controlling physiological signals is an efficient method of preventing laziness and providing support.

## SOCIAL ROBOTS WITH APPLICATIONS IN HEALTHCARE

Numerous terms that are associated with social robots have been found in the literature In a sense, all of them fit inside our scope (1–3). Other related phrases are also frequently used, encompassing everything from small, portable gadgets as shown in Figure 1.

*Figure 1. Different types of social robot interactions*

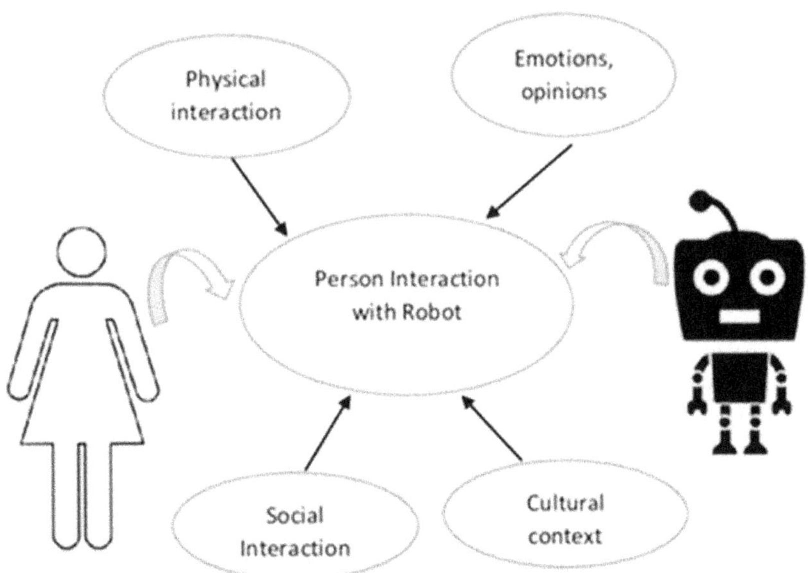

**RQ1.** In what way are communal automata created? Are they created morally?

The creation of social robots involves a number of algorithms as well as various implementations that must also be regarded as fundamental include facial recognition, speech recognition, cognitive and decision-making modules, emotional modules, and ergonomics(Ahn et al., 2019).

Regarding the design or functionality of robots meant for human use, Pleo, Paro, Parrot, Aibo, Huggable, or iCat are recommended pet-like SAR for youngsters or the elderly. Robotic emotive non-verbal behaviour can also be understood by humans. People who are sick when they are young and old appear to have less stress. According to, the fundamental issue with social robots is their cultural acceptance and aptitude. Users also favour robots with feminine genders. Understanding what the human is doing was another issue raised in this study; nevertheless, the appearance of the robot relies on its use, the user's age, and a number of other aspects(Banthia et al., 2016; Nergui, Komekine, Nagai, & Otake, 2014; Ono et al., 2015; Shayan et al., 2016).

It has been determined that the person-centered process is the optimum design methodology for developing and implementing AAL facilities because it enables interactions between participants and healthcare professionals. To enable the robot to identify some diseases, such as ASD, discovered that extremely industrialized procedures were required to mix more over-all reasoning features in the robot. Socio-Cognitive Engineering (SCE), a methodology that permits the integration of various theories, models, and patient and carer visions, is additional practice that is helpful in health care settings. The interaction can be enhanced by including additional components like feedback and prediction. When NAO is used in heart disease settings, users are given a fresh perspective on the significance of their vital signs through human-robot interaction(Meghdari, Alemi, Khamooshi, Amoozandeh, Shariati, & Mozafari, 2016).

Even though patient safety, dignity, and welfare can all be promoted by social robots, some researchers feel that using children as test subjects for such investigations is unethical . Additionally, care services are subject to strict regulation, and robotic care workers must be covered by specific legislation.

The CCVSD (Care-Centered Value-Sensitive Design) approach is a framework comprising ethically significant elements, according to the International Federation of Robotics (IFR). The following factors should be taken into account when evaluating a care robot: the use context, the care practise, the actors involved, the type of care robot (its capabilities, appearance, etc.), and the list of values involved for the practise in question in the stated context (i.e., the interpretation and prioritisation of care values)(Goris, Saldien, Vanderniepen, & Lefeber, 2009; Lewis & Canamero, 2013; Robinson, MacDonald, Kerse, & Broadbent, 2013; Sequeira & Ferreira, 2016) .

**RQ2.** What are the major functions and purposes of social robots in the medical field?

The elderly and children are the two patient populations for whom social robot applications are most common in the healthcare setting.

## Seniors

These assistive technology can be useful for senior citizens. According to the researchers' classification in, they fall under the headings of ICT. They discovered that the research focused on eight issues that affect older people The management of (1), (5), and (6) was assisted by social robots. Hospitals have also employed social robots for dementia rehabilitation. Elderly people's quality of life and mood can be enhanced while also feeling less lonely thanks to robots (Chung & Park, 2019; Cooper, Fava, Vivas, Marchionni, & Ferro, 2020; Goncalves & Arsenio, 2015; Logan et al., 2019; Moerman & Jansens, 2020).

Emotions are linked to social interactions, which is true for senior people and patients with dementia . It presents both advantages and obstacles for the PARO robot and is a promising field. The psychological needs of the aged in terms of inclusion, identity, attachment, occupation, and comfort can be supported through PARO . ARI uses several physical clues simultaneously to convey emotion and elicit empathy, primarily facial expressions, body movement, posture, and verbal cues . SARs (Social Assistive Robot and Companion) include AIBO, PARO, AIBO, and iCat. In the year 2017, 6,400 helping robots were sold globally for use in healthcare. The difficulty is in defining the function of robots in healthcare and enforcing standards and moral guidelines for the services they render (Chen et al., 2018; Cooper, Fava, Vivas, Marchionni, & Ferro, 2020; Diaz-Boladeras et al., 2016; Larriba et al., 2016; Moerman et al., 2019).

The condition of our homes has an impact on our wellbeing. They are a crucial quality in the most at-risk populations, such as the elderly, who frequently require medical support. For instance, older persons living independently at home received assistance from empathic coaches and virtual assistants.

For the elderly, robots are more than simply devices; they provide emotional support in a way similar to that of a communicating and coexisting buddy or companion.

## Youths

Children's quality of life can be enhanced by robots' interactions with them in schools.

Tele presence is a useful instrument, as we showed in the primary part, to luxury paediatric tumours over extended periods of isolation. Additionally, PLEO is employed in the kind scheme, anywhere the youngster's wellness is a priority, for long-term partnerships and bonding. Children found PLEO fascinating due to its appearance, which was evocative of dinosaurs or an electronic toy . Unfortunately, stress can also be a factor during prolonged periods of recuperation. Interactive plush animals can help lift a child's spirits and enhance their quality of life while they are in the hospital. They can also act as a distraction during a medical treatment and offer support when dealing with an illness. Social robots have been shown to have good benefits on kids, including better moods, engagement, trust, reduced discomfort or tension, more relaxed behaviour like smiling and being open, and improved communication or emotional connections with users (Alemi, Meghdari, Ghanbarzadeh, Moghadam, & Ghanbarzadeh, 2014; Sheridan, 2020).

The literature discusses (Kautish et. al 2022) automatic diagnosis in detail. It is used in ASD (Autism Spectrum Disorder), when it is still difficult to foresee how activities will turn out. Observing kinematics is not the same as comprehending interior mental states.

The phrase "in the wild" is typically used to refer to actual circumstances. Many tasks are carried out in controlled or laboratory settings, which are in no way realistic. Some studies, like those involving diabetic kids, approach the issue using a paradigm for a long-term care partnership between humans and robots (Burns et al., 2018). For instance, the Pleo robot, a robotic pet baby dinosaur, functions differently to help kids while they are in the hospital (Bartlett et al., 2020; Meghdari, Alemi, Khamooshi, Amoozandeh, Shariati, & Mozafari, 2016; Panico, Cordasco, Vogel, Trojano, & Esposito, 2020; Sequeira & Ferreira, 2016).

Other social robots were employed to test the urine of cancer-stricken youngsters, observe and assess therapy procedures, and save therapeutic treatments in databases (Beraldo et al., 2019; Neerincx et al., 2019; Sharif & Alsibai, 2017; Turja & Parviainen, 2020; Valles-Peris et al., 2018).

## USING SOCIAL ROBOTS AS AN ADDITIONAL KIND OF TREATMENT FOR CHRONIC, PROGRESSIVE DISEASES

### Social Robots can Assist Dementia Patients

Significant advancements in robotics have been made possible by technological innovations. There have been efforts to develop robots that can engage in social interactions with people. Because the robots that have been created up to this point are based on organisms that already exist in our environment, they either: features that are anthropomorphic or zoomorphic. New perspectives, as well as chances and fields for their application, have been made feasible by the raging technological evolution. These days to do research on the socialization-promoting benefits of machinelike interference, evaluating the consequences on human cognition and behaviour as shown in Figure 2.

*Figure 2. Social robots assisting dementia patients*

Demenassisting demential disease in which cognitive ability is compromised; as a result, it is associated with altered communication and a propensity for more depressed moods. Patients may therefore be at danger since they may have a greater propensity to feel alone and isolated, which will ultimately worsen their mental health (Turja & Parviainen, 2020). This creates a window of opportunity for the use of alternative therapies, which are intended to reduce symptoms and slow the progression of the disease . Through the use of drawing, and other doings that also incorporate the use of robots, these therapies provide stimulation, interaction, and a change in routine. The incorporation of complementary therapies becomes more important in these patients' treatment plans because neurodegenerative illness is a condition that is frequently incurable.

Due to the long history of human interaction with other animals—the first fossilised evidence dates to the time of Homo erectus—we should emphasise therapy with domestic animals as one of the complementary treatments for dementia. Domestic animals are frequently chosen nowadays because of their emotive return and as facilitators of social connection between people, since they are able to lessen and contribute. This close association has been maintained (Goncalves & Arsenio, 2015; Lewis & Canamero, 2013; Ngo et al., 2020; Panico, Cordasco, Vogel, Trojano, & Esposito, 2020; Sheridan, 2020; Wynsberghe, 2013).

Animals may instill calmness in the people they live with, therefore they continue to draw attention from the outside world today. Contrarily, the Social Support Hypothesis asserts that companion animals serve as both a social support and a facilitator of human connection. This is due to the fact that companion animals make their owners feel happier and less lonely.

However, due to potential challenges including sanitary, allergenic, or institutional issues, studying therapeutic intervention with domestic animals can occasionally be challenging. As a result, it presents a chance for the creation of a zoomorphic SAR application that aims to get beyond this restriction.

The majority of zoomorphic robot experiments to date involved the robotic baby seal (PARO), a canine-like of the studies aimed to encourage multimodal contact through vision, hearing, and touch.

The model that is being utilised and researched the most is PARO. It is a robot in the form of a darling closure that was created in Japan in 2004 and was authorised as a medical device by the Food and Drug Administration in 2009. It is currently available everywhere, and there are various nations where research using this approach can be done. EEG monitoring in elderly people revealed that exposure to the robot improved the cortical neurons' ability to function. As patients' cognitive abilities varied, so did their levels of emotional and cognitive involvement. Patients expressed emotional involvement ("you are so dear"), interpretations of the PARO's will ("you are so clever! "), and his feelings ("Do not do this because you frighten him!").

## AUTISM SPECTRUM DISORDER OR SOCIAL ANXIETY

A noticeable and enduring apprehension is a defining feature of social anxiety disorder. Many social circumstances that might cause feelings of embarrassment or humiliation . Despite their desire to interact with others, people with social anxiety disorder either entirely avoid social situations or tolerate them with excruciating discomfort . This is because they are so afraid of being scrutinised by others. Various reasons of social anxiety is shown in Figure 3.

75% of people who are diagnosed with social anxiety disorder report that it first manifested. The likelihood of a social anxiety persisting from childhood until adulthood increases if it develops before age 11.

*Figure 3. Various reasons of social anxiety*

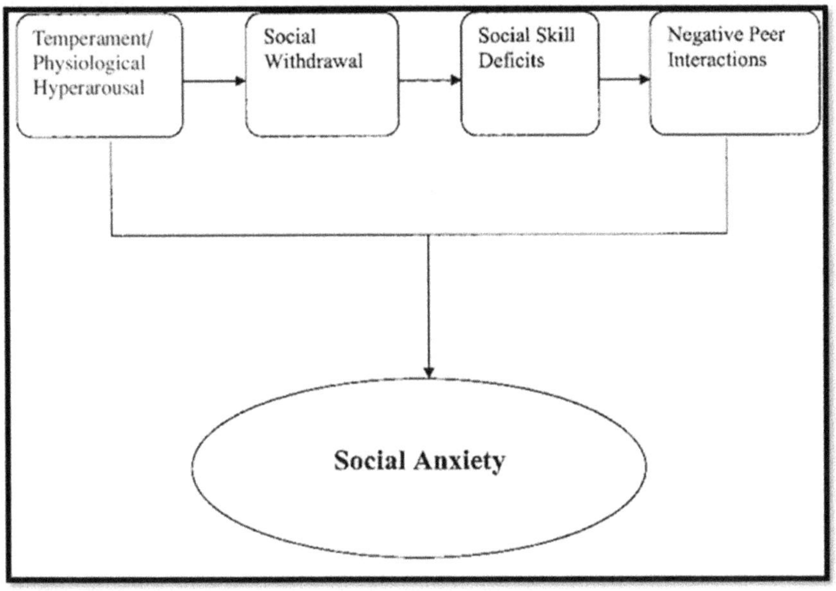


Children are predisposed to the risk of developing maladaptive social anxiety due to behavioural reticence, a constant propensity to react to unexpected situations with terror and retreat. Social anxiety disorder can have a chronic and lifelong history. Given the risk of morbidity and disability, early diagnosis and treatments are crucial for people with social anxiety disorder (Alemi, Meghdari, Ghanbarzadeh, Moghadam, & Ghanbarzadeh, 2014; Kang, 2020; Khosravi & Ghapanchi, 2016).

## Interviews Conducted by Robots

Numerous research in the area of robot-mediated interviews have shown that, regardless of the difficulty of the questions, children's interact in similar ways to rots as humans. The findings demonstrated that children gave both the humanoid and machinelike interviewers same gratified, i.e., comparable volumes and kinds of info. These findings suggest that kids engaged in the interviews and responded to the robotic interviewers. The findings of a qualitative analysis suggested that some special needs children may have shown greater interest in interviews conducted by robots as compared to human beings. According to a research by Bethel et al., children reported instances of bullying considerably more frequently while speaking with a robot than when speaking with a human interviewer. If the robots are created using Additionally, they noted that such systems might have particularly useful uses for kids who struggle to communicate and find it difficult to interact with people in general, let alone strangers, due to problems like ASD or anxiety. In these situations, a robot interviewer in the hands of mental health professionals or clinical psychologists could be a beneficial tool for offering counselling services (Dodds et al., 2018; Nergui, Komekine, Nagai, & Otake, 2014; Yu et al., 2015).

## Social Robots as Screening and Diagnosis Tools

The study of behaviours linked to social anxiety may benefit from the use of social robots by therapists. Social anxiety sufferers may display behavioural due to the stress and dread generated by social anxiety, people who suffer from these symptoms exhibit avoidance of eye contact, tight body postures, and incorrect speaking voices. Social robots are able to recognise social cues from people and respond appropriately. Example to show social robots in healthcare sector is shown in Figure 4.

*Figure 4. Social robots in healthcare sector*

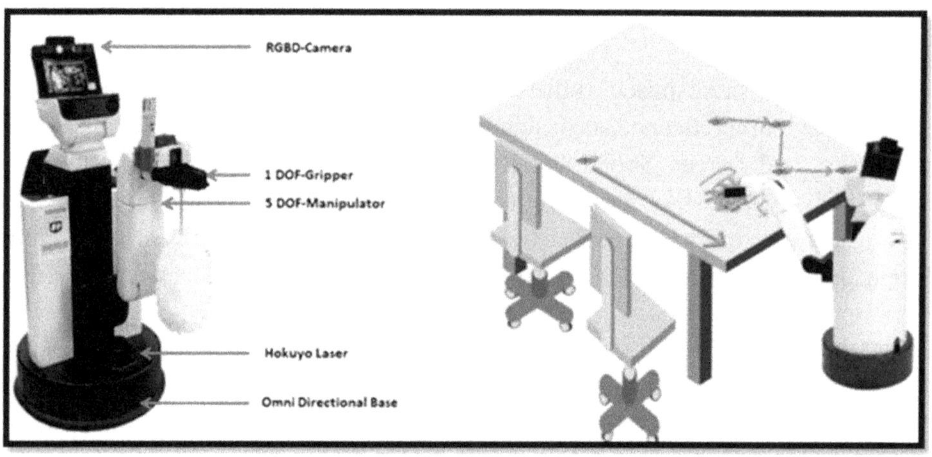


Additionally, human gaze patterns and behaviours including head movements and emotions linked to anxiety and sadness can be identified using machine learning algorithms. The therapist might be given access to this quantitative data in real-time to improve initial and ongoing assessments as well as diagnosis as the course of treatment advances (Eriksson, 2018; Hung et al., 2019; Robinson, MacDonald, Kerse, & Broadbent, 2013; Rouaix et al., 2017; Sarabia et al., 2018).

## Robotic Therapy

For people with severe social anxiety, the use of interactive robots may serve as a "stepping stone." They may, for instance, practise social behaviours and abilities before using them in actual interactions. Facilitating the development of various social and communication skills may be a part of therapeutic sessions. Individuals can exercise these abilities in the interim relying on their capacities to be ready for real-life challenges. Existing treatment modalities, like as exposure, may benefit from the ability to confront fearful social circumstances without the pressure of engaging with another person example of robotic therapy is shown in Figure 5.

*Figure 5. Example of robotic therapy*

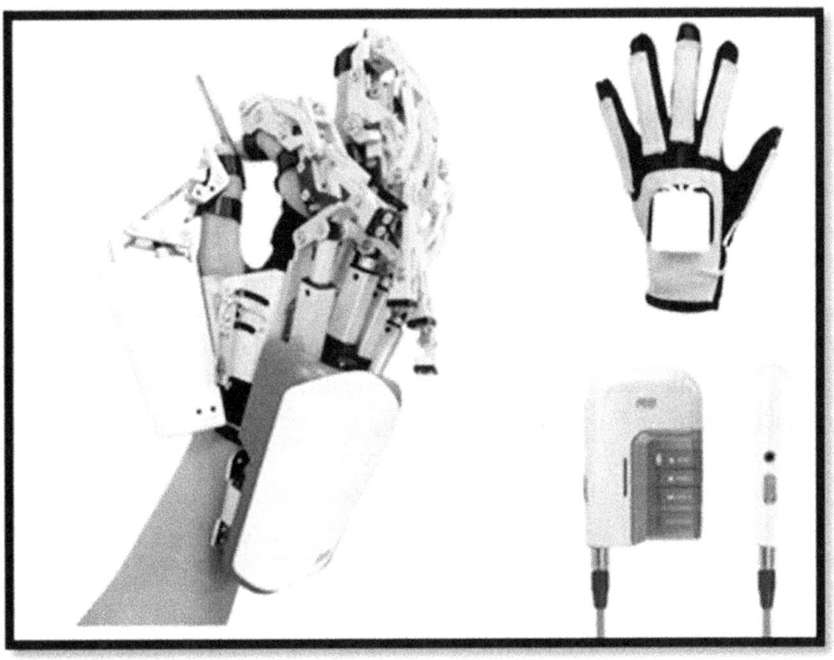

The idea is for the person to eventually confront social settings, including other people, so that the anxiety symptoms they feel during exposure may be more controllable as a result of this additional layer. Additionally, the use of robots in a therapeutic setting might add a comforting and enjoyable element to the therapy session, which might help with emotional support and stress release during trying times in the therapeutic process. When a robot participates in therapy sessions, the environment becomes friendlier and less intimidating, and clients may experience a calming effect.

## Interactive Social Companions in the Form of Social Robots

People who have social anxiety say they have fewer friends and have more trouble maintaining their relationships and friendships. To mimic the effects of therapy animals, a variety of robot pets are currently on the market. The benefits of interacting with robotic dogs on mental health have been highlighted in several studies, including improved mood and socialising as well as decreased depression, stress, and anxiety. Many of these research concentrated particularly on elderly people and those suffering from dementia. Another popular interactive companion robot, Paro is shaped like a adorable closure and has many different sensors. With the aid of these sensors, Paro can identify its surroundings and its inhabitants (Cooper, Fava, Vivas, Marchionni, & Ferro, 2020).

## Peer/Interactive Playmates: Social Robots

The physical presence of social robots as well as their abilities to communicate with which train the robot in particular skills. A further benefit of social robots over humans is their behavior's repeatability and predictability. Robots offer a way to investigate and practise communal services without being intimidated by the intricacy and variety of human-human communication. In order to impart the essential social skills and involve users in interactions that are therapeutically pertinent, social robots can engage in fun dyadic human-robot play activities. The robot will focus solely on one individual during this form of contact. Additionally, play activities can be created that are tailored to a person's unique requirements, interests, and abilities. For instance, based on how each user is doing, the level of social skills instruction could be incrementally changed (Turja & Parviainen, 2020).

## The Role of Social Robots as Social Mediators

Due to communication challenges and worries about being scrutinised by others, people who suffer from social anxiety sometimes find it difficult to participate in social situations. Werry et al. demonstrated in an observational research that situations with a robot as a very exciting social situation for viewing diverse social and non-social interaction patterns can be created by a media- tor and pairs of kids. The ability of kids with ASD in social situations can be determined as well as specific issues by analysing these behavioural patterns.

In three case studies, Robins et al. found that interacting with the minimally expressive robot Kaspar helped low-functioning autistic youngsters. A humanoid robot named Kaspar was created to assist kids with ASD. It has also been demonstrated that playing with Kaspar has a good impact on some kids' behaviour in specific domains, such as mimicry, prompted speech, focus, and communication.

### Social Robots as instructors or Coaches

The creation of social robots that may serve as a teacher or a guide is gaining popularity user's social, physical, or mental health by monitoring and involving them in a variety of healing. By tracking each participant's weight reduction and offering tailored feedback, Members stuck with it for a lot longer than those in the other two conditions after the first six weeks. Additionally, it was said that the working alliance with the robot had become much tighter. In a different study, Fasola and Mataric created Bandit, a robotic coach system, to encourage elderly users to engage in physical activity. The robotic coach methodology used in this study was created based on psychological studies on users' intrinsic motivation. In order to investigate the significance of physical embodiment, the researchers also contrasted the real-world Bandit robot with its virtual counterpart. The findings demonstrated that participants' assessments of the robot and the interaction were improved by physical embodiment.

## ELEMENTS INFLUENCING WHETHER SOCIAL ROBOTS ARE ACCEPTED

Social robots are those that are useful and have the social intelligence and abilities necessary to interact with people in a way that is acceptable to society. This implies that they must be able to interact with the user and be recognised as a social entity by the user (Panico, Cordasco, Vogel, Trojano, & Esposito, 2020). This term incorporates robots that are primarily used as companions.

The objective is to improve users' psychological and mental health, as well as service robots that assist humans in performing daily tasks.

Acceptability is characterised as "the robot being voluntarily absorbed into the life of the older person", which suggests long-term use.

### Beliefs and Concerns About Technology

Users develop a mental picture of robots before they have their first firsthand experience with one, which shapes how they react to the robot. Science fiction and the media are examples of secondhand sources of knowledge outside the individual that can affect mental models. For instance, zoomorphic robots like Paro, who assumes the form of a baby harp seal, may pique patient' interest and will also help them reminisce their past by conjuring up positive, nurturing feelings that were once produced while interacting with pets (Kang, 2020). The Flow of Factor affecting trust in human SAR is shown in Figure 6.

Expectations regarding what a robot can and cannot do influence prior experiences and a person's attitude toward it. This is related to anthropomorphism, which is the tendency for people to think of both robotic and non-robotic objects as having human-like mental capacities. We go into more detail about how this happens below.

*Figure 6. The flow of factors affecting trust in human SAR*

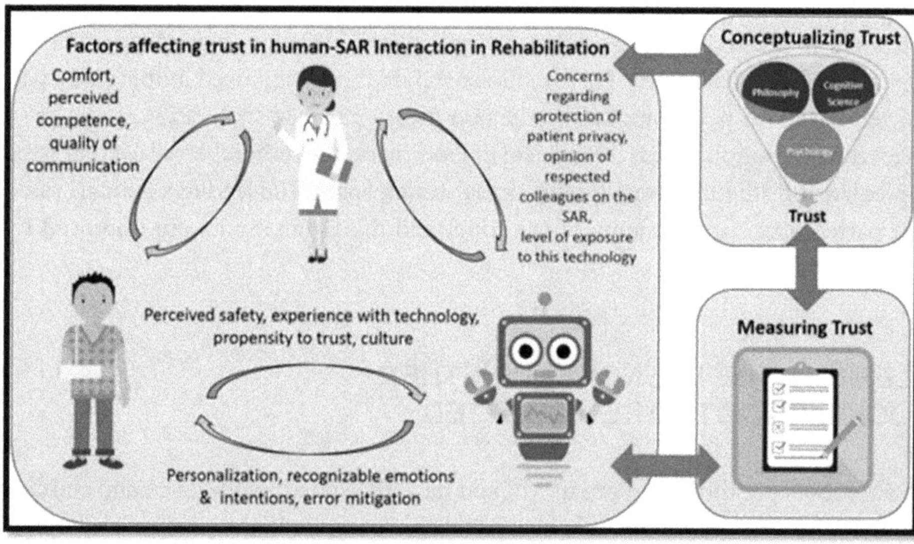


The way people feel about specific robots and the level of fear or emotion they cause depends on how much of a human's perception of a robot's capacity for feeling (mind experience) and action is shared by the robot (mind agency). The latter comprises opinions regarding its morality, memory, and self-control. According to Takayama, a robot that is thought to have human characteristics like its own thinking ability, moods, targets, wishes etc. A robot with a moderate level of perceived agency, is seen as a tool (Moerman & Jansens, 2020). Takayama makes a distinction between immediate insights and thoughtful insights that come from carefully considering a state of affairs. People tend to react to robots impulsively, which can cause them to be seen as highly agentic creatures in the here-and-now. This propensity could make it easier for people to connect emotionally with robots and evoke responses from others. It has been suggested that loneliness and thankfulness for a device that aids them may both increase a person's propensity to anthropomorphize.

It has been proposed that robots that are known to have little agency but lots of familiarity with similar situations will perform better as companion robots. This condition is met by Paro, a very successful robot that exhibits a great deal of emotion but minimal agency.

A complex picture is painted by the research about the effects of gender, age, and experience specific to computer, education in general on fear and attitude toward robots.

According to Arras and Cerqui, 35% of males had a additional favourable perception of automation than did 8% of females, who were shown to be more dubious about every facet of robot technology. Additionally, compared to 22% of those under the age of 18, 39% of older adults had a more favourable perception of robots. Although they rejected the notion that robots could replace human social interaction, Older Adults thought that robots could improve their quality of life and overall satisfaction (Ferreira & Sequeira, 2014).

The robot is a movable, cylinderical shape device with a screen face that of a female that can function independently and link to smart IOT technologies at home, according to the authors. According to data gathered through surveys, participants with higher levels of education were less likely to view the robot as a social being. Additionally, those with more CE said it was simpler to use (PEOU). Correlations between CE and PEOU and gender differences revealed that men had higher CE, which raised their PEOU. But this research also discovered that degrees of apprehension about the robot were massive and linked with education levels, CE, and age. Ten Factors influencing and accessibility of social robots are shown in Figure 7.

*Figure 7. Ten factors influencing and accessibility of social robots*

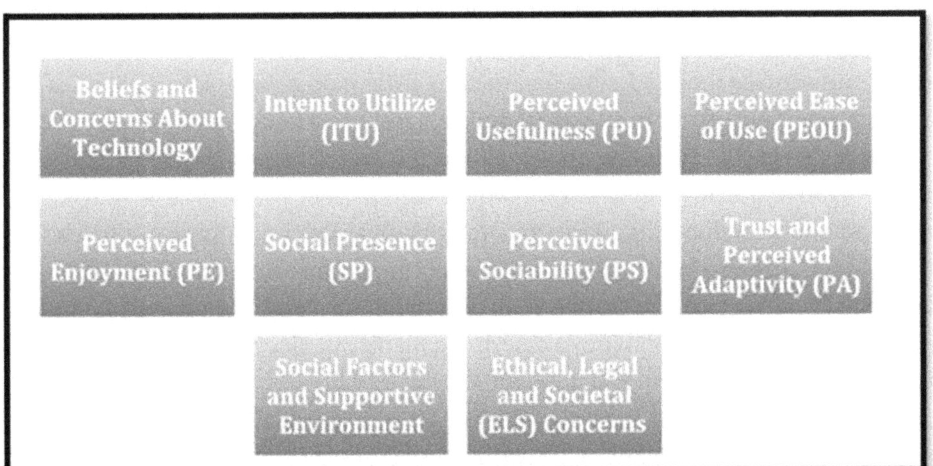

Normura et al. performed a poll to examine the relationship between age and apprehension regarding robots. They discovered that younger persons, who had direct or indirect exposure to humanoid robots, expressed higher degrees of apprehension toward robots than people in their 50s and 60s. However, older adults had a much higher scepticism of technology than did younger individuals. The former also had less grasp of the capabilities of technology and considered it more challenging to use. Compared to males, women were less open to the idea of using robots. It's interesting to note that the age groups adopted various learning methods while using new technologies: young people experimented, adults read manuals, and older adults preferred to seek assistance. The study also discovered that older adults preferred that robots not be able to move around the home freely and that just very few of older adults, compared to half of younger adults, said that having a robot perform duties in their home would make them feel entirely safe and at ease (Goris, Saldien, Vanderniepen, & Lefeber, 2009). The conclusion that older adults might react to technologies differently than younger persons was corroborated by Scopelliti et al. Three generations from six families (n = 23) participated in their pilot qualitative survey, which revealed that older adults had a favourable opinion of robotic technology. However, participants in other age groups voiced various goals, whereas older adults were focused on the harmonious integration of robots into the home environment.

## Intent to Utilize (ITU)

According to the data, using a robot and being more familiar with it, as opposed to only hearing about it from a third party, might change aspects that affect acceptance. Stafford et al. used a robot attitude measure to record attitudes toward the robot Cafero before and after employees and older adults residents in a retirement community interacted with it for 30 minutes. Both participant groups exhibited fewer hostile feelings toward the robots after the interaction. In their observational qualitative field trial that was carried out in a "smart" house, Gross and Schroeter et al. discovered a comparable improvement in attitude. They discovered that some older adults with mild cognitive impairment (MCI) and their care partner partners (n = 4 dyads) at first had unfavourable initial reactions to and considered the companion robot as frightening. However, after using it for one day, they began to recognise its advantages and found it to be more acceptable. With older adult residents played with an iCat for 3 min and filled a questionnaire to determine their ITU, Heerink examined whether ITU predicted actual robot usage (Kautish et al., 2022).

Following the session, iCat was left in a lounge where attendees could use it if they feel like. This subsequent use was captured on camera. They discovered that whereas ITU did not usually anticipate real consumption, it occasionally did.

The use of Steffie, a fictional screen persona who helped participants with online computer tasks, was observed in a second experiment with older adults (n = 30). The PCs belonging to the participants have this application installed. ITU, according to Heerink, is influenced by additional acceptance criteria that can predict the perceived value and attitude towards a robot.

ITU may present challenges for examining robot acceptability with older adults and PWD, according to Stafford. This is due to the fact that when participants are aware that they won't have access to a robot after the study is over they don't feel interested to give insights about their intentions to utilise them (Jupalle et al., 2022; Kaur & Kautish, 2022; Kautish et al., 2021; Madhu et al., 2022; Sampathkumar et al., 2020).

The work of Pfadenhauer and Dukat sheds light on the significance of contextually investigating acceptability parameters. The use of member remark and audiovisual explicit certification of roughly three group action sessions each calendar month, over a period of one year, allowed them to conduct an anthropological examination of the implementation of Paro in a German residential care environment for PWD. They discovered that Paro was utilised in a variety of capacities, such as a conversationalist, an observation tool, and a communication aid. Users' views of their unmet needs and how well they believe a specific robot would fulfil them are factors that impact their judgements.

## Perceived Usefulness (PU)

Users need to see social robots as relevant to their current unmet needs and as useful. In three 10-day stints spread over five months, De Graaf] investigated acceptance in the households of older adults (n = 6) with the help of the rabbit-like health promotion robot Karotz. The robot was programmed to welcome users, give weather updates, give advice on timely activity reminders, talk about daily routines, and prompt users check their weights. Semi-structured interviews were done and interactions were videotaped. Researchers discovered that during each phase of usage, participants mostly discussed whether or not the robot was beneficial to them. Various role of Role of social robots for children with ASD is shown in Figure 8.

This shows that precisely defining needs could increase human acceptance of robots. Determining the perceived demands of older adults and PWD, however, can be challenging and is influenced by a variety of circumstances. For instance, it might be difficult to recognise unmet requirements if older adults have a diminished awareness of their own needs as a result of habit or if they are reluctant to acknowledge their impairment out of fear of being stigmatised or losing their independence . Additionally, PWD may lack the cognitive capacity to recognise or articulate their requirements or they may think that social robots are useless if they are already being taken good care of by carers . Indeed, according to a number of studies, PWD and their carers may not agree on the specifics of their unmet requirements or on potential robot-based solutions. This fact has an effect on people's acceptance of robots, and societal influences are examined below in more detail.

*Figure 8. Role of social robots for children with ASD*

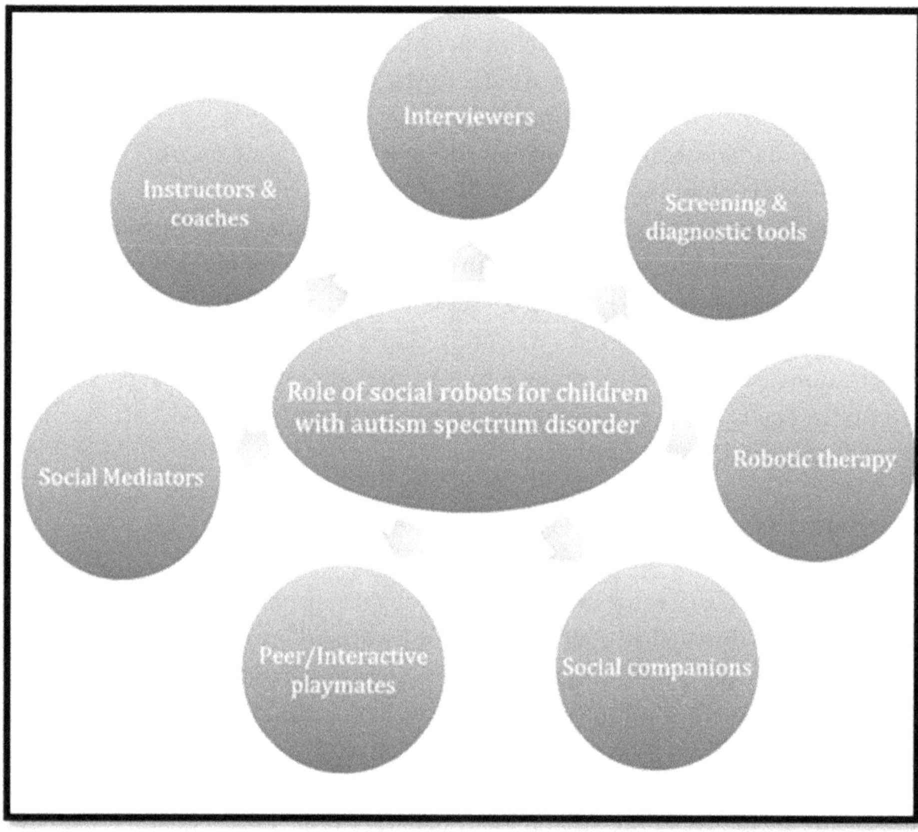


## Ease of Use

Robot design must be tailored to environmental considerations, user type, and individual requirements, in order to increase the acceptance and usability of robots. As a result, all social robots must be simple to clean. Those intended for use in people's homes must be durable, must require very less upkeep, and be able to manoeuvre through situations with moving and stationary obstructions, irregular floors and perhaps staircase, varying lighting situations, and varying door thresholds. Different designs are available in residential care because of the longer halls, potential static floor plans, and identical furniture (Jupalle et al., 2022; Whig & Ahmad, 2014d; Whig, Kouser, Velu et al, 2022; Whig, Nadikattu, & Velu, 2022).

Robots can be more user-friendly for older adults with impaired hearing, visual impairment, or cognitive deficit if they support a variety of interacting modalities. Khosala et al. discovered that during a game of Hoy with a robot. The numbers were called out vocally, but the robot's visual display helped participants recall and visualise them. It should be mentioned, too, that although humans prefer direct speech from robots over other acoustic and visual modalities of communication.

Longer use, habituation, and learning may result in a higher PEOU rating. In a usability lab modelled to real-life user apartments, Torta et al.'s study evaluated the appropriateness of a robot used with an integrated smart home system as a communication interface.

Some older adults had two sessions over a two-week period, while few older adults received eight sessions over a three-month period. Later sessions saw an improvement in the system's usability, with participants reflecting in particular on how they had grown accustomed to the robot's speed and behaviour.

In conclusion, it's critical that robots are matched to end users' demands and skills. PEOU can be enhanced with time, practise, and education. The following, however, have been noted in the literature as PEOU-related factors: the robot's voice and visual communication, button usability, and monitor adaptability.

## Perceived Enjoyment (PE)

When users have the option to utilise robot(s) in a home setting voluntarily, motivational variables like PE come into play since acceptability rises if the robot is seen as entertaining and fun. In a study utilising an iCat robot that was made talkative by using a hidden opera- tor, Heerink et al. discovered that PE substantially linked with minutes of actual usage (0.625, p 0.01) and minutes of intention to use (0.420, p 0.05). Semi-independent older adults were the participants (n = 30). Participants responded to surveys on their interactions with iCat during which they inquired about the weather, tell a joke, televisison schedules. De Graaf's and Torta et al. findings, however, indicated that perceived enjoyment diminished over 6 and 8 months, respectively.

## Social Presence (SP)

Users need to feel as though they are with their regular social companion when they are with robots whose aim it is to motivate and stimulate them, hence these robots must exhibit a level of social presence (SP) appropriate to that objective. In fact, it would seem that the ability of robots to possess SP is what sets them apart from non-robotic technology. By deploying personified robots, SP can be improved.

In an 8-month study with PWD (n = 9), Tapus and Tapus investigated the use of a robot as a tool to monitor and promote cognitive activities for PWD. Through playing music and interactive activities with the user, the robot provided customised cognitive stimulation. Researchers contrasted reactions to a simulation on a large computer screen with those to a humanoid torso design on a mobile platform. They discovered that participants consistently preferred the embodied robot over the computer.

However, the robot's size is also crucial since if it is too small, people may not notice it, resulting in SP that is less than ideal. In a usability lab designed to seem like an actual apartment, Torta et al. tested robot in a smart home environment as a communication interface. Robot acceptability was tested by older adults (n = 8) using scenarios including as calling a friend to arrange a meeting, enquiring about the weather, listening to music, exercising, and getting environmental warnings. Participants went through 2–8 sessions spread out over varying time frames of 2–12 weeks. Participants reported little nervousness and appreciated the robot, although its diminutive size resulted in a very low SP score.

The size of robots must also be considered. The acceptance of two robots, Guide and Paro, for person with dementia was examined by Robin-son et al.

While Paro is roughly 55 cm long, Guide is 1.6 m tall. A 1-hour interactive session with the robots was recorded, transcribed, and evaluated. Staff and family members were also the subjects of semi-structured interviews. The results showed that Paro (n = 6) received responses and conversations from residents more frequently than Guide (n = 2). In contrast to 40% (n = 4) who touched Guide, all residents touched Paro. Paro received better reviews from staff and family members than Guide. Paro, they reasoned, would be more practical in their environment because had aesthetic eyes and it promoted demonstrative touch. But some family members (n = 5) and employees (n = 3) felt it was overly huge and advised that its size should be reduced. Although most participants believed that PWD would not be able to utilise it on their own, it was noted that Guide had the ability to support activities and stimulate residents. Participants' opinions on the size of the Guide in particular were divided. Others agreed that its scale allowed unlikely to be noticed, while some thought it was too enormous and intimidating. The results of some studies, however, contend that big robots can evoke feelings of companionship, fear, and unease, are consistent with this conclusion.

## Perceived Sociability (PS)

Aspects of robot look, behaviour, and communication methods have an impact on PS.

People have different views about the components comprising the robots of and the color of their bodies, according to Scopelliti et al.'s research. A 40-inch-tall prototype robot (Ed), capable of delivering verbal cues to aid PWD in executing household sequences of actions like preparing a cup of tea, underwent an acceptance and feasibility evaluation by Begum et al. in a home simulation laboratory. Interviews with carers and PWD (n = 5) were recorded by researchers. They noted a lack of agreement over the gender a robot should portray and if it should have a soft or powerful voice.

How realistic a robot should appear and user preferences for a human- or machine-like appearance are other factors that affect robot design. These inquiries are related to the uncanny valley theory, that contends that users accept robots much more as they become similar to humans, but get uneasy with them when they approach becoming fully human. In the Stafford et al. study mentioned above, perceived human similarity was linked to higher levels of anxiety and heart rates in older adults volunteers than in their official carers. This implies that the idea of the uncanny valley differs between people and societies and may be connected to fear.

Zoomorphic robots, on the other hand, may experience a different effect of realism on acceptability. Paro was evaluated by Heerink et al. alongside various zoomorphic robots, including a newborn seal, dog, cat, dinosaur, and bear. They watched the reactions of 15 patients with intermediate dementia and 36 professional carers being interviewed. Each PWD was exposed to a different robot over the hour-long sessions, and their reactions were recorded. The infant seal received the greatest marks for its ease of use, softness, and because it was more portable and lighter than Paro. Second choice was the cat since it was more believable. The dinosaur Pleo received the lowest score because it was thought to be strange and reptilian.

According to research on face traits, people's preferences for certain aspects and whether or not they should be human-like vary. Some robot designers have investigated whether people might accept human-like robots with simple face features. Matilda is a little social robot with a minimalist baby like face that has undergone successful field trials, according to Khosla et al. She features a facial expression detection system and can incorporate user choices to tailor her services. Trials with PWD (n = 7) and their carers were carried out over the course of six months in seven Australian houses. Data was analysed to determine the emotional response of participants and the calibre of their interactions with the robot while conducting interviews and videotaping exchanges. According to the research, PWD seemed to appreciate their one-on-one time with Matilda. In response to the question, "Matilda makes me smile," every participant agreed or strongly agreed, stating that "Matilda is a friend" and "Matilda does not worry me."

It is obvious that there is no general agreement on what social robots should look like. Users react to a package that includes the robot's expressions and communicative behaviour rather than a robot's looks, which does not affect acceptance when viewed in isolation.

Robots may need to learn "robo-tiquette," according to some. Being perceived as warm, open, inventive, calm, spontaneous, effective, systemic, cooperative, polite, and cheerful must be part of this. User and robot personalities are also significant issues. A mobile robot that can remind users to take their medications and keep their appointments was used in two experiments by Brandon on relatively healthy and capable older adults (n = 22). The purpose of the studies was to determine what effect personality compatibility between the user and robot had. However, a robot's personality and actions must match its intended use and the expectations of its users. In a laboratory environment, Amirabdol-lahain et al. looked at how older adults (n = 41) responded to robots performing particular activities, and roles. They wanted to know if people's preferences for robots varied depending on the situation and their preconceived notions of what particular jobs entailed. They discovered that having a personality that matches the user's expectations for the specific activity and context—rather than having a complementing or similar personality to the robot—is what increases user acceptance of robots.

## Trust and Perceived Adaptivity (PA)

The significance of trust is examined, arguing that it influences and explains the demand for a sense of control over a social robot and its adaptivity. Users are said to need to have confidence in the robot and feel at ease with a certain amount of perceived control, but they also need communally astute robots to have some self-sufficiency. With regard to robot goal and deployment scenario, an optimal balance between these elements presumably differs between different users.

To support these claims, additional study with larger samples is necessary.

Heerink examined the effects of PA acceptance under the same experimental circumstances as those mentioned above (Heerink, 2011). The more adaptable robot was favoured by the participants, who also gave it higher ratings for ITU, perceived enjoyment, and perceived usefulness. The authors hypothesised that this was because they had less control over the robot's behaviour because they felt more uneasiness toward it.

Users must have faith in robots' dependability and safety, and this faith must be gained. A series of workshops were held with older adults who were housebound and had mild sensory and movement problems by Frennert et al. Responding to sketches of several robots, participants were invited to express their ideal robot preferences. Additionally, they conducted interviews with older adults (n = 5) and one couple who spent one week residing with foam polystine mock-ups of these ideas. They discovered that control sentiments were significant and linked to issues of privacy and trust.

A survey questionnaire with older adults (n = 292) was used to examine the factors influencing user trust and ITU with assisted living robots. In this study, participants were given two crisis situtations with robot assistance in a fire and during a time of extreme illness. Naturally, users indicated that they would prefer to employ the robots in these circumstances and that their level of trust in the robot was substantially correlated with ITU (0.51). A correlation between trust levels and PEOU (0.49), PU (0.50), and predicted reliability was also found (0.63). Scopelliti also discovered that responses from older adults about the advantages, downsides, and distrust of robots were influenced by trust in the capabilities of robots for usage in a home environment.

Eighty-five percent of individuals demonstrated mistrust by refusing to allow a robot free rein in the home, while 82% expressed concern over potential damage.

## Social Factors and Supportive Environment

Social influences cover broader cultural concerns as well, although few research in this study appear to account for these difficulties, and none provided information about the cultural background of their samples. Two studies were carried out across multiple nations.

Users in care facilities in Germany and UK and accepted PARO and PLEO to a similar extent, according to Klein and Cook. When asked about robot design, older adults in the UK and French Republic expressed stronger anxieties around the need for confidentiality than older adults in the Netherlands, according to Amirabdollahian et al. The former did not want pictures of their house to be shown to others. Social Factors and Supportive Environment is shown in Figure 9.

Negative ageist attitudes are a further cultural and societal problem that may hinder the acceptance of robots. Neven observed researchers interviewing older adults (n = 6) and 30–60-minute exchanges between them and an unnamed robot in order to investigate how images of older adults influence technological development. They discovered that older adults may have various interpretations of what being older implies and that ageist presumptions influenced robot design and deployment.

Additionally, older adults may be unwilling to associate with potential robot users if they are viewed by older adults as lonely, reliant, and isolated. This could be because employing the robot would go against their perception of themselves and the independent, healthy selves they desire to portray.

*Figure 9. Factors comprising the social environment of a human being*

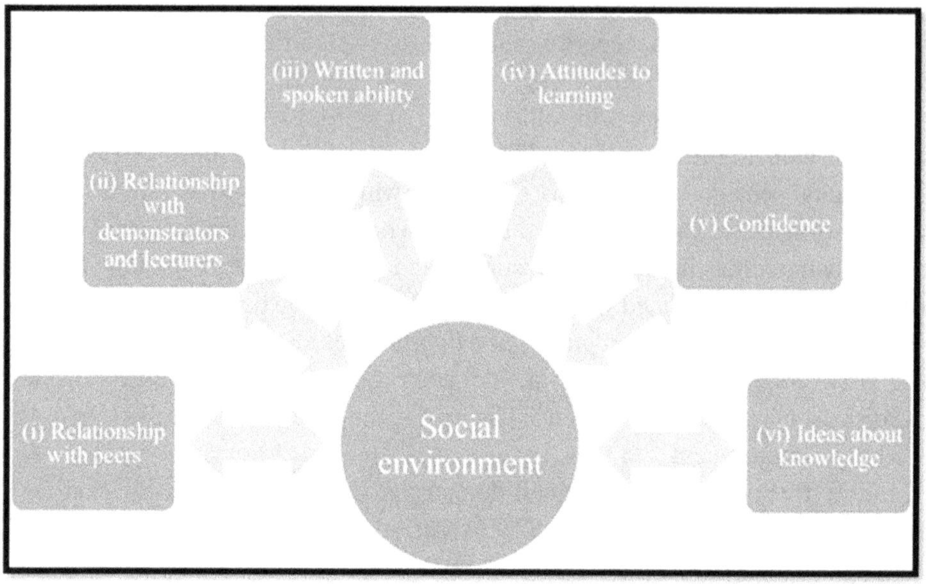


Stakeholder views on the morality of robot use also have an impact on acceptability. Three videotaped focus groups were done by Wu et al. with older adults and older adults with MCI who had a range of opinions concerning the what the robots should look like, but all participants and expressed anxiety about robots displacing or limiting human interaction and also highlighted the ethical problems associated with the robots.

## Ethical, Legal, and Societal (ELS) Concerns

In contrast to modern technologies, social robots have a different impact on people's privacy. This is crucial to separate the confidentiality features or types that apply in this situation in order to comprehend the privacy implications of social robots. We can distinguish between physical privacy and informational privacy in that sense (concerns Due to their ability to access private areas, social robots pose a threat to physical privacy. Informational self-determination, a key principle of European data protection law, is more in line with the concept of informational self-determination, which is referred to as informational privacy. Social robots make the acquisition of personal information "murky" because they are constantly processing data in the background and because the processing and dissemination of that data are difficult for a single user to understand. The ability to comprehend a social robot's data processing and disclosure capabilities is complicated by the anthropomorphic effect of social robots. Informational privacy issues apply to both direct interactions between people and robots as well as interactions between people mediated by robots, for instance, when a robot is compromised or a telepresence robot is used for spying.

It's time for legislation on robots. The European Union is working to provide legal certainty for robot legal compliance as a result of EP Resolution 2015/2103 (INL), as "doing otherwise would negatively impede the development and uptake of robotics." To better account for new developments in machinery types like robotics and digitization, the machinery directive is likely to be updated in the future.

We don't yet have any hard data on the lasting effects of health care and treatment robots due to the limited usage of social robots at this time. It is so challenging to provide an answer to the key question of whether prolonged involvement with robots decreases human-human interaction. Studies on therapeutic robots and carer robots like Paro, however, point to beneficial benefits on users, particularly for certain populations like young people with autism spectrum disorder and older people with dementia.

Last but not least, the difficulties in the field of human-robot interaction mainly concerned the societal effects of healthcare robots, such as generational differences in robot adoption or possibly reduced human-human connection.

## CONCLUSION AND FUTURE SCOPE

The ability to perform the function of an interactive social communicator and, therefore, to be a social robot is the most recent evolution of collaborative robots (historically suggested for collaboration with human beings). This new position has a lot of potential in terms of rehabilitation and assisting people with impairments, especially the elderly and disabled. SRs have shown significant promise in the care of the elderly and youngsters with communication disorders such as autism. We have recently seen increased activity in both scientific and therapeutic uses of SRs as a result of the COVID-19 epidemic. Once we think about SRs, and if we are concerned about the problems listed above, we must also consider the other aspect of the coin, which is that, during this pandemic season, a robot of this type might provide solutions to many issues faced in hospital environments, such as a lack of personnel. During a lockdown, many elderly and disabled persons are left totally alone in their houses, with no access to health care. Furthermore, even before the COVID-19 epidemic, there existed an issue of help for the old, fragile, crippled, sick, solitary, and ego. It is little question that robotics will play a major role in future health and care. The robots will aid in surgical operations, rehabilitation, building automation, hospital cleaning, serve meals and prescriptions, and provide general assistance. While it is true that robots cannot now communicate emotional states, they can do tasks in a precise and effective manner and might be of great assistance in dealing with handicap issues and other health-care issues. From just an economical perspective, it is highly intriguing for insurance firms in a variety of ways, including the prospect of establishing new insurance formulae centred on the use of care-robots, as well as the introduction of new policies that cover the risks associated with the use of robots. Similar to other artificial intelligence applications, a major element for the spread of SRs will undoubtedly be the opinion and acceptance of all concerned actors, ranging from physicians, nurses, and carers to clients and their companions. This book chapter is very useful for the researchers working in the same field.

## REFERENCES

Agrawal, S. (2018). *The encyclopedia of medical robotics.* World Scientific. doi:10.1142/10770

Ahn, H. S., Yep, W., Lim, J., Ahn, B. K., Johanson, D. L., Hwang, E. J., Lee, M. H., Broadbent, E., & MacDonald, B. A. (2019). Hospital Receptionist Robot v2: Design for Enhancing Verbal Interaction with Social Skills. In *Proceedings of the 2019 28th IEEE International Conference on Robot and Human Interactive Communication (RO-MAN),* New Delhi, India. 10.1109/RO-MAN46459.2019.8956300

Alemi, M., Meghdari, A., Ghanbarzadeh, A., Moghadam, L. J., & Ghanbarzadeh, A. (2014). Effect of Utilizing a Humanoid Robot as a Therapy-Assistant in Reducing Anger, Anxiety, and Depression. In *Proceedings of the 2014 Second RSI/ISM International Conference on Robotics and Mechatronics,* Tehran, Iran. 10.1109/ICRoM.2014.6990993

Babaiasl, M., Hamed, S., Poorya, J., & Mojtaba, Y. (2016). A review of technological and clinical aspects of robotaided rehabilitation of upper-extremity after stroke. *Disabil Rehabil Assist Technol Taylor & Francis, 11*(4), 263–280. doi:10.3109/17483107.2014.1002539 PMID:25600057

Banthia, V., Maddahi, Y., May, M., Blakley, D., Chang, Z., Gbur, A., Tu, C., & Sepehri, N. (2016). *Development of a Graphical User Interface for a Socially Interactive Robot: A Case Study Evaluation.* IEEE.

Bartlett, M. E., Costescu, C., Baxter, P., & Thill, S. (2020). Requirements for Robotic Interpretation of Social Signals "in the Wild": Insights from Diagnostic Criteria of Autism Spectrum Disorder. *Information (Basel), 11*(2), 81. doi:10.3390/info11020081

Belda-Lois, J. M., Mena-del Horno, S., Bermejo-Bosch, I., Moreno, J. C., Pons, J. L., Farina, D., Iosa, M., Molinari, M., Tamburella, F., Ramos, A., Caria, A., Solis-Escalante, T., Brunner, C., & Rea, M. (2011). Rehabilitation of gait after stroke: A review towards a top-down approach. *J Neuroeng Rehabil BioMed Central Ltd*, *8*(1), 66. doi:10.1186/1743-0003-8-66 PMID:22165907

Beraldo, G., Menegatti, E., Tommasi, V. D., Mancin, R., & Benini, F. (2019). A Preliminary Investigation of Using Humanoid Social Robots as Non-Pharmacological Techniques with Children. In *Proceedings of the IEEE International Conference on Advanced Robotics and Its Social Impacts (ARSO)*, Beijing, China. 10.1109/ARSO46408.2019.8948760

Burns, R., Jeon, M., & Park, C. H. (2018). Robotic Motion Learning Framework to Promote Social Engagement. *Applied Sciences (Basel, Switzerland)*, *8*(2), 241. doi:10.3390/app8020241 PMID:35582004

Bustamante Valles, K., Montes, S., Madrigal, M. J., Burciaga, A., Martínez, M. E., & Johnson, M. J. (2016). Technologyassisted stroke rehabilitation in Mexico: A pilot randomized trial comparing traditional therapy to circuit training in a robot/technology-assisted therapy gym. *Journal of Neuroengineering and Rehabilitation*, *13*(1), 1–15. doi:10.118612984-016-0190-1 PMID:27634471

Carrillo, F. M. (2018) Physiotherapists' acceptance of a socially assistive robot in ongoing clinical deployment. In: *Proceedings of 27th IEEE international symposium on Robot and Human Interactive Communication, RO-MAN'18*, Nanjing China. 10.1109/ROMAN.2018.8525508

Chen, C., Garrod, O. G. B., Zhan, J., Beskow, J., Schyns, P. G., & Jack, R. E. (2018). Reverse Engineering Psychologically Valid Facial Expressions of Emotion into Social Robots. In Proceedings of *the 2018 13th IEEE International Conference on Automatic Face & Gesture Recognition,* Xi'an, China. 10.1109/FG.2018.00072

Chung, K., & Park, R. C. (2019). Chatbot-Based Heathcare Service with a Knowledge Base for Cloud Computing. *Clust. Comput. J. Netw. Softw. Tools Appl.*, *22*(S1), 1925–1937. doi:10.100710586-018-2334-5

Cooper, S., Fava, A. D., Vivas, C., Marchionni, L., & Ferro, F. (2020). ARI: The Social Assistive Robot and Companion. In *Proceedings of the 2020 29th IEEE International Conference on Robot and Human Interactive Communication (RO-MAN)*, Naples, Italy. 10.1109/RO-MAN47096.2020.9223470

Diaz-Boladeras, M., Angulo, C., Domenech, M., Albo-Canals, J., Serrallonga, N., Raya, C., & Barco, A. (2016). Assessing Pediatrics Patients' Psychological States from Biomedical Signals in a Cloud of Social Robots. In *Proceedings of the XIV Mediterranean Conference on Medical and Biological Engineering and Computing,* (pp. 1179–1184). Springer. 10.1007/978-3-319-32703-7_230

Dodds, P., Martyn, K., & Brown, M. (2018). Infection Prevention and Control Challenges of Using a Therapeutic Robot. *Nursing Older People*, *30*(3), 34–40. doi:10.7748/nop.2018.e994 PMID:29569863

Eriksson, Y. (2018). The Perception of Aging and Use of Robots. In *International Conference on Human Aspects of IT for the Aged Population,* (pp. 30–39). Springer: Cham, Switzerland. 10.1007/978-3-319-92034-4_3

Ferreira, I., & Sequeira, J. When Children Interact with Robots: Ethics in the Monarch Project; IEEE: Piscatway, NJ, USA, 2014; ISBN 978- 981-4623-34-6.

Gnjatovic, M. (2018) Pilot corpus of childrobot interaction in therapeutic settings. In: *Proceedings of 8th IEEE international conference on Cognitive Infocommunications, CogInfoCom 2018*, Debrecen, Hungary. 10.1109/CogInfoCom.2017.8268252

Goncalves, D., & Arsenio, A. (2015). *Human-Driven Multi-Robot Design Process for Social Interactions with Children on Complex Environments*. IEEE. doi:10.1109/ICARA.2015.7081127

Goris, K., Saldien, J., Vanderniepen, I., & Lefeber, D. (2009). The Huggable Robot Probo, a Multi-Disciplinary Research Platform. In *International Conference on Research and Education in Robotics*. Springer. 10.1007/978-3-642-03558-6_4

Hung, L., Gregorio, M., Mann, J., Wallsworth, C., Horne, N., Berndt, A., & Liu, C. (2019). Exploring the Perceptions of People with Dementia about the Social Robot PARO in a Hospital Setting. Dement. *Int. J. Soc. Res. Pract.*

Jupalle, H., Kouser, S., Bhatia, A. B., Alam, N., Nadikattu, R. R., & Whig, P. (2022). Automation of human behaviors and its prediction using machine learning. *Microsystem Technologies*, *28*(8), 1–9. doi:10.100700542-022-05326-4

Kang, S. (2020). A Study on Smart Homecare for Daily Living Ability and Safety Management of the Elderly. *Inf. Sci. Appl.*, *621*, 707–710. doi:10.1007/978-981-15-1465-4_72

Kaur, R., & Kautish, S. (2022). Multimodal sentiment analysis: A survey and comparison. *Research Anthology on Implementing Sentiment Analysis Across Multiple Disciplines*, 1846-1870.

Kautish, S., Peng, S. L., & Obaid, A. J. (Eds.). (2021). Computational intelligence techniques for combating COVID-19. Springer International Publishing. doi:10.1007/978-3-030-68936-0

Kautish, S., Reyana, A., & Vidyarthi, A. (2022). SDMTA: Attack detection and mitigation mechanism for DDoS vulnerabilities in hybrid cloud environment. *IEEE Transactions on Industrial Informatics*, *18*(9), 6455–6463. doi:10.1109/TII.2022.3146290

Khan, M. J., & Hong, K.-S. (2017). Hybrid EEG–fNIRSbased eight-command decoding for BCI: Application to quadcopter control. *Frontiers in Neurorobotics*, *11*, 6. doi:10.3389/fnbot.2017.00006 PMID:28261084

Khosravi, P., & Ghapanchi, A. H. (2016). Investigating the Effectiveness of Technologies Applied to Assist Seniors: A Systematic Literature Review. *International Journal of Medical Informatics*, *85*(1), 17–26. doi:10.1016/j.ijmedinf.2015.05.014 PMID:26216463

Larriba, F., Raya, C., Angulo, C., Albo-Canals, J., Diaz, M., & Boldu, R. (2016). Externalising Moods and Psychological States in a Cloud Based System to Enhance a Pet-Robot and Child's Interaction. *Biomedical Engineering Online*, *15*(S1), S72. doi:10.118612938-016-0180-3 PMID:27454772

Law, P. (2015). Social robotics in health-care service: the case of rehabilitation programmes in Hong Kong. In J. Vincent, (Ed.), *Social robots from a human perspective* (pp. 55–65). Springer International Publishing. doi:10.1007/978-3-319-15672-9_5

Lewis, M., & Canamero, L. (2013). Are Discrete Emotions Useful in Human-Robot Interaction? Feedback from Motion Capture Analysis. In *Proceedings of the 2013 Humaine Association Conference on Affective Computing and Intelligent Interaction*, Geneva, Switzerland. 10.1109/ACII.2013.23

Li, J., Wu, T., Xu, Z., & Gu, X. (2014). A pilot study of post-total knee replacement gait rehabilitation using lower limbs robot-assisted training system. *European Journal of Orthopaedic Surgery & Traumatology*, *24*(2), 203–208. doi:10.100700590-012-1159-9 PMID:23412304

Logan, D. E., Breazeal, C., Goodwin, M. S., Jeong, S., O'Connell, B., Smith-Freedman, D., Heathers, J., & Weinstock, P. (2019). Social Robots for Hospitalized Children. *Pediatrics*, *144*(1), e20181511. doi:10.1542/peds.2018-1511 PMID:31243158

Madhu, G., Govardhan, A., Ravi, V., Kautish, S., Srinivas, B. S., Chaudhary, T., & Kumar, M. (2022). DSCN-net: A deep Siamese capsule neural network model for automatic diagnosis of malaria parasites detection. *Multimedia Tools and Applications*, *81*(23), 34105–34127. doi:10.100711042-022-13008-6

Meghdari, A., Alemi, M., Khamooshi, M., Amoozandeh, A., Shariati, A., & Mozafari, B. (2016). Conceptual Design of a Social Robot for Pediatric Hospitals. In *Proceedings of the 2016 4th International Conference on Robotics and Mechatronics (ICROM),* Tehran, Iran. 10.1109/ICRoM.2016.7886804

Miao, Q., Zhang, M., Cao, J., & Xie, S. Q. (2018). Reviewing high-level control techniques on robot-assisted upper-limb rehabilitation. *Advanced Robotics*, *32*(24), 1253–1268. doi:10.1080/01691864.2018.1546617

Moerman, C. J., & Jansens, R. M. L. (2020). Using Social Robot PLEO to Enhance the Wellbeing of Hospitalised Children. *Journal of Child Health Care*. PMID:32840383

Moerman, C. J., van der Heide, L., & Heerink, M. (2019). Social Robots to Support Children's Wellbeing under Medical Treatment: A Systematic State-of-the-Art Review. *Journal of Child Health Care*, *23*(4), 596–612. doi:10.1177/1367493518803031 PMID:30394806

Neerincx, M. A., van Vught, W., Blanson Henkemans, O., Oleari, E., Broekens, J., Peters, R., Kaptein, F., Demiris, Y., Kiefer, B., Fumagalli, D., & Bierman, B. (2019). Socio-Cognitive Engineering of a Robotic Partner for Child's Diabetes Self-Management. *Frontiers in Robotics and AI*, *6*, 118. doi:10.3389/frobt.2019.00118 PMID:33501133

Nergui, M., Komekine, K., Nagai, H., & Otake, M. (2014). *Development of Laughter Motion on the Cognitive Robot "Bono-02" Assisting Group Conversation*. IEEE. doi:10.1109/U-MEDIA.2014.30

Ngo, H. Q. T., Le, V. N., Thien, V. D. N., Nguyen, T. P., & Nguyen, H. (2020). Develop the Socially Human-Aware Navigation System Using Dynamic Window Approach and Optimize Cost Function for Autonomous Medical Robot. *Advances in Mechanical Engineering*, *12*(12), 12. doi:10.1177/1687814020979430

Niyetkaliyev, A. S., Hussain, S., Ghayesh, M. H., & Alici, G. (2017). Review on design and control aspects of robotic shoulder rehabilitation orthoses. *IEEE Transactions on Human-Machine Systems*, *47*(6), 1134–1145. doi:10.1109/THMS.2017.2700634

Ono, S., Obo, T., Kiong, L. C., & Kubota, N. (2015). Robot Communication Based on Relational Trust Model. In *IECON 2015-41st Annual Conference of the IEEE Industrial Electronics Society*. IEEE. 10.1109/IECON.2015.7392941

Panico, F., Cordasco, G., Vogel, C., Trojano, L., & Esposito, A. (2020). Ethical Issues in Assistive Ambient Living Technologies for Ageing Well. *Multimedia Tools and Applications*, *79*(47-48), 36077–36089. doi:10.100711042-020-09313-7

Piau, A. (2019) Use of a robotic walking aid in rehabilitation to reduce fear of falling is feasible and acceptable from the end user's perspective: a randomised comparative study. Maturitas Elsevier 120:40–46. doi:.maturitas.2018.11.008 doi:10.1016/j

Robinson, H., MacDonald, B., Kerse, N., & Broadbent, E. (2013). The Psychosocial Effects of a Companion Robot: A Randomized Controlled Trial. *Journal of the American Medical Directors Association*, *14*(9), 661–667. doi:10.1016/j. jamda.2013.02.007 PMID:23545466

Rouaix, N., Retru-Chavastel, L., Rigaud, A.-S., Monnet, C., Lenoir, H., & Pino, M. (2017). Affective and Engagement Issues in the Conception and Assessment of a Robot-Assisted Psychomotor Therapy for Persons with Dementia. *Frontiers in Psychology*, *8*, 950. doi:10.3389/fpsyg.2017.00950 PMID:28713296

Sampathkumar, A., Rastogi, R., Arukonda, S., Shankar, A., Kautish, S., & Sivaram, M. (2020). An efficient hybrid methodology for detection of cancer-causing gene using CSC for micro array data. *Journal of Ambient Intelligence and Humanized Computing*, *11*(11), 4743–4751. doi:10.100712652-020-01731-7

Sarabia, M., Young, N., Canavan, K., Edginton, T., Demiris, Y., & Vizcaychipi, M. P. (2018). Assistive Robotic Technology to Combat Social Isolation in Acute Hospital Settings. *International Journal of Social Robotics*, *10*(5), 607–620. doi:10.100712369-017-0421-z

Schröder, J., Truijen, S., Van Criekinge, T. A., & Saeys, W. (2019). Feasibility and effectiveness of repetitive gait training early after stroke: A systematic review and meta-analysis. *Journal of Rehabilitation Medicine*, *51*(2), 78–88. doi:10.2340/16501977-2505 PMID:30516821

Sequeira, J. S., & Ferreira, I. A. (2016). *Lessons from the MOnarCH Project*. IEEE. doi:10.5220/0005998102410248

Shamsuddin, S. (2017) Preliminary study on the use of therapeutic seal robot for patients with depression. In: *Proceedings of IEEE 4th International Symposium on Robotics and Intelligent Sensors: empowering robots with smart sensors*, Tokyo, Japan. 10.1109/IRIS.2016.8066065

Sharif, M. S., & Alsibai, M. H. (2017). *Medical Data Analysis Based on Nao Robot: An Automated Approach Towards Robotic Real-Time Interaction with Human Body*. IEEE.

Shayan, A. M., Sarmadi, A., Pirastehzad, A., Moradi, H., & Soleiman, P. (2016). RoboParrot 2.0: A Multi-Purpose Social Robot. In *Proceedings of the 2016 4th International Conference on Robotics and Mechatronics*, Tehran, Iran. 10.1109/ICRoM.2016.7886775

Sheridan, T. B. (2020). A Review of Recent Research in Social Robotics. *Current Opinion in Psychology, 36*, 7–12. doi:10.1016/j.copsyc.2020.01.003 PMID:32294577

Tang, J., Zhou, Z., & Yu, Y. (2017) A hybrid computer interface for robot arm control. In: *Proceedings of 8th International Conference on Information Technology in Medicine and Education, ITME 2016*, Fuzhou, China. 10.1109/ITME.2016.0088

Tuisku, O., Pekkarinen, S., Hennala, L., & Melkas, H. (2018). Robots do not replace a nurse with a beating heart. *Information Technology & People, 32*(1), 47–67. doi:10.1108/ITP-06-2018-0277

Turja, T., & Parviainen, J.(2020). The Use of Affective Care Robots Calls Forth Value-Based Consideration. In *Proceedings of the 2020 29th IEEE International Conference on Robot and Human Interactive Communication (RO-MAN),* Naples, Italy.

Valles-Peris, N., Angulo, C., & Domenech, M. (2018). Children's Imaginaries of Human-Robot Interaction in Healthcare. *International Journal of Environmental Research and Public Health, 15*(5), 970. doi:10.3390/ijerph15050970 PMID:29757221

Whig, P., & Ahmad, S. N. (2014d). Simulation of linear dynamic macro model of photo catalytic sensor in SPICE. *COMPEL: The International Journal for Computation and Mathematics in Electrical and Electronic Engineering.*

Whig, P., Kouser, S., Velu, A., & Nadikattu, R. R. (2022). Fog-IoT-Assisted-Based Smart Agriculture Application. In *Demystifying Federated Learning for Blockchain and Industrial Internet of Things* (pp. 74–93). IGI Global. doi:10.4018/978-1-6684-3733-9.ch005

Whig, P., Nadikattu, R. R., & Velu, A. (2022). COVID-19 pandemic analysis using application of AI. *Healthcare Monitoring and Data Analysis Using IoT: Technologies and Applications*, 1.

Whig, P., Velu, A., & Bhatia, A. B. (2022). Protect Nature and Reduce the Carbon Footprint With an Application of Blockchain for IIoT. In *Demystifying Federated Learning for Blockchain and Industrial Internet of Things* (pp. 123–142). IGI Global. doi:10.4018/978-1-6684-3733-9.ch007

Whig, P., Velu, A., & Naddikatu, R. R. (2022). The Economic Impact of AI-Enabled Blockchain in 6G-Based Industry. In *AI and Blockchain Technology in 6G Wireless Network* (pp. 205–224). Springer. doi:10.1007/978-981-19-2868-0_10

Whig, P., Velu, A., & Nadikattu, R. R. (2022). Blockchain Platform to Resolve Security Issues in IoT and Smart Networks. In *AI-Enabled Agile Internet of Things for Sustainable FinTech Ecosystems* (pp. 46–65). IGI Global. doi:10.4018/978-1-6684-4176-3.ch003

Whig, P., Velu, A., & Ready, R. (2022). Demystifying Federated Learning in Artificial Intelligence With Human-Computer Interaction. In *Demystifying Federated Learning for Blockchain and Industrial Internet of Things* (pp. 94–122). IGI Global. doi:10.4018/978-1-6684-3733-9.ch006

Whig, P., Velu, A., & Sharma, P. (2022). Demystifying Federated Learning for Blockchain: A Case Study. In Demystifying Federated Learning for Blockchain and Industrial Internet of Things (pp. 143–165). IGI Global. doi:10.4018/978-1-6684-3733-9.ch008

Wynsberghe, A. (2013). van A Method for Integrating Ethics into the Design of Robots. *Ind. Robot. Int. J. Robot. Res. Appl.*, *40*(5), 433–440. doi:10.1108/IR-12-2012-451

Yu, R., Hui, E., Lee, J., Poon, D., Ng, A., Sit, K., Ip, K., Yeung, F., Wong, M., Shibata, T., & Woo, J. (2015). Use of a Therapeutic, Socially Assistive Pet Robot (PARO) in Improving Mood and Stimulating Social Interaction and Communication for People With Dementia: Study Protocol for a Randomized Controlled Trial. *JMIR Research Protocols*, *4*(2), e45. doi:10.2196/resprot.4189 PMID:25934173

Zhang, M., Davies, T. C., & Xie, S. (2013). Effectiveness of robot-assisted therapy on ankle rehabilitation – a systematic review. *Journal of Neuroengineering and Rehabilitation*, *10*(1), 30. Advance online publication. doi:10.1186/1743-0003-10-30 PMID:23517734

## ADDITIONAL READING

Breazeal, C. (2011). Social robots for health applications. In *Proceedings of the 33rd Annual International Conference of the IEEE EMBS,* Boston, Massachusetts, USA.

Fosch-Villaronga, E., Lutz, C., & Tamò-Larrieux, A. (2019). Gathering expert opinions for social robots' ethical, legal, and societal concerns: Findings from four international workshops. *International Journal of Social Robotics*, *11*, 281–297. doi:10.100712369-019-00605-z

Frennert, S., & Östlund, B. (2014). Review: Seven matters of concern of social robots and older people. *International Journal of Social Robotics*, *6*(2), 299–310. doi:10.100712369-013-0225-8

Góngora Alonso, S., Hamrioui, S., de la Torre Díez, I., Motta Cruz, E., López-Coronado, M., & Franco, M. (2018). Social robots for people with aging and dementia: A systematic review of literature. *Journal of Medical Systems*, *42*, 253. doi:10.100710916-018-1080-8 PMID:30136901

González-González, C. S., Violant-Holz, V., & Gil-Iranzo, R. M. (2021). Social robots in hospitals: A systematic review. *Applied Sciences (Basel, Switzerland)*, *11*(13), 5976. doi:10.3390/app11135976

Heerink, M., Vanderborght, B., Broekens, J., & Albó-Canals, J. (2016). New friends: Social robots in therapy and education. *International Journal of Social Robotics*, *8*(4), 443–444. doi:10.100712369-016-0374-7

Rasouli, S., Gupta, G., Nilsen, E., & Dautenhahn, K. (2022). Potential applications of social robots in robot-assisted interventions for social anxiety. *International Journal of Social Robotics*, *14*(5), 1167–1198. doi:10.100712369-021-00851-0 PMID:35096198

Scassellati, B., Boccanfuso, L., Huang, C.-M., Mademtzi, M., Qin, M., Salomons, N., Ventola, P., & Shic, F. (2018). Improving social skills in children with ASD using a long-term, in-home social robot. *Science Robotics*, *3*(21), eaat7544. Advance online publication. doi:10.1126cirobotics.aat7544 PMID:33141724

## KEY TERMS AND DEFINITIONS

**Alzheimer's Disease:** It is a brain ailment that gradually damages memory and thinking skills, as well as the capacity to do the most basic tasks.

**Anorexia Nervosa:** It is a serious and potentially life-threatening psychological and eating illness that is curable. It is distinguished by severe food restriction and an acute dread of gaining weight.

**Autism Spectrum Disorder:** It is a developmental impairment caused by brain differences due to which patients struggle with social communication and interaction, as well as confined or repetitive habits or hobbies.

**Cerebral Palsy:** It is a collection of disorders that impair a person's ability to move, balance, and maintain posture

**Dementia:** It is a condition characterized by progressive or persistent loss of intellectual ability, particularly impairment of memory and abstract reasoning, and frequently by personality change, as a result of organic brain disease.

**Humanoid:** A humanoid is a nonhuman creature that resembles or acts like a human being, such as a robot.

**Rehabilitation:** It is the process of returning someone to health or a regular life following incarceration, addiction, or sickness through training and counseling.

**Social anxiety:** It is characterized by an intense and persistent fear of being observed and judged by others.

**Social Psychology:** The scientific study of how ideas, feelings, and behaviours are impacted by the real or imagined presence of other people or by social standards is known as social psychology.

**Social Robot:** A social robot is a self-aware robot that interacts and communicates with people or other autonomous physical agents by adhering to social behaviours and rules specific to its job.

# Chapter 6
# Training a Legged Robot to Walk Using Machine Learning and Trajectory Control for High Positional Accuracy

**Amit Biswas**
*Triassic Robotics, India*

**Neha N. Chaubey**
*Dharmsinh Desai University, India*

**Nirbhay Kumar Chaubey**
*Ganpat University, India*

## ABSTRACT

*Legged robots are a class of biologically inspired robots that use articulated leg mechanisms for locomotion. Legged motion is very complex and requires specialized actuation mechanisms and complicated motion control systems to operate. Traditional legged robots were controlled by purely physics-based, however, recent developments of artificial intelligence (AI), and machine learning (ML) techniques have opened new opportunities to train locomotion skills in a legged robot in a much more efficient way than the traditional physics-based controllers. In this chapter, the authors study how machine learning techniques are used to train quadruped robots in basic locomotion skills, evaluate training accuracy, training speed, and also discussed performance, simulation environment, trajectory control, and how the authors achieved accurate tracking of trajectories. Furthermore, this chapter delves into the details of the actual quadruped robot that the authors built to evaluate locomotion policies and some challenges that were faced in building the real robot.*

DOI: 10.4018/978-1-6684-8171-4.ch006

## INTRODUCTION

Legged locomotion is a challenging task for robots. Locomotion in unstructured or uneven terrain is one of the most challenging tasks for legged robots. Legged motion is more complicated and requires specialized control systems that are difficult to develop. Recent advancements in artificial intelligence and machine learning techniques have given rise to novel methods of training and controlling legged locomotion in robots. Walking on legs by itself is a computationally intensive task. Several aspects of the locomotion must be simultaneously managed amongst a very large set of possible selections. The agent typically uses limbs/legs to support a body. Sequences of leg movements must be synchronized to achieve a state wherein the body is set in motion. Leg movements have to be carefully orchestrated to maintain the body in motion while also keeping it supported on the legs. This problem by itself is challenging given the number of possible ways the legs can be arranged. Once set in motion, several other physical aspects must be considered such as the momentum of the body, reactionary forces from the ground, speed and position control, and so on. Maintaining a fine balance between all these aspects is the challenging aspect of legged locomotion.

Further complications arise from irregularities in the terrain being traversed. Uneven surfaces create an element of unknown support point for the legs. For example, an expected footfall may not be achievable due to the unevenness of the terrain. The footfall position may not be reachable as well. The unpredictable availability of the surface makes it even more difficult to plan for footfall. Reactionary forces from the ground vary in direction and magnitude due to the unevenness of the surface. This introduces further complications in the walking process.

Traditional methods of achieving legged locomotion in robots relied on purely physics-based controllers. Purely physics-based controllers took a very rudimentary approach to calculating the physical aspects of the agent and attempted to work out a solution that would balance the agent on the legs. It would then be possible to walk on legs by planning to move the agent's body forward. While these approaches work well to a degree of success in an ideal set of environments, they fail quickly in any type of challenging environment such as on uneven terrain. Even minor deviations from the expected environment are difficult to accommodate in such methods of walking. Moreover, traditional physics-based controllers are extremely difficult to build and maintain. They require extreme fine-tuning to achieve the desired results.

Using artificial intelligence and machine learning techniques to solve the problem of robotic locomotion introduces a new frontier of work. Machine learning techniques can be used to train the robot to walk instead of pre-training it based on physical aspects of the environment. Using machine learning techniques, the robot can start with a blank slate and learn from the environment what works best to achieve the desired goals. As a result, the challenges of the environment need not necessarily be defined by the robot. As the robot attempts to walk in an environment, it observes reactions and responses from the environment, these feedbacks are then fed into the machine learning technique, which then factors in all these inputs to deice a further course of action. This process repeats itself for several iterations, gradually improving over time. After a predetermined amount of iterating over the feedback-correction loop, the robot would have worked out multiple combinations of control policies that have already worked well in the training period. This collection of solutions that are known to be good is then referred to by the algorithm as the robot continues to walk.

## RELATED WORK

Several methods of controlling a robot's motion have been tried on several different robots. For example, the MIT Cheetah robot generates simple reference trajectories and performs model predictive control to get desired contact forces and then uses Jacobian transpose control to realize them, as discussed by Tuomas Haarnoja in 2018. The ANYmal robot, described by M. Hutter in 2016, plans footholds based on the inverted pendulum model. These methods of control work well but require considerable knowledge of the locomotion task to perform and also a deep understanding of the robot's dynamics. These requirements are often a limiting factor. A much better approach will be to use a method that can work well without prior knowledge of the robot dynamics or the locomotion task. Reinforcement learning-based training promises to achieve that. Tuomas Haarnoja in 2018 demonstrated that using reinforcement learning it is possible to learn locomotion skills without an explicit model of the robot dynamics.

In 2021, A Kumar demonstrated that using their Rapid Motor Adaptation (RMA) algorithm, a legged robot was able to walk on terrain that was never encountered during training. Jens Kober in 2013 discussed many different possibilities of using Reinforcement learning for robotic locomotion. Related literature shows numerous applications of machine learning techniques for robotic locomotion as observed by N. Kohl in 2004, Tuomas Haarnoja in 2018, R. Hafner in 2020 and L. Smith in 2021. Several different approaches to reinforcement learning have been proposed with varying degrees of success and performance. D Jain in 2019 explored a hierarchical reinforcement learning method that decomposes complex tasks into simpler sub-tasks. Reinforcement learning-based training methods often lead to robust locomotion gaits that can be transferred to real robots with higher performance accuracy than other traditional methods. Deep reinforcement learning models have been successfully used to train robots in simulation. However, transferring the trained policies to a real robot remains a challenging task. Discrepancies between the real world and the simulation environment create a gap between training and actual performance. This leads to a loss of performance in the real world as compared to simulation.

Several approaches have been proposed to bridge the gap between simulation and the real world. One approach would be to train the robot in a real-world physical environment. While this eliminates the simulation gap altogether, in most scenarios this is not practically possible due to challenges like resetting the experimental setup after every cycle, collecting data from the robot, controlling environmental factors, changing the robot's dynamic parameters, and so on. A much better approach would be to train the robot in a simulated environment and then use the trained policies in the real world. Training robots in simulation, although much faster than training in the real world, is still a very time-consuming and computationally intensive process. Augmented Random Search (ARS) is a much faster algorithm compared to other existing algorithms. Authors claim about 15 times more efficient than the fastest competing methods. Trajectory control is an aspect that requires special attention. Several factors such as hardware constraints, motor controller limitations, and other real-world physical aspects make it difficult to realize the trajectories in the robot hardware. Related literature discusses multiple approaches to achieve accurate trajectory control and trajectory optimization. Using reference trajectories has been shown to be useful in multi-joint systems such as articulated robot legs. In 2022, Nguyen Chuong showed the use of reference trajectories and contact timings. One method proposes using simplified rigid body dynamics to solve for contact timings and a reference trajectory of the robot body. However, David DeFazio in 2021 showed that human knowledge in the form of Linear Temporal Logic formulas can be applied to speed up the RL-based locomotion learning process without using any reference trajectories. Several recent works have shown the importance of trajectory optimization as well. Other methods discuss how joint positions are mapped to the output of high-level trajectory generators.

## USING REINFORCEMENT LEARNING TECHNIQUES

Using reinforcement Learning techniques to train locomotion skills in a robot is a promising approach for multiple reasons. Using these techniques, we no longer need detailed information about the physical aspects of the robot or the environment (Kautish et. al 2022).

This makes the trained policies more robust and more resilient from any discrepancies in modelling. Discrepancies between the simulated and the real physical system are the main reason why many controllers fail in the real world although they perform well in simulation. Every aspect of the physical world cannot be accurately modelled. Dynamic characteristics are extremely difficult to model accurately. These unmodelled parameters contribute to the poor performance of the controllers in the real world. Jie Tan in 2018 showed that locomotion tasks are particularly vulnerable to discrepancies in the model. Deep reinforcement learning methods have seen a lot of progress in recent times, they can solve locomotion problems from scratch but suffer from the limitations arising out of model discrepancies. In our experiments, we try to eliminate model discrepancies as much as possible through accurate modelling. We also try to accommodate for any residual discrepancies through robust training and experimental setup.

Control policies trained on reinforcement learning-based methods have been shown to be more robust to external disturbances and new unseen environments. We experimented with two different approaches to training the robot: using joint space to map joint positions to the output of a trajectory generator. Another approach is to use positional mapping directly in the cartesian space. As shown by Guillaume Bellegarda in 2021, using cartesian space is as effective as joint space. Accurate trajectory control was crucial in both these methods. John Schulman in 2015 proposed new methods that yielded strong empirical results on highly challenging 3D locomotion tasks, learning running gaits for bipedal and quadrupedal simulated robots.

## EXPERIMENTAL SETUP

Our experimental setup consists of a simulation environment as well as a real quadruped robot. The simulation is an accurately modeled representation of the real robot. Control policies are developed and tested in the simulation environment. Any modifications and refinements are first tried out on the simulated environment. This saves development time and allows us to experiment with multiple variations of the setup. The real robot leg is then used to validate the policies. Testing on real hardware also uncovers some discrepancies that could not be modeled accurately in the simulation environment. For experiments related to motion control, trajectory control, and tracking accuracy, testing on the real robot has frequently uncovered issues that were difficult to uncover in simulation. The robot we built is a quadrupedal-legged robot. It has four legs, with three joints in each leg, providing 3 degrees of freedom namely the 'Abad' joint, the 'Femur' joint, and the 'Tibia' joint. Each joint is actuated by an electric motor which is attached through a set of gear reducers. Each joint has a high-resolution position sensor that provides positional feedback to the motor controller.

*Figure 1. Robot leg configuration and "Stego" the real quadrupedal robot*

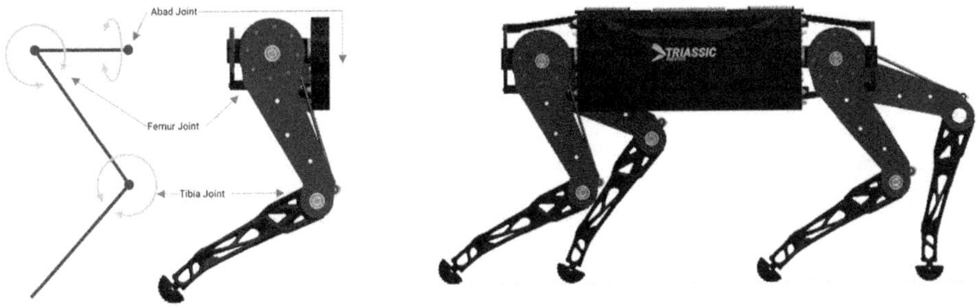


Each of the 12 motors is controlled by an onboard motor controller that receives positional feedback through the high-resolution position sensor attached to the joint. This positional feedback is then used to precisely control the position and velocity of the joint. We used FOC-based motor controllers for this experiment. The motor controller receives high-level control instructions from an onboard computer that executes a locomotion gait. The onboard computer is responsible for synchronizing the leg movements in order to achieve the desired gait. The computer continuously receives positional feedback from the controller and uses it to update its own internal model of the robot.

The onboard computer determines the leg movements based on the trained locomotion policy. It tries to match the actual state of the robot to the desired state of the robot. This is achieved by sending high-level instructions to the motor controller. The motor controller then computes low-level instructions that are required to reach the desired state. These low-level instructions are executed on the motor and simultaneously being monitored in a closed loop. Our robot has many design similarities to a quadruped robot named MIT Cheetah 3 as described by G. Bledt in 2018.

*Figure 2. Joint control and feedback loop*

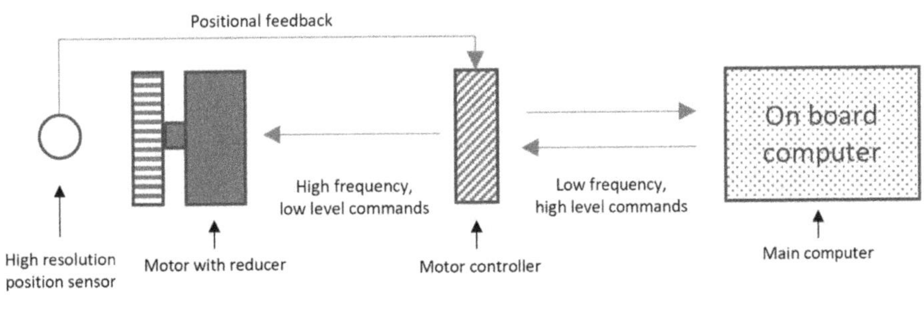

The real robot leg is mounted on a rig that supports the leg and allows for movement along the vertical (z) axis and along the abad (y) axis. All the joints of the leg are free to move. The leg is connected to a computer for command and control. This computer acts as the on-board computer that sends high-level commands to match the robot's state to the desired state.

*Figure 3. Experimental setup of the robot leg and the simulation environment*

## TRAJECTORY CONTROL

While the trained model may plan joint trajectories, the controller must ensure those trajectories are being tracked accurately. Inaccurate tracking of trajectories introduces errors that offset the accuracy of training. So, it becomes necessary that commanded trajectories are tracked as accurately as possible.

Accurate trajectory tracking becomes a challenge due to several factors. One of them is the inaccuracies in modeling and some physical characteristics that are not even considered while modeling. It is difficult to model factors that affect the dynamics of the robot such as motor parameters, and real-world details like friction, material properties, mass distribution, etc. Since these factors affect the dynamics of the robots significantly, any inaccuracy in modeling them results in inaccurate tracking of trajectories.

Another reason for inaccurate trajectory tracking is due to inaccurate execution of tracking commands. This happens because of small nuances in the way the main computer, motor controller and the motor interact with each other.

Without proper trajectory control, we noticed a significant amount of tracking error (Fig 4). With proper trajectory control in place, we were able to achieve much better tracking (Fig 5). We realized that the manner in which the motor controller was trying to execute tracking points did not compensate for the momentum the leg acquired once it is in motion.

At higher velocities and with higher mass, the momentum was significant enough to introduce a lot of tracking errors. Further, as the tracking error increases, the controller tries to correct the error by moving fast in the opposite direction, thereby introducing even more momentum to handle in the next tracking point.

*Figure 4. Poor trajectory control, shows tracking error in position and velocity*

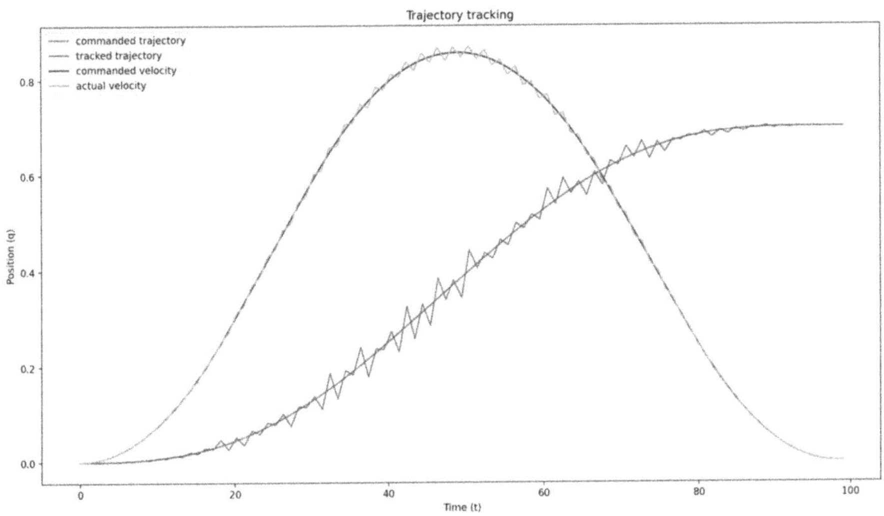


As a result of this controlling behavior, we found that tracking error increases significantly when the velocities are high. Also when the moving mass is high, such as in the lower joints of the leg, the momentum is high at lower speeds, again introducing tracking error. As a solution to this problem, we found that leading the position by a factor of $n$ steps ahead of the velocity and acceleration curve leads to much better tracking accuracy. Since velocity and acceleration determine the position reached, the controller adjusts these values to reach the desired position. If the desired position is always kept ahead of the current step, the controller never reaches the position until the last $n$ steps and hence the velocity and acceleration do not change abruptly.

*Figure 5. Improved tracking by using trajectory control with positional lead*

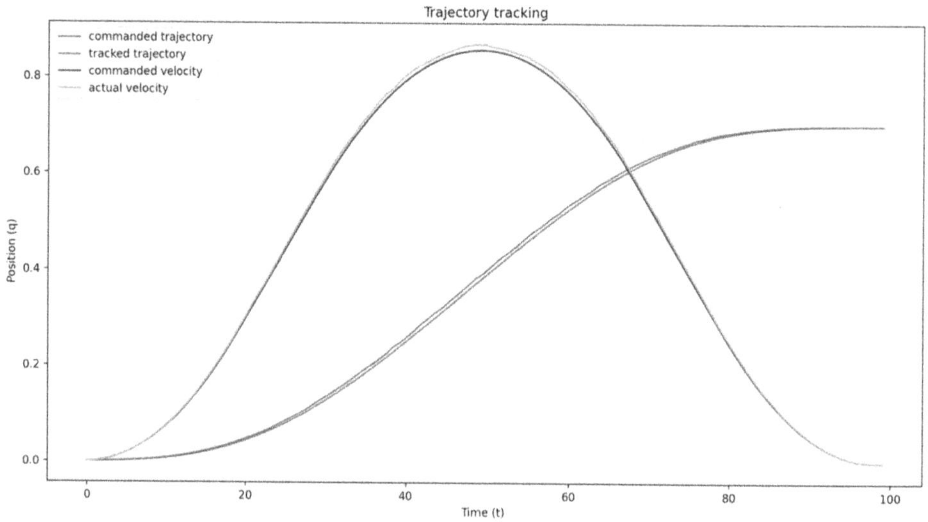


*Figure 6. Trajectory control with positional lead*

**Algorithm 1:** Trajectory Control with positional lead in every step

```
//Parameters:
pos_lead        //Factor of positional lead
s_list          //List of all steps in the trajectory
step            //One step in the list of steps s_list
acc             //Acceleration value at the current step
vel             //Velocity value at the current step
pos             //Position value at the current step

While s_list has steps, loop through each step
       step = current step
       acc = step.acc
       vel = step.vel
       if current step is less than total steps - pos_lead
              pos = (current step + pos_lead).pos
       else
              pos = step.pos
       Update motor controller(acc, vel, pos)
       state = Read state from motor controller
       update local state
End while
```

## SIMULATION ENVIRONMENT SETUP

To test the locomotion policies in the simulation we set up an environment that closely mimics real-world scenarios. We used PyBullet as the simulation environment. We created accurate models of the real robot from the original CAD files that were used to make the real robot. These models were loaded in the simulation environment using URDF files. A similar environment was set up in a Linux environment using Gazebo as the simulation engine.

*Figure 7. Simulation environment setup using PyBulllet along with real robot leg supported on a test rig*

## USING AUGMENTED RANDOM SEARCH FOR TRAINING

Dynamic movements that require intricate balancing like walking, running, jumping, etc., are extremely difficult for robots to perform. Traditional methods of control require extensive experience and tedious manual tuning. Several approaches to machine learning have been proposed to solve this problem.

John Schulman in 2017 proposed Proximal Policy Optimization (PPO) which is much simpler to implement and experiments by the author have shown the algorithm to outperform various benchmarks including simulated robotic locomotion tasks, In our experiments, we used "Augmented Random Search" to train our robot's basic locomotion skills such as walking.

A. Rajeswaran in 2017 showed that tasks that required continuous control can be solved by training even simple policies with linear and RBF parameterizations. Timothy P Lillicrap showed in 2015 that reinforcement learning can be effective for tasks requiring continuous control. Augmented Random Search (ARS), as proposed by Horia Mania in 2018, is a random search method for training static, linear policies for continuous control problems. It is an improvement over Basic Random Search (BRS). Yan Duan in 2016 benchmarked several approaches to solving continuous control tasks.

*Figure 8. Algorithm of augmented random search*

**Algorithm 2** Augmented Random Search (ARS): four versions **V1**, **V1-t**, **V2** and **V2-t**

**Hyperparameters:** step-size $\alpha$, number of directions sampled per iteration $N$, standard deviation of the exploration noise $\nu$, number of top-performing directions to use $b$ ($b < N$ is allowed only for **V1-t** and **V2-t**)

**Initialize:** $M_0 = \mathbf{0} \in \mathbb{R}^{p \times n}$, $\mu_0 = \mathbf{0} \in \mathbb{R}^n$, and $\Sigma_0 = \mathbf{I}_n \in \mathbb{R}^{n \times n}$, $j = 0$.

**while** ending condition not satisfied **do**

Sample $\delta_1, \delta_2, \dots, \delta_N$ in $\mathbb{R}^{p \times n}$ with i.i.d. standard normal entries.

Collect $2N$ rollouts of horizon $H$ and their corresponding rewards using the $2N$ policies

$$\textbf{V1:} \begin{cases} \pi_{j,k,+}(x) = (M_j + \nu\delta_k)x \\ \pi_{j,k,-}(x) = (M_j - \nu\delta_k)x \end{cases}$$

$$\textbf{V2:} \begin{cases} \pi_{j,k,+}(x) = (M_j + \nu\delta_k)\,\mathrm{diag}\,(\Sigma_j)^{-1/2}\,(x - \mu_j) \\ \pi_{j,k,-}(x) = (M_j - \nu\delta_k)\,\mathrm{diag}(\Sigma_j)^{-1/2}(x - \mu_j) \end{cases}$$

for $k \in \{1, 2, \dots, N\}$.

Sort the directions $\delta_k$ by $\max\{r(\pi_{j,k,+}), r(\pi_{j,k,-})\}$, denote by $\delta_{(k)}$ the $k$-th largest direction, and by $\pi_{j,(k),+}$ and $\pi_{j,(k),-}$ the corresponding policies.

Make the update step:

$$M_{j+1} = M_j + \tfrac{\alpha}{b\sigma_R} \sum_{k=1}^{b} \left[r(\pi_{j,(k),+}) - r(\pi_{j,(k),-})\right] \delta_{(k)},$$

where $\sigma_R$ is the standard deviation of the $2b$ rewards used in the update step.

**V2** : Set $\mu_{j+1}$, $\Sigma_{j+1}$ to be the mean and covariance of the $2NH(j+1)$ states encountered from the start of training.[2]

$j \leftarrow j + 1$

**end while**

## RESULTS

Using our method of trajectory control, we were able to achieve a very high level of positional accuracy. We used a method of positional lead where the position leads the current step by a factor $n$. This prevents the controller from adjusting acceleration and velocity too abruptly. As a result, we were able to get smooth motion and high positional accuracy.

We were also able to achieve very close trajectory tracking performance in simulation and in real robot legs. This demonstrates the advantages of proper use of trajectory control in legged robots. Joint space control and cartesian space control were both equally effective although there were some differences. Joint space control makes it difficult to achieve pre-determined trajectories, as a result, the gait may vary from what the controller instructed. With cartesian space control, it is much easier to track trajectories. Gait sequences are also executed accurately. We have successfully trained a legged robot with basic locomotion skills using reinforcement learning methods. We trained the robot's simple walking gaits without extensive use of the robot's dynamics and without prior knowledge of the dynamics involved in the locomotion task. Our robot was trained in a simulated environment that closely mimics real-world scenarios. After successful training, the robot was able to perform simple walking gaits based on the trained policies. We used the algorithm Augmented Random Search. We have built a quadrupedal-legged robot that was used for the experiments. The robot was able to perform walking gaits accurately although transferring trained policies from simulation to real robot remains a challenging task. Limitations of hardware and limitations in the control electronics make it difficult to match the performance from simulation to real robots.

*Figure 9. Sequence of gait patterns, the robot is executing a simple walking gait*

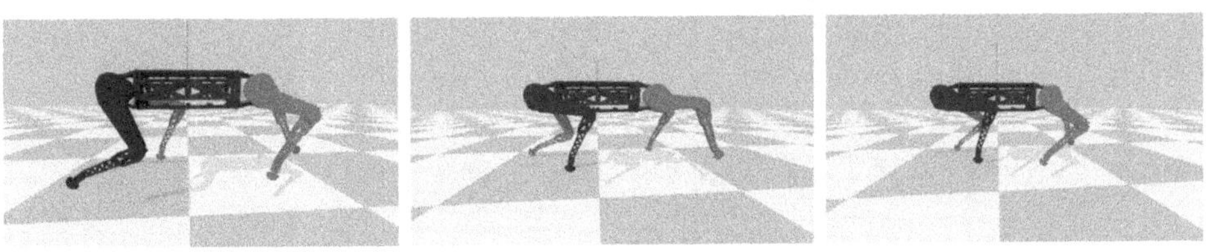

## CONCLUSION

Using machine learning techniques to train our robot has resulted in significant performance improvements. We were able to train basic locomotion tasks like walking in a much more efficient way in terms of training time and performance. We were able to achieve dynamic walking gaits in simulation that are difficult to achieve without using machine learning techniques. Augmented Random Search (ARS) is a significantly faster algorithm. It performs especially well in tasks that require continuous control. The challenges we faced mainly originated from discrepancies in modeling. There are numerous physical characteristics that could not be modeled accurately.

*Figure 10. Simultaneous trajectory planning for all the joints of the robot. Shows a frame of walking gait*

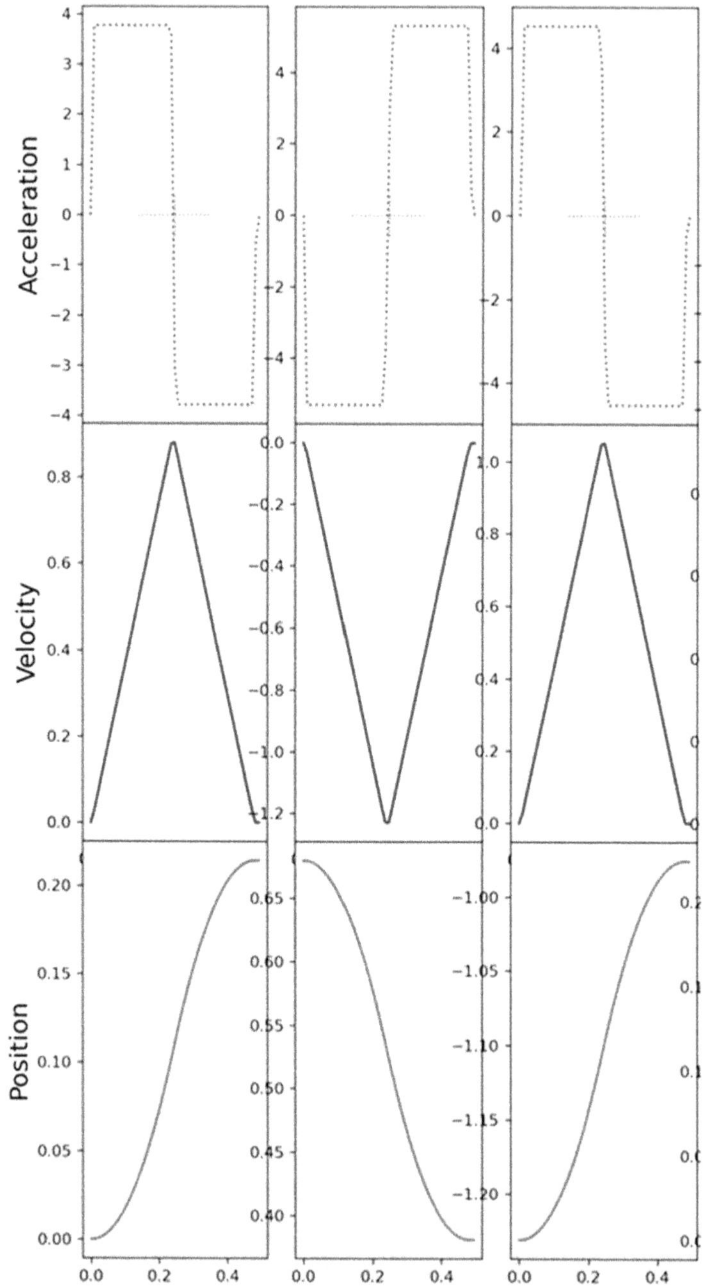

*Figure 11. Locomotion policies being tested on Stego*

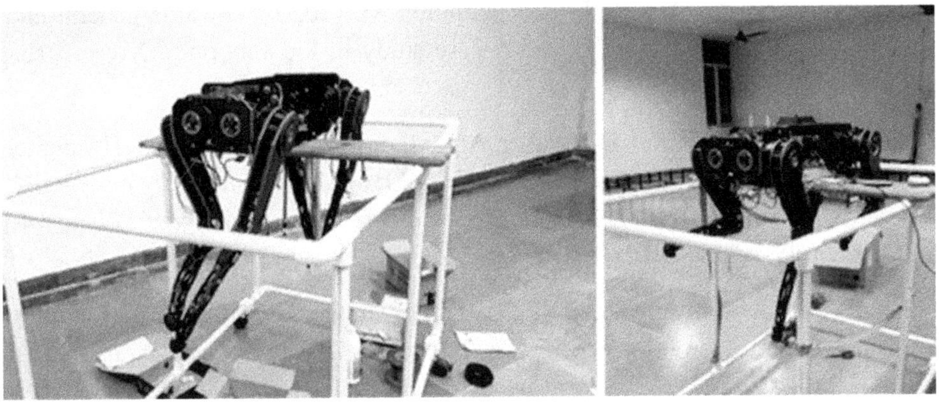

Tasks like robotic locomotion are particularly vulnerable to discrepancies in the model. Using reinforcement learning for training the policies turned out to be a good choice especially because we had to deal with discrepancies. We also found that proper trajectory control was necessary to achieve performance accuracy in real-world scenarios. Trajectory control was poorly affected by limitations in hardware in interfacing between the main computer, motor controller, and motor. We found that accurate trajectory tracking can be achieved by leading the positional value by a factor of n while keeping the acceleration and velocity the same.

## REFERENCES


Bellegarda, G., & Nguyen, Q. (2021). Robust High-speed Running for Quadruped Robots via Deep Reinforcement Learning. arXiv:2103.06484

Bledt, G., Powell, M. J., Katz, B., & Di Carlo, J. (2018). MIT cheetah 3: Design and control of a robust, dynamic quadruped robot. *International Conference on Intelligent Robots and Systems (IROS)*. IEEE.

DeFazio, D., & Zhang, S. (2021). Leveraging Human Knowledge to Learn Quadruped Locomotion Policies. arXiv:2107.10969

Duan, Y., & Chen, X. (2016). *Rein Houthooft, John Schulman, and Pieter Abbeel. "Benchmarking deep reinforcement learning for continuous control.* Proceedings of the 33rd International Conference on International Conference on Machine Learning (pp. 1329–1338). *JMLR.org*, .

Hafner, R., Hertweck, T., Kloppner, P., Bloesch, M., Neunert, N., Wulfmeier, W., Tunyasuyunakool, S., Hees, N., & Riedmiller, M. (2020). Towards general and autonomous learning of core skills: A case study in locomotion. arXiv preprint arXiv:2008.12228

Hutter, M., Gehring, C., Jud, D., Lauber, A., Bellicoso, C. D., Tsounis, V., Hwangbo, J., Bodie, K., Fankhauser, P., & Bloesch, M. (2016). *Anymal-a highly mobile and dynamic quadrupedal robot. International Conference on Intelligent Robots and Systems (IROS).* IEEE. 10.1109/IROS.2016.7758092

Jain, D., Iscen, A., & Caluwaerts, K. (2019 ). Hierarchical Reinforcement Learning for Quadruped Locomotion. *IEEE/RSJ International Conference on Intelligent Robots and Systems (IROS),* Macau.

Kober, J. & Peters, J. (2012). Reinforcement learning in robotics: A survey. In *Reinforcement Learning", pages* (pp. 579–610). Springer.

Schulman, J., Moritz, P., Levine, S., Jordan, M., & Abbeel, P. (2015). High-dimensional continuous control using generalized advantage estimation. arXiv preprint arXiv:1506.02438.

Kaur, R., & Kautish, S. (2022). Multimodal sentiment analysis: A survey and comparison. *Research Anthology on Implementing Sentiment Analysis Across Multiple Disciplines*, pp. 1846-1870.

Kautish, S., Peng, S. L., & Obaid, A. J. (Eds.). (2021). *Computational intelligence techniques for combating COVID-19.* Springer International Publishing. doi:10.1007/978-3-030-68936-0

Kautish, S., Reyana, A., & Vidyarthi, A. (2022). SDMTA: Attack detection and mitigation mechanism for DDoS vulnerabilities in hybrid cloud environment. *IEEE Transactions on Industrial Informatics*, *18*(9), 6455–6463. doi:10.1109/TII.2022.3146290

Kumar, A., Fu, Z., Pathak, D., & Malik, J. (2021). Rma: Rapid motor adaptation for legged robots. arXiv preprint arXiv:2107.04034. doi:10.15607/RSS.2021.XVII.011

Madhu, G., Govardhan, A., Ravi, V., Kautish, S., Srinivas, B. S., Chaudhary, T., & Kumar, M. (2022). DSCN-net: A deep Siamese capsule neural network model for automatic diagnosis of malaria parasites detection. *Multimedia Tools and Applications*, *81*(23), 34105–34127. doi:10.100711042-022-13008-6

Mania, H., Guy, A., & Recht, B. (2018). Simple random search provides a competitive approach to reinforcement learning. arXiv:1803.07055v1

Nate. Kohl and Peter. Stone, "Policy gradient reinforcement learning for fast quadrupedal locomotion," *IEEE International Conference on Robotics and Automation, 2004. Proceedings. ICRA'04.*

Nguyen, C., & Nguyen, Q. (2022). Contact-timing and Trajectory Optimization for 3D Jumping on Quadruped Robots. arXiv:2110.06764. doi:10.1109/IROS47612.2022.9981284

Rajeswaran, A., Lowrey, K., Todorov, E., & Kakade, S. (2017). Towards generalization and simplicity in continuous control. arXiv preprint arXiv:1703.02660

Sampathkumar, A., Rastogi, R., Arukonda, S., Shankar, A., Kautish, S., & Sivaram, M. (2020). An efficient hybrid methodology for detection of cancer-causing gene using CSC for micro array data. *Journal of Ambient Intelligence and Humanized Computing*, *11*(11), 4743–4751. doi:10.100712652-020-01731-7

Schulman, J., Wolski, F., Dhariwal, P., Radford, A., & Klimov, O. (2017). Proximal policy optimization. *Algorithms. CoRR, abs/1707.06347.*

Smith, L., Kew, J. C., Peng, X. B., Ha, S., Tan, J., & Levine, S. (2021). Legged robots that keep on learning: Fine-tuning locomotion policies in the real world. arXiv preprint arXiv:2110.05457.

Tan, J., Zhang, T., Coumans, E., Iscen, A., Bai, Y., Hafner, D., Bohez, S., & Vanhoucke, V. (2018). Sim-to-Real: Learning Agile Locomotion For Quadruped Robots. arXiv:1804.10332v2

Timothy, P., Lillicrap, J., Hunt, J., Pritzel, A., Heess, N., Erez, T., Tassa, Y., Silver, D., & Wierstra, D. (2015). Continuous control with deep reinforcement learning. arXiv preprint arXiv:1509.02971, 2015

Haarnoja, T., Zhou, A., Abbeel, P., & Levine, S. (2018). Soft actor-critic: Off-policy maximum entropy deep reinforcement learning with a stochastic actor. arXiv preprint arXiv:1801.01290

Haarnoja, T., Ha, S., Zhou, A., Tan, J., Tucker, G., & Levine, S. (2018). Learning to walk via deep reinforcement learning," arXiv:1812.11103

# Chapter 7
# Cloud Computing Principles for Optimizing Robot Task Offloading Processes

**U. Rahamathunnisa**
*Vellore Institute of Technology, India*

**K. Sudhakar**
*Madanapalle Institute of Technology and Science, India*

**Tamilarasi K. Murugan**
*Vellore Institute of Technology, Chennai, India*

**Thivaharan S.**
*PSG Institute of Technology and Applied Research, India*

**Rajkumar M.**
*Vellore Institute of Technology, India*

**S. Boopathi**
*Mythayammal Engineering College (Autonomous), India*

## ABSTRACT

*In this chapter, the integration of two rapidly evolving concepts for integration of robotics and cloud computing has been illustrated. Intelligent robots should be preferred over low-cost conventional robots for dynamic and sophisticated applications, the internet, cloud computing, and artificial intelligence principles. "Task offloading" is one of the popular cloud computing techniques for extending the constrained capabilities of robots. The various task features of cloud robotics have also been explained. Literature on various optimization techniques used for task offloading (task offloading, path planning, and access points) in cloud-enhanced robots has been included. The data handling framework and genetic algorithm-based task offloading mechanism of cloud robotic systems have been illustrated. The system architecture and task-flow chart for smart city and manufacturing control applications have been graphically illustrated.*

DOI: 10.4018/978-1-6684-8171-4.ch007

## INTRODUCTION

The capacity to incorporate self-governing robotic intelligence has increased thanks to the introduction and rapid development of cloud innovation and the Internet of Things. More study has recently been concentrated on creating collective robots for certain task-based applications in order to make up for the significant testing and changing demands of these applications. Any device that depends on a system for support is regarded as a part of organized robotics. The purpose of characterizing this in this way is to include both existing and potential frameworks, such as UAVs and warehouse robots (Kumar & Michael, 2012). The impressive chores of detecting, instigating, imparting, and reasoning are distributed among a group of robots in network improved robotics. Such frameworks can be used for a wide range of jobs, including mechanical assistance, planetary meanderer control, clinical medical treatments, administrative-based duties, and so on. Even with all the advancements, designing autonomous robots with limitless capabilities is still not conceivable. Whether a framework is based on hardware or software, all automated systems have their own set of restrictions(Samikannu et al., 2022; Vanitha et al., 2023; Vennila et al., 2022). These include asset restrictions, communication constraints, educational standards, and so forth. Robot design and development still have a lot of room for improvement. Expanding the battery's capacity could increase power usage (by an octuplet). The method of task-offloading involves moving a mechanical task to a distant asset in order to conserve energy, speed up response, and prolong the battery life of the robot. The concept of "cloud enhanced robotics" perfectly combines mechanical autonomy with additional support from the cloud foundation.

**Cloud Enhanced Robotics (CER):** CER is a distinct area that has combined the two rapidly evolving concepts of robotics and cloud computing. It initially advises turning more heavily to the omnipresent virtual resources of the cloud and less to human information. However, networked robots have a variety of requirements, such as assets, learning capacities, and communication. The ubiquitous and on-demand benefits provided by "the cloud" enable the robots to maximize their potential and identify resources in order to overcome their own obstacles. Through virtualization, the robot offloads work by utilizing these cloud benefits (Mell & Grance, 2011). The creation of dynamic and sophisticated applications now has a greater capacity for coordinating self-governing detection and incitation due to the growth of CER and the Internet of Things (IoT). For the combination of mechanical advancements, particularly in robotizing applications, CER is thought to be a significant enabling agent. In essence, CER represents the standard for system assets that have increased operational capabilities and mechanical assets that are combined with lower registration costs. The various applications of CER is illustrated in Figure 1.

The main guideline for the sequencing of exams is dependent on the components of the problem. These highlights have led to several recent readings on resource administration and distribution for cloud computing. Sending computationally expensive jobs to the repeated, affordable, and adaptive cloud servers is known as "task offloading," and it has become a popular technique for extending the constrained capabilities of robots. Task offloading and the variables (such as cost, and trade-off) associated with its dynamic decisions need to be better understood.

*Figure 1. Multiple domains of cloud enhanced robotics*

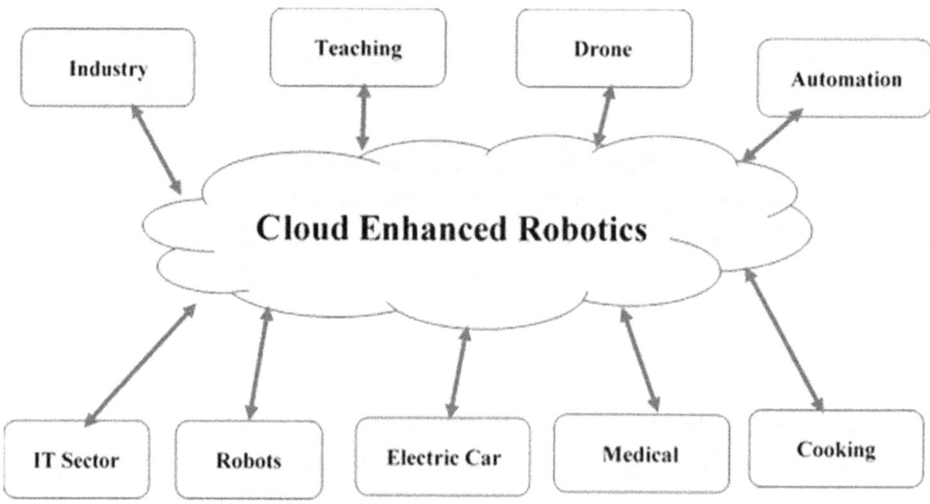


**Task Offloading:** Similar to how they do it when applying computation aids, gadgets could offload jobs to speed up task execution. According to NIST, cloud computing is a concept that enables ubiquitous and on-demand access to a shared pool of storage, servers, networks, and other resources(Mell & Grance, 2011). Task offloading is a potential strategy for preserving nearby resources (such as device energy, task completion time, battery life, and so on) while also enhancing the performance of the framework. When performing such operations, the key objective is to minimize energy use and delay while maintaining the level of service (QoS). In order to find energy-saving models and increase the battery life of applications, some logical tests are being conducted. The need for additional transistors is implied by the development of robots and the inventive creation of smaller-measured semiconductors for low energy reasons. Unfortunately, this leads to an increase in energy use. Energy can be conserved by programming autonomous robot components or the complete machine to go into rest or reserve mode when not in use. The task offloading in cloud robotics have been used for the controlling the different task is illustrated in Figure 2.

With task offloading in cloud robotics, the robot can save energy while also lowering the cost of the framework's equipment. When weighing the pros and downsides of work offshoring, a trade-off is necessary. Additional aspects include the network's configuration, robot development, the mechanical limit, and others. Although there is no assurance that job offloading will result in energy savings, it will also speed up work completion. There are potential mixes, needs, and seriality for many jobs (task flow) that should be maintained. Furthermore, because there are numerous equal jobs, unloading allows the assets to be used more frequently. The additional features of the robotic framework, such as development decision and communication link determination, are advantageous to a robot's on-demand portability and offloading of choice operations. With these factors in mind, a specific and more thorough analysis is needed to focus on the crucial factors that affect the task offloading dynamic in various kinds of cloud robotic applications.

*Figure 2. Features of task offloading in cloud robotics*

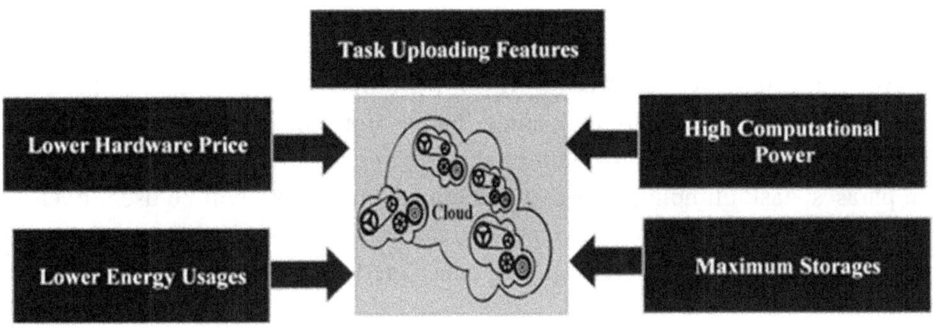


**Motivational Scopes:** It is well accepted that task offloading in CER reduces execution delay and robot energy consumption. Numerous challenges could significantly impede the development of a framework for QoS in the future. In a CER application, the following basic factors should be taken into account when making task offloading decisions:

**Applicability**: Each program contains a variety of tasks, not all of which should be offloaded. Some tasks should be completed locally because cloud connectivity can be time and energy consuming. Some tasks may be more appropriate for offloading than others (considering their presentation trade-off).

**Accessibility and Consumption of assets**: In applications involving autonomous robots, the workload's task component may need to be split between the robot and the cloud virtual machine. The availability of assets is another consideration for dumping. There can be instances where the robots or other resources are inaccessible throughout the application. Offloading duties might occasionally be advantageous in order to more effectively utilize the resources available at a given time. Offloading decisions may be influenced by the availability or selection of assets, resulting in the best possible outcome for the application.

**Availability of cloud and communication interface**: The accessibility of APs makes the cloud foundation available to the client since the robots choose the access point (AP) to communicate with the cloud (robot). Large-scale activities might have to be conducted locally due to AP accessibility issues. The AP's assessment of the openness of the relevant communication channels serves as a key paradigm when deciding how a robot should move. The client has access to the cloud foundation thanks to the accessibility of APs because the robots choose the access point (AP) to communicate with the cloud (robot)(Boopathi, Jeyakumar, et al., 2022; Domakonda et al., 2022; Vanitha et al., 2023). The inaccessibility of APs may necessitate local completion of duties in the case of large-scale activity. An important paradigm for offloading decisions about how a robot should move is the openness of the proper communication channels (as decided by the AP).

**Robotic Energy Consumption Minimization:** To compare and contrast the energy usage of a robot with offloading the jobs for remote execution in order to determine the best task technique for the entire application. In the opposite sense, the phrases "task dumping," "task distribution," and "task" will be used. Even if development itself consumes more energy, a suitable trade-off between movement energy and data transmission should be considered when making decisions about job offloading.

**Time/Delay Minimization**: Reducing application delay or reaction time is turning into a crucial issue, especially for robotic applications that require lots of computing. The robot recognizes the optimal task offloading choices and makes use of its resources to do tasks more quickly. The adaptability of the robot and the AP's choice also have an impact on the dynamic execution by reducing the use of time delay. In order to send jobs to the cloud and get results as soon as possible, it is important to offload the work using the optimal communication channels.

**Combined Minimization of Distance and Energy**: The robots' on-demand portability allows them to obtain reasonable APs and conserve energy. Even when there is development, it requires more energy. One of our current priorities is to strike a proper balance between energy and separation while offloading dynamic decision-making.

## OPTIMIZATION TECHNIQUES FOR TASK OFFLOADING PROCESS

The task offloading, path planning, and access point (AP) are selected for CER applications. In order to plan our mobility-driven and communication-aware offloading for a cloud-enhanced robotic framework, the associated link between these three factors is helpful. The act of selecting the optimal element from a set of available options is known as optimization. The term "optimization problem" refers to an NP-hard, non-raised quadratic programme that is quadratically forced. In the discipline of engineering, there are several optimization studies on numerous topics and areas on Mobile Cloud computing (MCC) devices. In order to overcome the barriers associated with a mobile device's performance, MCC integrates the cloud registering foundation into the portable condition. Studies on task offloading optimization techniques have gained attention since the offloading process has a significant impact on the relationship between mobile and cloud. Prior to 2000, the main objective of offloading research was to make it feasible. After 2000, the focus shifted to improving the algorithms that identify the critical offloading choices. Additionally, this led to the division of applications into different jobs and the distribution of the remaining workload across clients, just like in the cloud. At that time, improvements in virtualization led to a better offloading architecture with greater QoS. In order to introduce computing resources adjacent to the base station of a cellular network, technologies like mobile edge computing were developed. Regular mobile systems include clients that communicate with cloud servers across a considerable delay(Boopathi, Balasubramani, et al., 2021; Boopathi, Siva Kumar, et al., 2023; Domakonda et al., 2022; Kumara et al., 2023). The use of the offloading system on mobile devices (to the cloud) for the single client situation is specifically emphasized by several offloading techniques. There are then methods that expedite the process by saving time. The many optimization techniques and their subtler manifestations will be examined in this section. The offloading of computation at the strategy level into the framework of the Internet of Things (IoT) platform is made possible by a number of technologies (Kansal & Chana, 2015). These technologies make it simpler for engineers to move mobile applications to the cloud.

Some contend that it is advantageous to concentrate on offloading strategies at the "task level." These tactics include, which created a mobile application with the "LARAC" task offloading computation as the optimal solution(Zhang et al., 2017). Surprisingly, its applicability only applies to simple scenarios since it perceives tasks as having a direct topology. A task graph is a diagram of an application that incorporates conditions (Guo et al., 2018). Other studies actually put task offloading tactics for programmed with condition-filled task graphs into practice. Each activity in the task diagram is offered as a potential offloading option in our plan, whether it should be completed and ready-to-go or on the cloud. than a generic strategy, still. Explicit studies are required because task assignments in multi-client applications will become more general and sophisticated (Kautish et. al 2022). Multi-user offloading is presented as joint optimization issues by several modern approaches. Multiple autonomous tasks are handled by each mobile client, and they are resolved using techniques like straight programming and stochastic demonstration. Decision-making in difficult settings is designed to be delegated using transformative algorithms. For some difficult problem sets, they compete with the so-called "conventional methods," such as artificial neural networks (ANNs), heuristics, TOPSIS, VIKOR, AI, and fuzzy systems (FSs)(Boopathi, 2022b)(Boopathi, 2019; Boopathi, Myilsamy, et al., 2021; Boopathi, 2022c, 2022a; Boopathi & Sivakumar, 2013; Kavitha et al., 2022; Myilsamy & Sampath, 2021). These methods can survive changing conditions and provide instantaneous responses in a variety of settings. Workload distribution among cloud hubs and dynamic job planning have both been accomplished using ant colony optimization (ACO). An ACO-based offloading technique while also introducing a computerized application offloading middleware for the ACO decision-making process. It has also been shown that scheduling algorithms on the cloud use bee colony province techniques(Boopathi, Thillaivanan, et al., 2021). BCA has been designed to conserve energy while using resources. Another type of developing computation used for load scheduling is Particle Swarm Optimization (PSO). A well-known, game-changing approach for cloud computing models like the Firefly Algorithm and the Harmony Search Algorithm is the Cuckoo Algorithm (CAA). Even while there is evidence that resource management problem solutions exist, the same can be stated of the CSA in regards to scheduling issues. This is where the Hybrid Intelligent Generic Algorithm (HIGA) leaves its mark in relation to the models(Domakonda et al., 2022; Kumara et al., 2023; Samikannu et al., 2022; Vanitha et al., 2023; Vennila et al., 2022; Yupapin et al., 2022).

Natural selection and hereditary traits are evolutionary notions that inform the Genetic Tool (GA), an adaptive heuristic search algorithm. It illustrates clever use of arbitrary searches, which are used to address optimization problems. In GA, weak and unsuitable species face eradication through the process of routine determination. In order to coordinate the search into the area of better execution inside the search space, GAs use real data. Due of its scalability and flexibility, genetic algorithms (GA) have been employed in hybrid cloud computing techniques for mobile clients. Numerous ongoing projects concentrate on mobile cloud computing applications to enhance decision-making for offloading(Boopathi, 2021). GA is comparable to other evolutionary techniques in that it has been designed to balance load. GA is a realistic alternative that might perform better in an unstructured setting than traditional techniques and other evolutionary methodologies. The most important aspect is that it cannot develop further to become CER. Given the circumstances surrounding robotics applications, GA is a potentially viable alternative to conventional strategies. Considerations of the robot's position (movement) and accessible communication links are necessary when mapping the task to a robot (bandwidth). Two characteristics, movement and bandwidth, affect one another and work together have been taken. The design process for the task offloading technique for mobile cloud computing, infused life into the GA-based offloading approach. Although comparative technique is used for finding the additional mobility factors necessitate a multi-targeted CER decision-making procedure(Zhang et al., 2017).

## TASK OFFLOADING IN CLOUD ENHANCED ROBOTICS

Robots can choose how they wish to go to more effective communication links thanks to on-demand mobility. Task offloading is a key component of mobile cloud computing and CER. In addition, robots can do activation tasks, which demand more processing and complexity. For instance, Center uses the Software-as-a-Service (SaaS) model for a robust and scalable cloud-enabled object recognition system, while uses the SaaS model for synchronous restriction and mapping (SLAM) (Arumugam et al., 2010). The goal of Robot-as-a-Service (RaaS) is to concur with standard assistance thresholds for robotics activity. Rapyuta uses the PaaS model for cloud-based robot information sharing applications like RoboEarth. The potential of a virtual machine being managed via cloud computing is presented (Xu et al., 2013). In our approach, on-request task offloading is implemented via the cloud rather than storage alone. Data obtained from geographically isolated robots is used to use the cloud as a capacity vault. Another important topic for projects using cloud robotics is quality of service (QoS).

The task offloading together with the mobility and communication viewpoints will improve the success of the decision-making process. Instead, in order to improve the quality of operations. The supports the system has integrated with best task offloading. A task offloading decision process is converted into a multi-target optimization problem through the process of optimization. Recently, a few optimization techniques have been developed with single- and multi-operator cloud robotics applications like material management and robot development in mind.

There are numerous offloading optimization algorithms available that can explicitly be adjusted for perfect distribution in CER. Data retrieval framework of cloud robotic system is shown in Figure 3. The study that considered task mapping in a submerged mobile sensor array by executing a dynamic and coordinated effort between an autonomous underwater vehicle (AUV) and the cloud server is one of the researches that can be found in the literature (Pandey et al., 2015). The GA-based architecture with a focus on task offloading selections is taken rather than cloud resource allocation (Figure 4).

*Figure 3. Data retrieval framework of cloud robotic system*

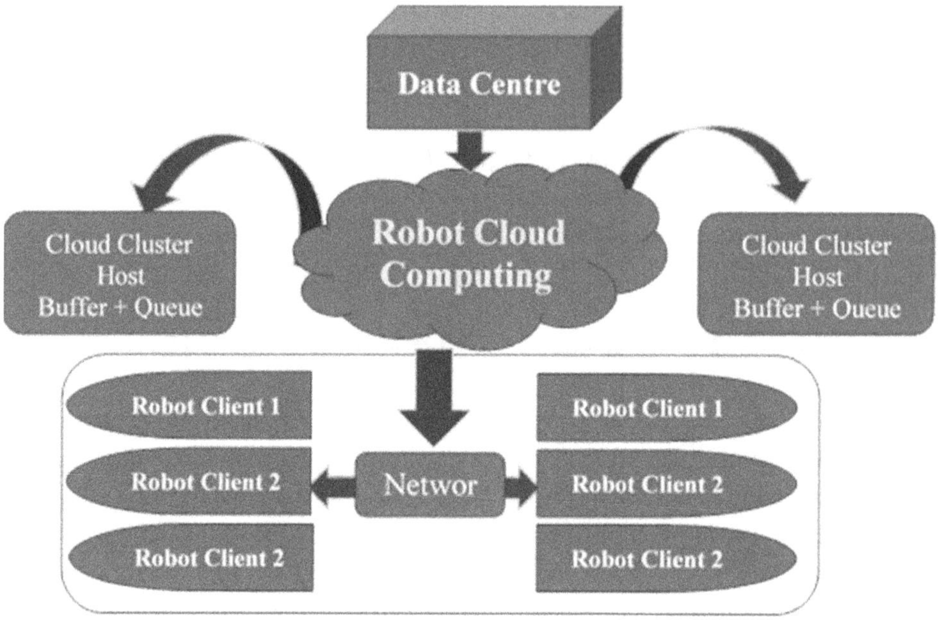


*Figure 4. GA-based task offloading mechanism*

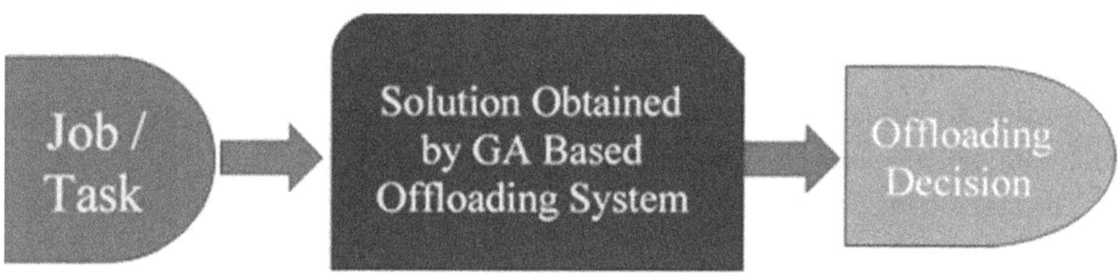


The portioning and planning for a 3D domain mapping in their continuous multi-sensor information recovery structure for multi-operator robotics frameworks(Wang et al., 2018). In complex scenarios, using evolutionary computation for optimization may result in better execution. One such tactic that offers significant advantages over traditional and other transformative techniques mentioned in the literature is the genetic algorithm. When adequate information from earlier sources is not available, GA can quickly fulfil agreements. It is useful in robotics for unstructured scenarios with various destinations in eerie surroundings in this way. GA-based approaches are limited to mobile cloud computing scenarios and applications in writing. Three crucial parameters-offloading, movement, and communication-depend on the robot because of its on-demand mobility. This both broadens the scope of the problem and increases the complexity of the decision-making process. The GA approach is ideal for our CER application area. Guo et al. provide the most relevant and recent example of the GA-based approach in cloud robotics. These three decision-making configurations to a modified GA-based technique are mapped due to increased complexity.

Due to elements like parameter dependencies, mapping difficulties, and other issues, working with dynamic and multi-functional frameworks in software development is becoming significantly more complex. As a result, more dependent parameters are added to describe and address problems with joint optimization. Examining the two constrained and related factors mobility and communication and how they relate to task offloading for CER is necessary. These three a significant factor in the decision-making processes has been built.

## IMPLICATIONS OF COMMUNICATION AND MOBILITY

For CER, the ability to proceed on-demand means the robot may be able to choose its connection to accommodate or influence the creation of its offloading decision. The MCC frameworks often have inactive mobility and communication angles, thus this is out of the ordinary (client subordinate). A growing trend in distributed computing for networked robotics is task offloading to the cloud. The CER design transfers capacity, communication, and computing between the robot and the cloud. Serious jobs entail computing that uses a lot of energy and introduces latency while being carried out by the robot.

Remote offloading of tasks to cloud servers is possible, and exceptionally equipped virtual machines can do the work. The communication's data transmission is the crucial element here. For cloud communication, a rapid communication link (high bandwidth) reduces idle time, which saves the robot energy. Additionally, WiFi is available, which balances out variations in accessibility, energy use, and delay. A list of comparisons between WiFi and cellular networks for mobile clients (Xia et al., 2014).

The communication interface or portal that links the robot and the cloud is crucial for the WiFi to operate as intended. The decision to unload is what drives the selection of the communication channel. Several studies considered the accessibility of communication links and suggested techniques based on this. Several optimizations have simultaneously shown that the communication channel chosen for offloading creates a framework that is unmatched for client and cloud-based communication.

The ability of access points (APs) to considerably enhance work offloading to the cloud execution is mostly obvious. As the clients are currently progressing, this creates the opportunity for them to select the appropriate channel to meet their unloading requirements. The selection of the AP was presented as a key component of an optimization study (Xu et al., 2013), where a calculation was suggested that was dependent on the Analytic Hierarchy Process (AHP), which encouraged dominating framework execution. In this design, a combined optimization problem that includes offloading and AP determination as a decision-making factor. Robots can choose locations for offloading based on their identification of the network gateway thanks to on-request mobility. A few inquiries on communication for cloud robots have not utilized this approach yet. The communication perspective may also be used by networked robotics to offload data to the cloud. The offloading and system selection decisions (via the AP decision) is coordinated, resulting in a joint optimization problem. In any case, the extra component of on-demand development that makes our structure communication- and mobility-driven. It is for task offloading to use a communication interface. The robot's capacity to "reach" the appropriate locations and identify the appropriate APs for cloud communication serves as its driving force. Since on-demand mobility enables robots to take advantage of these benefits, it is obvious that communication and mobility are intertwined for delegating decision-making. Task offloading heavily relies on mobile cloud computing. The ability to transfer data after a person moves is referred to as mobility. A system for mobility-aware service designation was proposed by Rahimi et al., in which the application was shown as task area time work processes. Online location-aware offloading was taken into consideration for a two-layered framework (consisting of cloudlets and remote clouds) with the goal of utilising cloudlets, clouds, and mobile clients (Stenmark et al., 2015). An important piece of research looked into the problem of mobility-assisted deft computation offloading by abusing the contact designs controlled by these devices.

An adaptive evolutionary algorithm was used to make the offloading decisions in proposal for a mobility-aware offloading decision technique for communicated mobile cloud computing for cloudlet unwavering quality estimation. The joint optimization that includes the concepts of offloading, communication, and mobility. Our application area is cloud-enhanced robotics, where portable robots create on demand. The robot can actually plan its direction and enforce offloading rather than relying on client portability models. Numerous researches have attempted to examine offloading tactics that involved mobility. For remote code execution for a QoS and mobility-aware asset assignment engineering is particularly noteworthy(Akter et al., 2017; S. et al., 2022; Sampath et al., 2022). For communication, they looked at 5G ultra-thick cell technologies. In contrast to earlier task offloading research, our study emphasizes communication and mobility rather than QoS and storage. Researchers have created a robot that can handle all development-related decisions while delegating the decision-making for communication connections to a customer.

The client (robot) organizes all development-related decisions, and the robot has amazing yet difficult qualities of on-request mobility.

The key mobility benefit of CER over MCC is its on-demand capacity. Offloading is further improved by the ability to transfer to a better system upon request. Although many studies have tried to describe cloud computing in networked robotics structures. Recently, path planning algorithms that are solely focused on movement choices have been presented in a few robotic studies that have been published in the literature. In addition, our decisions on path planning and mobility are a component of multi-target decision-making that essentially aids in the selection of work offloading strategies for CER applications. One of the main goals of the study is to consider work offloading, communication, and mobility as multi-target optimization problems for CER.

## POSITIONING OF THE CONTRIBUTIONS

In cloud-enhanced robotics, task-offloading decision-making is a different topic that requires separate research. Task offloading, mobility, and communication are fundamentally altered as a result of the robot's on-demand services. Most analyses have been conducted from the standpoint of mobile cloud computing.

When shifting a task to the cloud may compensate for the robot's local computing, task offloading is beneficial. It might be possible to conserve energy by moving point jobs to the cloud. According to CER, AP selection (communication) and way planning (mobility) should be interdependent.

### An Integrated Optimal Task Offloading Framework for CER Applications

It is necessary to have a task offloading framework capable of handling such layers of complexity. Our suggested topology then coordinates wireless sensor networks, cloud infrastructures, and robotic systems. It affects CER applications where the combination of sensing, actuation (robots), and computation sustainment (cloud) will supply several independent forms of support. The best task offloading choices for distinct CER framework types are to be identified using a HIGA-based task offloading strategy. Two delegate applications, smart manufacturing and smart cities are selected from a variety of examples. The effects of mobility and communication are evaluated through leisure thanks to the proposed structure. Integration of task offloading, path planning and access point selection concept is depicted in Figure 5.

*Figure 5. Integration of task offloading, path planning and access point selection*

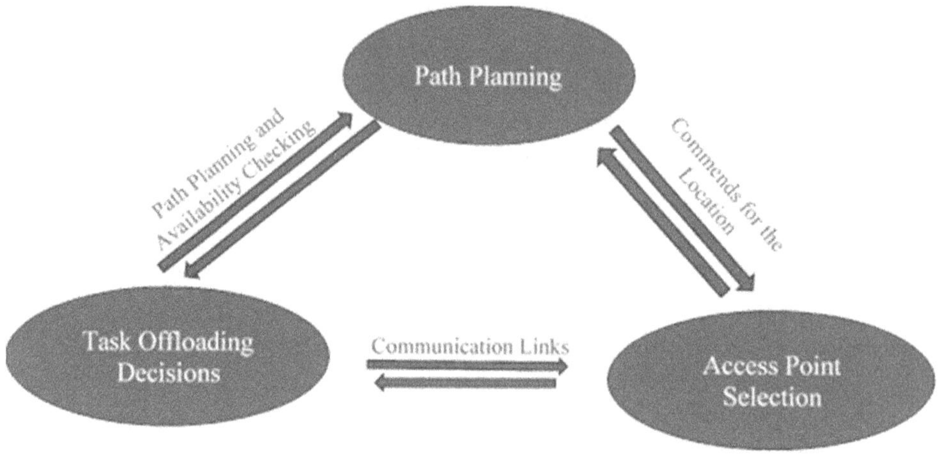


## Optimization for Cloud Enhanced Robot

"Path planning" addresses the problem of optimal robot movement between several focuses, including decision-making with a specific goal in mind. The AP selection, in contrast, associates with data transmission appraisal in which robots do not have their own communication channels. The amount of data transfer bandwidth allocated to robots is not the maximum a specific AP will supply. To have the best availability and transmission capacity, clients can select from a variety of systems. For offloading decision-making, both path planning and AP determination have been taken into consideration separately. Using task offloading and path planning to determine APs for CER is an innovative technique to increase framework performance while using less energy. Path planning represents the robot's ability to choose movement strategies that are appropriate for unloading at this time. The robot might be forced to move to particular locations close to APs as a result of offloading decisions at the same time.

We design a 3-layer decision set for a HIGA-based approach (offloading, path planning, and AP choice). Nobody has exploited the interdependent relationship between AP determination (communication) and route arranging (mobility) for a combined report up until this point, despite the fact that movement or communication has only recently been considered in deciding on offloading decisions.

## TASK OFFLOADING OPTIMIZATION FOR CLOUD ENHANCED ROBOTICS

For usage in the industrial and commercial sectors, an application that combines IoT-enabled sensors, the cloud, and the robot is currently being developed. To demonstrate how this technology might be applied in the real world, we give study situations that are consistent with the focus of our work. The benefit of job offloading using our suggested HIGA-based strategy appears to be present in an itemized recreation that provides a nearly flawless answer. In contrast to conventional mobile cloud computing approaches, our system analyses how communication and mobility perspectives affect decision-making.

A growing standard called cloud robotics makes it possible for autonomous robotic nodes to collaborate and communicate with a cloud computing organization. By creating a comprehensive architecture where robots transfer data and complex processing to the inevitable cloud, it improves the Internet of Things (IoT). In the literature, task offloading in robotics has received a lot of attention. However, the astounding achievements of networked robotics are at the center of our work. They aid in enhancing the task offloading decision-making process in various CER applications through the use of their offloading systems(Chakravarthi et al., 2022; Jeevanantham et al., 2022; Samikannu et al., 2022).

Cloud robotics may be used to address problems with localization and mapping. A resilient cloud-robot system that is appropriate for diverse situations was developed (Osunmakinde & Ramharuk, 2014). I suggested a resource allocation model that would be strengthened by the suggested structure. However, none of them spent any time designing a system with task offloading in mind. The secret is to create a sophisticated system that facilitates robot work offloading and provides the benefit of on-demand mobility.

## SYSTEM ARCHITECTURE FOR CLOUD ENHANCED RBOTICS

To cloud computing architectures that incorporate machine-to-cloud collaboration/communication in CER in Figure. Several elements are considered for QoS, interoperability, and flexibility to influence our plan decision because each model has advantages and disadvantages of its own. System architecture for cloud enhanced robotics is illustrated in Figure 6. With the recent introduction of the Internet of Things, a variety of heterogeneous sensors are delivered through a smart foundation that recognizes fundamentally important events and monitors physical magnitudes. The best barrier is the absence of mobility and actuation given the integrated structure of detecting (sensing) and cloud computing assets. Robots are a crucial ICT component

*Figure 6. System architecture for cloud enhanced robotics*

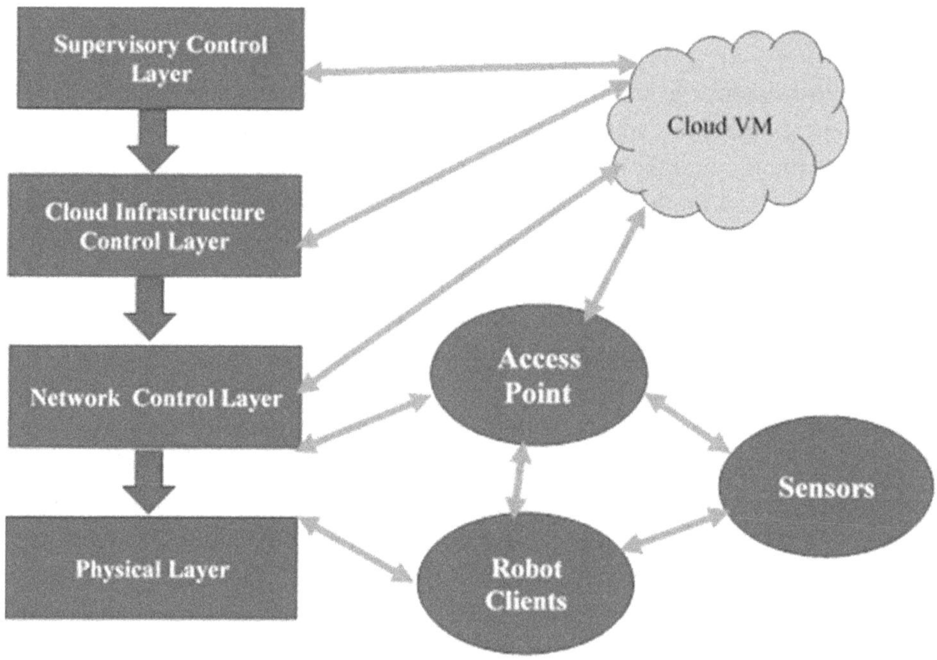


of the smart city framework and are an essential part of it. Robotic nodes may be useful since they can carry out mobile and intelligent services(Babu et al., 2022; Boopathi, Arigela, et al., 2022; Boopathi, Khare, et al., 2023; Boopathi, Siva Kumar, et al., 2023; Chakravarthi et al., 2022; Domakonda et al., 2022; Jeevanantham et al., 2022; Kumara et al., 2023; Palaniappan et al., 2023; S. et al., 2022; Samikannu et al., 2022; Sampath et al., 2022; Senthil et al., 2023; Trojovský et al., 2023; Vanitha et al., 2023; Vennila et al., 2022).

**Physical Layer**: The WSN is composed of low-effort and low-energy smart sensor devices that are integrated into machinery, allowing them to communicate with one another, collect raw data, and analyses it for crucial application-specific activities. By utilizing APs, the AI in sensors transmits required data to the robot over the cloud. Robots have the distinctive ability to move on demand; they plot their path as necessary to travel to the cloud while in motion. Additionally, robots may use the cloud to receive instructions from the supervisory control center. This further enhances the offloading process to guarantee that they always have access to additional aid.

**Network Layer**: AP is described as a smart device with internet functionality that enables the robot to reach the cloud's core. Access points in this layer allow the sensor and robotic framework to communicate with one another. Additionally, it resolves any problems that can arise between physical layer components for upstream data collecting and cloud services. The tangible network layer known as the AP is what allows insubstantial data to stream freely. Due to the variety of APs, the robots can choose to increase their own stream rates for communication depending on their environment and the AP's choice, in accordance with IEEE 802.11 WLAN conventions.

**Cloud framework Layer:** In order to address the needs of testing and dynamic structures, cloud computing has supplied inexpensive, flexible, and readily available assets. The cloud system implicitly includes hardware and software components, such as servers, storage, and virtualization software. It includes a reflection layer that virtualizes resources and exposes them to clients via API interfaces. Virtual machines (VM) with shared power that supply the necessary usefulness to execute all high-thickness working frameworks are typically included in the cloud association. The cloud infrastructure also needs a lot of computational power to handle a variety of complicated and unpredictable client (robot) requests.

**Supervisory Control Layer:** Through the cloud system, networked robots can be remotely controlled and supervised by people thanks to this supervisory control layer. Customers can view data in the cloud that has been collected from sensors and movement reports produced by robots through control terminals. Designers can filter robots and continue execution of uses (as needed) for large-scale tasks thanks to this two-way communication.

## Smart Manufacturing and Smart Cities Application

A self-governing robot would develop its computing capabilities to carry out increasingly challenging but lucrative service tasks. The usage of this architecture for service jobs in the travel sector (help robots), safety services (watch robots, crowd management robots), and fire-fighting robots has been recognized in a number of explicit application areas.

## Smart Manufacturing

Self-sufficient innovations are transforming traditional processing factories into futuristic smart production lines (Stenmark et al., 2015). According to Shrouf and his co-authors, the rise of "smart manufacturing" is the fourth revolution now affecting the industrial industry. The solution is in creating a framework for astute monitoring and the usage of pooled resources as needed. The coordination of cloud computing (CC) and remote sensor networks has resulted in increased effectiveness for natural observation since structure and activity in modern activities contain diverse assortments of decision-making. Robotics has also had a significant impact on modern applications at the same time. Later, it advanced to computerization, and as time passed, it finally reached the era of cloud robots. Similar to IoT, robotics cloud computing enhances robot activities through computation support, storage, robotic learning, and crowd sourcing. Material handling, mechanical production systems, warehousing process maintenance, cooperative navigation, and other popular models are used in Industry 4.0 applications(Lee & Shin, 2015).

### Smart City

The idea of a smart city is one of the most potential applications for cloud computing. This is a city that uses data and communication innovations to make services increasingly aware, understandable, and useful. Examples of such techniques are a completely remote-controlled street lighting corridor of lampposts, an urban automation network (UAN), and a city-wide wireless sensor network. The integrated cloud-enhanced robotics system provides its occupants with numerous forms of help as well.

Our application's scope is distinct and propose an intelligent video observation framework designed to restrict video processing and transmission in a camera network. The breadth of our application, which uses a robotic system and WSN for surveillance, is very broad. The cloud-based technique for a smart city robotic framework has been applied, however their application is focused on the use of unmanned aerial vehicles (UAVs).

## OPTIMIZATION FOR SINGLE TASKFLOW PROCESS

The "cloud robotic group control" is applied in smart cities. The entire action can be divided into three crucial phases or moments that call for the robot to travel to three different locations. In a 20-hub task flow, the numerous activities where the framework will rely on the execution of these tasks.

**Identification Event**: The application's foundational phase combines robotic PRO and camera WSN surveillance. The camera sensor network compiles surveillance information to find and recognize any individual of interest who might be a threat or need help. This starts the first-on-scene event, which sends the robotic administrator wireless position information.

**Investigation Event**: It will then determine whether to alert the relevant authorities or direct the person to a nearby safe spot. The investigative event doesn't involve much mobility, but it does include a few challenging activities that must be completed successfully.

**Guide Event:** The ability to manage or accompany a person of interest at a certain location is one of the fundamental components of a robot-enabled tour operator. This feature is known as the "guide event." A robotic system is a three-tiered architecture made up of a top-level cloud framework, a wireless sensor network, and robotic experts. The middle tier of the framework's basic segment, which provides collaboration, movement, and actuation, is presented by the robot specialist. The cloud framework provides a computational stage to the robotic system to aid the robotic operator in offloading duties to the cloud.

## SUMMARY

This chapter provides an illustration of how two fast expanding concepts—robotics and cloud computing— can be combined. For dynamic and advanced applications, the internet, cloud computing, and AI principles, intelligent robots should be favored over inexpensive conventional robots. One of the well-liked cloud computing methods for expanding robots' limited capabilities is "Task Offloading." It has also been detailed how cloud robotics' various task aspects work. The literature on different path planning, access point, and task offloading optimization approaches utilized in cloud-enhanced robots is also included. The cloud robotic systems' data handling architecture and duty offloading mechanism based on genetic algorithms have been demonstrated., the system architecture and task-flow diagram for smart city and manufacturing control applications have been demonstrated.

## REFERENCES

Akter, M., Zohra, F. T., & Das, A. K. (2017). Q-MAC: QoS and mobility aware optimal resource allocation for dynamic application offloading in mobile cloud computing. *ECCE 2017 - International Conference on Electrical, Computer and Communication Engineering,* (pp. 803–808). IEEE. 10.1109/ECACE.2017.7913013

Arumugam, R., Enti, V. R., Bingbing, L., Xiaojun, W., Baskaran, K., Kong, F. F., Kumar, A. S., Meng, K. D., & Kit, G. W. (2010). DAvinCi: A cloud computing framework for service robots. *Proceedings - IEEE International Conference on Robotics and Automation,* (pp. 3084–3089). IEEE. 10.1109/ROBOT.2010.5509469

Babu, B. S., Kamalakannan, J., Meenatchi, N., Karthik, S., & Boopathi, S. (2022). Economic impacts and reliability evaluation of battery by adopting Electric Vehicle. *IEEE Explore*, (pp. 1–6). IEEE.

Boopathi, S. (2019). Experimental investigation and parameter analysis of LPG refrigeration system using Taguchi method. *SN Applied Sciences*, *1*(8), 892. doi:10.100742452-019-0925-2

Boopathi, S. (2021). Improving of Green Sand-Mould Quality using Taguchi Technique. *Journal of Engineering Research*. doi:10.36909/jer.14079

Boopathi, S. (2022a). An Extensive Review on Sustainable Developments of Dry and Near-Dry Electrical Discharge Machining Processes. *Journal of Manufacturing Science and Engineering*, *144*(5), 1–37. doi:10.1115/1.4052527

Boopathi, S. (2022b). An investigation on gas emission concentration and relative emission rate of the near-dry wire-cut electrical discharge machining process. *Environmental Science and Pollution Research International*, *29*(57), 86237–86246. doi:10.100711356-021-17658-1 PMID:34837614

Boopathi, S. (2022c). Experimental investigation and multi-objective optimization of cryogenic Friction-stir-welding of AA2014 and AZ31B alloys using MOORA technique. *Materials Today. Communications*, *33*, 104937. doi:10.1016/j.mtcomm.2022.104937

Boopathi, S., Arigela, S. H., Raman, R., Indhumathi, C., Kavitha, V., & Bhatt, B. C. (2022). Prominent Rule Control-based Internet of Things: Poultry Farm Management System. *IEEE Explore*, (pp. 1–6).

Boopathi, S., Balasubramani, V., Kumar, R. S., & Singh, G. R. (2021). The influence of human hair on kenaf and Grewia fiber-based hybrid natural composite material: An experimental study. *Functional Composites and Structures*, *3*(4), 045011–045021. doi:10.1088/2631-6331/ac3afc

Boopathi, S., Jeyakumar, M., Singh, G. R., King, F. L., Pandian, M., Subbiah, R., & Haribalaji, V. (2022). An experimental study on friction stir processing of aluminium alloy (AA-2024) and boron nitride (BNp) surface composite. *Materials Today: Proceedings*, *59*, 1094–1099. doi:10.1016/j.matpr.2022.02.435

Boopathi, S., & Khare, R. K. G., J. C., Muni, T. V., & Khare, S. (2023). Additive Manufacturing Developments in the Medical Engineering Field. In Development, Properties, and Industrial Applications of 3D Printed Polymer Composites (pp. 86–106). IGI Global.

Boopathi, S., Myilsamy, S., & Sukkasamy, S. (2021). (Preprint). Experimental Investigation and Multi-Objective Optimization of Cryogenically Cooled Near-Dry Wire-Cut EDM Using TOPSIS Technique. *Ijamt*. Advance online publication. doi:10.21203/rs.3.rs-254117/v1

Boopathi, S., Siva Kumar, P. K., & Meena, R. S. J., S. I., P., S. K., & Sudhakar, M. (2023). Sustainable Developments of Modern Soil-Less Agro-Cultivation Systems. In Human Agro-Energy Optimization for Business and Industry (pp. 69–87). IGI Global. doi:10.4018/978-1-6684-4118-3.ch004

Boopathi, S., & Sivakumar, K. (2013). Experimental investigation and parameter optimization of near-dry wire-cut electrical discharge machining using multi-objective evolutionary algorithm. *International Journal of Advanced Manufacturing Technology*, *67*(9–12), 2639–2655. doi:10.100700170-012-4680-4

Boopathi, S., Thillaivanan, A., Pandian, M., Subbiah, R., & Shanmugam, P. (2021). Friction stir processing of boron carbide reinforced aluminium surface (Al-B4C) composite: Mechanical characteristics analysis. *Materials Today: Proceedings*, *50*(5), 2430–2435. doi:10.1016/j.matpr.2021.10.261

Chakravarthi, P. K., Yuvaraj, D., & Venkataramanan, V. (2022). IoT-based smart energy meter for smart grids. *ICDCS 2022 - 2022 6th International Conference on Devices, Circuits and Systems*, (pp. 360–363). 10.1109/ICDCS54290.2022.9780714

Domakonda, V. K., Farooq, S., Chinthamreddy, S., Puviarasi, R., Sudhakar, M., & Boopathi, S. (2022). Sustainable Developments of Hybrid Floating Solar Power Plants: Photovoltaic System. In Human Agro-Energy Optimization for Business and Industry (pp. 148–167). IGI Global.

Guo, Y., Mi, Z., Yang, Y., & Obaidat, M. S. (2018). An energy sensitive computation offloading strategy in cloud robotic network based on ga. *IEEE Systems Journal*, *13*(3), 3513–3523. doi:10.1109/JSYST.2018.2830395

Jeevanantham, Y. A., Saravanan, A., Vanitha, V., Boopathi, S., & Kumar, D. P. (2022). Implementation of Internet-of Things (IoT) in Soil Irrigation System. *IEEE Explore*, (pp. 1–5). IEEE.

Kansal, N. J., & Chana, I. (2015). Artificial bee colony based energy-aware resource utilization technique for cloud computing. *Concurrency and Computation*, *27*(5), 1207–1225. doi:10.1002/cpe.3295

Kaur, R., & Kautish, S. (2022). Multimodal sentiment analysis: A survey and comparison. *Research Anthology on Implementing Sentiment Analysis Across Multiple Disciplines*, 1846-1870.

Kautish, S., Peng, S. L., & Obaid, A. J. (Eds.). (2021). *Computational intelligence techniques for combating COVID-19*. Springer International Publishing. doi:10.1007/978-3-030-68936-0

Kautish, S., Reyana, A., & Vidyarthi, A. (2022). SDMTA: Attack detection and mitigation mechanism for DDoS vulnerabilities in hybrid cloud environment. *IEEE Transactions on Industrial Informatics, 18*(9), 6455–6463. doi:10.1109/TII.2022.3146290

Kavitha, C., Malini, P. S. G., Charan, V., Manoj, N., Verma, A., & Boopathi, S. (2022). (in press). An experimental study on the hardness and wear rate of carbonitride coated stainless steel. *Materials Today: Proceedings*. doi:10.1016/j.matpr.2022.09.524

Kumar, V. Y., & Michae, L. N. (2012). Opportunities and challenges with autonomous micro aerial vehicles. *The International Journal of Robotics Research, 31*(11), 1279–1291. doi:10.1177/0278364912455954

Kumara, V., Mohanaprakash, T. A., Fairooz, S., Jamal, K., Babu, T., & B., S. (2023). Experimental Study on a Reliable Smart Hydroponics System. In *Human Agro-Energy Optimization for Business and Industry* (pp. 27–45). IGI Global. doi:10.4018/978-1-6684-4118-3.ch002

Lee, K., & Shin, I. (2015). User mobility model based computation offloading decision for mobile cloud. *Journal of Computing Science and Engineering : JCSE, 9*(3), 155–162. doi:10.5626/JCSE.2015.9.3.155

Madhu, G., Govardhan, A., Ravi, V., Kautish, S., Srinivas, B. S., Chaudhary, T., & Kumar, M. (2022). DSCN-net: A deep Siamese capsule neural network model for automatic diagnosis of malaria parasites detection. *Multimedia Tools and Applications, 81*(23), 34105–34127. doi:10.100711042-022-13008-6

Mell, P., & Grance, T. (2011). The NIST-National Institute of Standars and Technology-Definition of Cloud Computing. *NIST Special Publication 800-145*, 1–3.

Myilsamy, S., & Sampath, B. (2021). Experimental comparison of near-dry and cryogenically cooled near-dry machining in wire-cut electrical discharge machining processes. *Surface Topography : Metrology and Properties, 9*(3), 035015. doi:10.1088/2051-672X/ac15e0

Osunmakinde, I., & Ramharuk, V. (2014). Development of a survivable cloud multi-robot framework for heterogeneous environments. *International Journal of Advanced Robotic Systems, 11*(10), 1–22. doi:10.5772/58891

Palaniappan, M., Tirlangi, S., Mohamed, M. J. S., Moorthy, R. M. S., Valeti, S. V., & Boopathi, S. (2023). Fused Deposition Modelling of Polylactic Acid (PLA)-Based Polymer Composites: A Case Study. In Development, Properties, and Industrial Applications of 3D Printed Polymer Composites (pp. 66–85). IGI Global.

Pandey, P., Pompili, D., & Yi, J. (2015). Dynamic Collaboration between Networked Robots and Clouds in Resource-Constrained Environments. *IEEE Transactions on Automation Science and Engineering*, *12*(2), 471–480. doi:10.1109/TASE.2015.2406115

S., P. K., Sampath, B., R., S. K., Babu, B. H., & N., A. (2022). Hydroponics, Aeroponics, and Aquaponics Technologies in Modern Agricultural Cultivation. In *Trends, Paradigms, and Advances in Mechatronics Engineering* (pp. 223–241). IGI Global. doi:10.4018/978-1-6684-5887-7.ch012

Samikannu, R., Koshariya, A. K., Poornima, E., Ramesh, S., Kumar, A., & Boopathi, S. (2022). Sustainable Development in Modern Aquaponics Cultivation Systems Using IoT Technologies. In *Human Agro-Energy Optimization for Business and Industry* (pp. 105–127). IGI Global.

Sampath, B. C. S., & Myilsamy, S. (2022). Application of TOPSIS Optimization Technique in the Micro-Machining Process. In Trends, Paradigms, and Advances in Mechatronics Engineering (pp. 162–187). IGI Global. doi:10.4018/978-1-6684-5887-7.ch009

Sampathkumar, A., Rastogi, R., Arukonda, S., Shankar, A., Kautish, S., & Sivaram, M. (2020). An efficient hybrid methodology for detection of cancer-causing gene using CSC for micro array data. *Journal of Ambient Intelligence and Humanized Computing*, *11*(11), 4743–4751. doi:10.100712652-020-01731-7

Senthil, T. S., Puviyarasan, M., Babu, S. R., Surakasi, R., Sampath, B., & Associates. (2023). Industrial Robot-Integrated Fused Deposition Modelling for the 3D Printing Process. In Development, Properties, and Industrial Applications of 3D Printed Polymer Composites (pp. 188–210). IGI Global.

Stenmark, M., Malec, J., Nilsson, K., & Robertsson, A. (2015). On distributed knowledge bases for robotized small-batch assembly. *IEEE Transactions on Automation Science and Engineering*, *12*(2), 519–528. doi:10.1109/TASE.2015.2408264

Trojovský, P., Dhasarathan, V., & Boopathi, S. (2023). Experimental investigations on cryogenic friction-stir welding of similar ZE42 magnesium alloys. *Alexandria Engineering Journal*, *66*(1), 1–14. doi:10.1016/j.aej.2022.12.007

Vanitha, S. K. R., & Boopathi, S. (2023). Artificial Intelligence Techniques in Water Purification and Utilization. In *Human Agro-Energy Optimization for Business and Industry* (pp. 202–218). IGI Global. doi:10.4018/978-1-6684-4118-3.ch010

Vennila, T., Karuna, M. S., Srivastava, B. K., Venugopal, J., Surakasi, R., & Sampath, B. (2022). New Strategies in Treatment and Enzymatic Processes: Ethanol Production From Sugarcane Bagasse. In Human Agro-Energy Optimization for Business and Industry (pp. 219–240). IGI Global.

Wang, Z., Zhao, Z., Min, G., Huang, X., Ni, Q., & Wang, R. (2018). User mobility aware task assignment for Mobile Edge Computing. *Future Generation Computer Systems*, *85*, 1–8. doi:10.1016/j.future.2018.02.014

Xia, Q., Liang, W., Xu, Z., & Zhou, B. (2014). Online algorithms for location-aware task offloading in two-tiered mobile cloud environments. *Proceedings - 2014 IEEE/ACM 7th International Conference on Utility and Cloud Computing, UCC 2014*, (pp. 109–116). IEEE. 10.1109/UCC.2014.19

Xu, F., Liu, F., Jin, H., & Vasilakos, A. V. (2013). Managing performance overhead of virtual machines in cloud computing: A survey, state of the art, and future directions. *Proceedings of the IEEE*, *102*(1), 11–31. doi:10.1109/JPROC.2013.2287711

Yupapin, P., Trabelsi, Y., Nattappan, A., & Boopathi, S. (2022). Performance Improvement of Wire-Cut Electrical Discharge Machining Process Using Cryogenically Treated Super-Conductive State of Monel-K500 Alloy. *Iranian Journal of Science and Technology - Transactions of Mechanical Engineering*, 1–17. doi:10.1007/s40997-022-00513-0

Zhang, L., Fu, D., Liu, J., Ngai, E. C. H., & Zhu, W. (2017). On Energy-Efficient Offloading in Mobile Cloud for Real-Time Video Applications. *IEEE Transactions on Circuits and Systems for Video Technology*, *27*(1), 170–181. doi:10.1109/TCSVT.2016.2539690

## APPENDIX: ABBREVIATIONS

ACO- Ant Colony Optimization
AHP- Analytic Hierarchy Process
AI- Artificial intelligence
ANN - Artificial Neural Networks
AP- Access Points
AUV - Autonomous Underwater Vehicle
CAA- Cuckoo Algorithm
CC - Cloud Computing
CER - Cloud Enhanced Robotics
GA- Genetic Algorithms
HIGA - Hybrid Intelligent Generic Algorithm
IoT Internet of Things
MCC- Mobile Cloud computing
PSO- Particle Swarm Optimization
QoS - Quality of Service
SLAM -SaaS model for synchronous restriction and mapping
TOPSIS-Technique for Order Preference by Similarity to Ideal Solution
VIKOR - VIekriterijumsko KOmpromisno Rangiranje
UAN - Urban Automation Network
UAVs - Unmanned Aerial Vehicles
VM - Virtual Machines

# Chapter 8
# Autonomous Navigation of Rovers Using ML and DL Techniques

**Megha Bhushan**
*DIT University, India*

**Maanas Singal**
*DIT University, India*

**Shirshendu Layek**
*DIT University, India*

**Arun Negi**
*Deliotte USI, India*

## ABSTRACT

*The growth in the sector of autonomous robots in the field of transportation has increased since the 2000s. Improving the dynamics of a robot is a continuous process to tailor its user experience. This field involves the use of planetary and extra-terrestrial robots' (called rovers) autonomous navigation capabilities. It allows analysis of terrain irregularities, climate and weather monitoring, sample collection, etc. As rovers require a significant investment, therefore, it is essential that the rover performs autonomously according to the expectations while ensuring its own safety. It is achieved by the use of complex mathematical models, image analysis techniques, machine learning (ML) and deep learning (DL) models, and allowing execution of the required tasks efficiently. Further, it provides insight on various aspects of rovers' navigation such as intended missions of rover, ML and DL models, comparison in terms of precision and accuracy, merits, and demerits along with future scope.*

DOI: 10.4018/978-1-6684-8171-4.ch008

## INTRODUCTION

The applications, and various use-cases of Machine Learning (ML) (Kholiya et al., 2021), Deep Learning (DL) (Pal et al., 2022), and Computer Vision (CV) has led to a revolution in the industrial and scientific applications. Today Artificial Intelligence (AI) (Verma et al., 2022) and ML techniques, and frameworks are used for a wide range of high priority tasks such as scientific computation, autonomous navigation, mechanical and civil engineering requirements, mining, and much more (Alzubi et al., 2018). This boom in the AI/ML frameworks has prompted researchers to use it for space exploration satellites and rovers (Mortlock & Kassas, 2021) to increase the efficiency, and the credibility of the exploration findings for further enhancements under human supervision. Also, quality of data (Bhushan et al., 2017; Bhushan et al., 2018a; Bhushan et al., 2018b; Bhushan & Goel, 2016; Negi & Kaur, 2017) is critical for obtaining better results using AI/ML techniques. It will increase the efficiency, and the credibility of the exploration findings for further enhancements under human supervision. It has expanded the unmanned space exploration missions which are conducted on a large scale. Also, it is not limited to geo-sensing satellites, solar and planetary orbiters (Mortlock & Kassas, 2021), unmanned space vehicles such as lunar and mars rovers, asteroid sample collection bots (Jiang et al., 2020), etc. which aids in the investigation of the immediate solar system. The existing autonomous systems are utilized by aerospace giants for accomplishing various mission parameters, eliminating or minimizing the need for human intervention while assuring the safety parameters.

This chapter describes the use cases of AI/ML in the navigation of rovers and autonomous vehicles. The rovers are used for a variety of purposes; such as planetary exploration, mineral scouting, sample collection, autonomous and remote testing etc. They can be categorized into several types such as mobile robots, agriculture rovers, planetary rovers, unmanned self-driving vehicles, and asteroid sample collectors. The use of a conglomerate of AI/ML frameworks that cover several elements of image recognition, forecasts, strategic planning, and reinforcement of current models instigates this field of research. The DL (Rana & Bhushan, 2022) and Neural Networks (NN) used in the existing works can be tailored and are scalable to suit the custom requirements of the mission or directive given to the rovers. This helps in making optimal measures for any tasks according to the constraints laid out by the limitation of resources which might be different for each and every rover (Alzubi et al., 2018; Daftry et al., 2022). However, it is important to note that while the standard frameworks with their existing parameters may not be much helpful for a more complex task such as one requiring more elaborative processing etc.

Subsequently, it provides various aspects and facilitates provisions for modification using a combination of techniques, and mathematical modelling such as Gaussian

Process (Takebayashi et al., 2021), Markov Decision Process (MDP) (Abcouwer et al., 2021)and so on to provide abridgement of data. ML techniques like Linear Regression (LR) (Arrouch et al., 2022)and Multiple Linear Regression (MLR) models can be used to predict limited computation involving problems like determination of the position of solar panels, etc. However, NNs play a significant role as the tasks become increasingly complex involving multiple iterative decisions based on several dynamic variables like object classification and traversal methods.

Designing various frameworks involve complex calculations, and annotations with an iterative decision-making process which is critical for rover operations using ANNs, CNNs or RNNs (Alzubi et al., 2018). The ANNs are able to perform with incomplete comprehension whereas the CNNs are extremely helpful for image processing problems. The RNNs work with multiple convolution layers incorporated in the same framework allowing it to perform more iterative decision-making tasks. The last step of the process includes CV with OpenCV (a versatile library for incorporating image recognition capabilities) which is useful in many aspects like object detection, and terrain mapping. For the testing, tailored models and publicly available datasets such as the ones provided by NASAs Martian Rovers Curiosity and Perseveranceare generally used for training and validation purposes. Along with the aforementioned datasets, some synthetic Earth traffic or terrain-based datasets are also sometimes used for further fine-tuning of the proposed models (Daftry et al., 2022).

The role of orbital satellites also has a major stake in this as the data accumulated by orbital imaging, refines the on-board data collection making the mission more effective, productive, and cost-efficient. The actual testing of the complete framework is usually done on analogue Lunar or Mars sites such as the Rio Tinto analogue site using test bots like NASA Athena Rover (Mortlock & Kassas, 2021). Mostly, complex, challenging actual Earth terrains resembling ground contours like the celestial body are also employed for the investigation in required aspects. Apart from the test rovers, miniature test mobile robot (Takebayashi et al., 2021) kits which resemble basic rover structure are also used as they are readily available.

## LITERATURE REVIEW

The existing studies related to autonomous navigation of rovers using ML and DL techniques studies have been segmented according to specific 'use-cases' as shown in Table 1. Also, these studies have been categorized as follows:

## Data Collection and Accounting

The quality of data is required for the accurate implementation of any type of model whether it is ML, DL, CV, Meta-Learning (Gaudet et al., 2020) or Neuromorphic Computing (Piñero-Fuentes et al., 2021). The data collection from the Low Earth Orbiting (LEO) satellites, and exoplanetary sources such as telescopes, drones, and rovers (Matthies et al., 2022) is reliable only up to the extent it is transmitted (Alimo et al., 2021). Often the data has errors due to unforeseen events such as diverse solar events which affect the throughput of data transmission making it more difficult to analyses (Abcouwer et al., 2021). Furthermore, reducing downlink time durations (when the systems are unable to exchange data due to the position of celestial bodies (Matthies et al., 2022)) is vital. Thus, the research works conducted in this aspect include the modelling of algorithms to either reduce the downlink time or reduce the data error redundancy (Abcouwer et al., 2021; Alimo et al., 2021).

## Aerial Assistance in Dataset Collection

The role of LEO, unmanned aerial vehicles, and other remote sensory equipment is indisputable as the on-board capabilities of the rovers are limited. Thus, to reduce the effect of this shortcoming, the data required for the training and validation of the AI/ML/DL/CV models have been presented in the works of Mortlock & Kassas (2021), Wang et al. (2022) and Sun et al. (2020).

## Implementation of ML for Navigation Systems

ML/DL models have been implemented as an ad-hoc to the pre-existing models as discussed by Daftry et al. (2022). It primarily focused on traversal path optimization where the most convenient, and efficient path for the traversal in the given set of tasks is required. In previous work of Mortlock & Kassas (2021), promising results on mobile robotic models as well as analogue rovers have been attained. Several emerging techniques involving state of the art DL algorithms along with emerging aspect technologies such as Meta-Learning (Gaudet et al., 2020), and Neuromorphic Computing have been presented in the research by authors Piñero-Fuentes et al. (2021). Also, special systems were used for enhancing ground clearance models like ACE (Abcouwer et al., 2021). The object avoidance and mitigation strategies consisting of DL, NN, and CV models were used in former works of Zhang et al. (2021), Ziya and KARAKOSE (2020), and Wang et al. (2022).

### Hazard Mitigation

The planetary rovers are especially prone to safety issues which can affect the mission parameters such as getting stuck in a pit of soft sand, or wheel slippage risks (Coloma et al., 2022). Various types of models employed in the mitigation of the aforementioned issue comprise of several DL, and image processing models (Ono et al., 2020) to test their precision in diverse conditions.

### Computational and Energy Resources

The on-board computational power of the rover equipment is very limited which creates issues while running heavy algorithms. Thus, various studies like by authors Chiodini et al. (2020) and Visca et al. (2022) have been conducted and analyzed to optimize or reduce the need of computational power per unit cycle along with path optimization models.

## OPEN CHALLENGES

Following are the challenges delineated into specific 'use-cases' (as illustrated in Table 1) after reviewing existing studies on achieving maximum accuracy or precision in the autonomous navigation of rovers by using AI and ML for maximizing throughput of the assigned mission (Daftry et al., 2022; Matthies et al., 2022).

### Localization Through Imaging

Better crater mappings can be obtained by exploring orbital trajectories or maneuvers (Matthies et al., 2022). These pre-trained DRL architectures can become a step in the development of hopping rovers for the autonomous exploration of asteroids. The results in dependable chaotic map propositions may be improved (Jiang et al., 2020). As discussed in the work of Duo et al. (2019), upgrades may include integrating an on-board mechanism to estimate solar power generation which can be coupled with the work demonstrated in Chiodini et al. (2020) to enrich datasets for fine tuning on ML/DL models. Also, the framework's application can be expanded for diverse tasks (Wang et al., 2022) such as object detecting, and tracking.

*Table 1. Existingliterature on autonomous navigation of rovers using ML and DL techniques*

| Reference | Year | Summary | Technique | Algorithm | Dataset | Results | Merits | Demerits |
|---|---|---|---|---|---|---|---|---|
| Mortlock & Kassas (2021) | 2021 | A solution to improve "GITL" conceptual cycle involved in the lunar rover navigation is presented to improve accuracy, decrease downlink time and costs. This optimizes the cycle in question to further increase the transpired range of the rovers. | Mapping the lunar craters from orbit, and parameterizing their landmarks with LIDAR, or Visual instrumentation on-board.<br>The craters were detected using 3D Point Cloud data from on-board instrumentation as well as using clues from earlier observations. | LIDAR and Image Processing, Computer Vision, TensorFlow, KPP, and CNN | Custom datasets were generated using data from Apollo missions, and an analogue crater site at Cinder Lake crater field near Flagstaff, AZ. | Crater Diameter(m) / Position Error(m):<br>5/1.02<br>10/3.22<br>15/4.48<br>20/4.35 | Low downlink/downtime for the "GITL" cycle which further improved the traversal time for the rover with higher accuracy of predicted +-5 to 10 meters using localized data. | Rigorous training was required with more diversified datasets.<br>There are a number of discrete loopholes or gaps in the presented model which marks way for a randomized arrival of errors while navigation which coupled with the absence of the live orbital.<br>Data feedback could lead to a chance of mis-navigation thereby jeopardizing the entire mission plan. |
| Abcouwer et al.(2021) | 2021 | The trailblazing DRL constant ad-mistering method for map-less navigation gave the experimental rover the exploration capabilities to have higher possibilities of reaching navigational targets produced perpetual actions. | The state of the art DRL model has been enriched by RL using Markov Process along with PPO (Policy Proximal Optimization) Algorithms. | RL (Markov Process), and PPO | No previous dataset used. | Continuous action training failed to approach the conclusion, and the oscillations go up and down for the reward within the span with no indication of reaching the target location. | With validation of perpetual actions allowed the agent to have richer investigations, supple movements and thus greater likelihood to reach the intended target.<br>In comparison to traditional discrete actions; this significantly improved autonomy in navigation missions. | The maples navigation problem can be addressed in a much broader aspect than discussed in the paper.<br>Although the model presented depicted a near ideal solution for resolution of the issue: however, the lateral dependencies on the critical hardware were intensive. |
| Wang et al.(2022) | 2022 | A terrain analysing method was developed for Martian Rover Navigation using CNN labelling utilizing the newly DeepLabv3+ pre-trained network for semantic segmentation in input image. It generated accurate 3D maps up to 12 metres distant from the camera. | It used a feed corrected stereo picture to estimate the depth map. The scene's point cloud was turned into a voxel grid for attaining a volumetric image with simultaneous labelling using DeepLabv3+ to get a 3D representation. | CNN, ImageNet, DeepLabv3+, and Bayes Models | ESA Katwijk Beach Planetary Rover Dataset | MLESAC has shown the best performance on Data 2 with accuracy of 0.97, and IoU of 0.58 | A voxel grid was used along with a DeepLabv3+ labeller to generate a volumetric representation of the environment for optimum navigational decisions.<br>It was successful on the ESA Planetary Rover dataset, thus, it was used for real rover navigation for autonomous locomotion. | It is restricted to testing on a single technique on a single ESA dataset which limits the potential credibility on various diversified terrain possibilities. |

*continues on following page*

*Table 1. Continued*

| Reference | Year | Summary | Technique | Algorithm | Dataset | Results | Merits | Demerits |
|---|---|---|---|---|---|---|---|---|
| Higa et al. (2019) | 2019 | A semi-supervised model pertaining to machine vision on Mars for resolving two specific tasks, i.e., classification and segmentationusing contrastive teachwas presented. It was a powerful representation learning technique.<br>For reconciling the contradiction arising due to data mismatch between the actual and synthetic data, annotations and unlabelled data were incorporated to improve performance. | A Semi Supervised Learning (SSL) model for Martian machine vision is expanded by contrastive learning to include supervised inter-class and unsupervised similarity-only variants for classification. Further, element-wise contrastive learning for segmentation and additional supervision through online pseudo labelling. | SICCL and USL | The MSL surface dataset along with the 34k unlabelled images from NASA's Planetary Data System were utilized for the study. | Segmentation performance on the AI4Mars Dataset indicated ImageNet pertaining to the proposed method with accuracy of 82.82%, and mIoU of 70.34%. | Annotations were used to resolve discrepancies caused by data abridgement and to further benefit from unlabelled data to enhance performance.<br>The experimental results demonstrated that the studied learning strategies can significantly outperform the far more complex approaches in terms of results. | It is restricted to SSL algorithms which may limit the experimental result readings, and precision.<br>It makes the validation of the framework on real-time scenarios questionable. |
| Prinz et al. (2022) | 2022 | An enhanced NN created on the DQN framework was used to decide the most appropriate course of action for the hopping rover, three different architecture modulesare merged to improve the effectiveness of the DRL algorithm. | Three different architecture modules; DDQN, Duelling Network, and Prioritized Experience Replay—are merged to improve the effectiveness of the DRL algorithm and boost the accuracy of the model's relearning. | DRL, DNN, DQN, DDQN, Duelling Network, and Prioritized Experience Replay | Custom datasets were tailored, and utilized for the study. | The proportion of perturbed cells: 4% \|<br>Indicator 1: 180504\|<br>Indicator 2: 450s\|<br>Indicator 3: 7665 | The robustness is assessed by considering asteroid landscape charts' static ambiguity.<br>It was done utilising map cells, which demonstrated that perturbed map scenarios, optimal pathways are effectively obtained. | The error margin of 4% is a disadvantage to this state-of-the-art model as it raises doubts on its validation as there is a gap or abridgement of data which may give variance in results. |
| Daftry et al. (2022) | 2022 | A ML improvised path planning while respecting safety constraints by training the model on perseverance data for actual Mars terrain simulation along with various synthetic complex terrain data. | "MLNav" model was trained on actual and synthetic terrain datasets along with implementation on the existing ENav Platform. | U-Net, SegNet, DeeplabV3+, and PSPNet | Custom generated maps, and data were utilized specific to the conducted study. | Accuracy:<br>U-Net:95.1%<br>SegNet:81.7%<br>DeeplabV3+:82.8%<br>PSPNet: 85.2% | Higher efficiency and throughput. The observations along with the inferring numbers are better up to 10% [+- 2%].<br>The overall cost of running the apparatus will take less amount than former apparatus. | The complexity of the presented model is relatively higher than the original model.<br>Even though the long run costs would be lower, however; the initial cost of implementing new model with its additional requirements will be more. |

*continues on following page*

*Table 1. Continued*

| Reference | Year | Summary | Technique | Algorithm | Dataset | Results | Merits | Demerits |
|---|---|---|---|---|---|---|---|---|
| Takebayashi et al. (2021) | 2021 | The reduction in computational time per unit cycle of enhanced navigation platform operating on-board the perseverance mars rover by optimizing the Approximate Clearance Estimator (ACE) system along with more successful traverses. | DCNN based approach using encoder/decoder architecture of U-Net algorithm which was further implemented on Tensor Flow using Monte Carlo simulation of the model on baseline ENav algorithm. | DCNN, U-Net, and TensorFlow | Custom data extracted from Height-Maps was utilized for training Heuristics. | The highest accuracy was obtained on the Learned Heuristic Model with no min ACE evaluations (border free) with 99.9% accuracy on benign terrains and 73.9% on complex ones. | Reduced computational time per cycle for the plan estimator along with significant increase in successful traverses.<br>Also, the reduction in ACE evaluations, and increased path efficiency elevating performance from the baseline ENav for robust throughput. | Need to implement a suite of other ML/DL algorithms for the improvement.<br>The function of the ENav system remained far too dependent on a single variable (tool) for path estimation which can result in problems. |
| Arrouch et al. (2022) | 2022 | For future Mars rover missions, every monitoring predictive framework has been developed using a vision-based framework to predict the driving energy usage parenthetically using ML.<br>A comparison between VeeGer-EnergyNet and Veeger-TerramechanicsNet ML models was also given. | ML techniques improved the slippage detection with fine tuning of hyper parameters.<br>Using the models, continuous values of data were predicted from images.<br>DL networks were used for classification which is utilized for regression. | DL, SVN, and CNN | A test dataset newly collected by the Athena rover at the JPL Mars Yard. | Estimated Power difference between Ground Truth Energy Consumption and VeerGerTetramechmics provided mean of 0.76, and standard deviation of 0.64, respectively. | Better, and more energy efficient way of rover navigation via the analysis of the tetra mechanical aspects of the rovers' environment allows the efficiency of the mission to rise up considerably. | The limited exposure of the training dataset despite being enormous in size presents a restriction to the use of the model.<br>Also, complex reprogramming required for sensor, and model recalibration for steady operation makes on the spot deployment difficult. |
| Furfaro et al.(2020) | 2020 | The precision landing on all sizes of planetary bodies requires highly intelligent systems for the task; thus, a pliant directing system was built on the classical ZEM/ZEV.<br>It is improved upon via the use of ML to overcome its' limitations, and created a closed loop directing model which is able to adapt to the given constraint scenario. | Using ML actor critical method based on strategy gradient incorporating advantage function estimation; the capabilities of the classical ZEM/ZEV have been expanded.<br>A developed guiding NN that may be used immediately aboard is provided by the actor-critic approach. | EML, MDP, RL, and PGT | No previous dataset used. | STM components are highest in the event of STM [12] (~22) and lowest in the event of STM [11] (~0) whereas the same is applicable in the metrics of ~23 and ~0.1 respectively. | The pliant strategy (A-ZEM/ZEV) has significant improvement of performance in parameters of fuel utilization and allows path restrictions to be used directly with the directing law.<br>The actor-critic algorithm outputs a competent directing neural network that can be directly used on-board which further improves its' effectiveness and efficiency. | This ELM-based actor-critic approach may be utilised in cases when the state space can be adequately discretized. The fact that the benefit function estimate is unbiased and the cost function is appropriately set up ensures method convergence.<br>Thus, work is required to improve the learning performance. |

*continues on following page*

*Table 1. Continued*

| Reference | Year | Summary | Technique | Algorithm | Dataset | Results | Merits | Demerits |
|---|---|---|---|---|---|---|---|---|
| Duo et al. (2019) | 2019 | As a step towards accelerating specificity by minimising search space and time, this method for perspective-to-satellite image localization of rover surfaces by ML improved the efficiency and accuracy of the same. It also served as a guideline for improving absolute localization on planetary surfaces. | Absolute localization accuracy can be improved by using ML models to enable robust systematisation, and optimized computation. It generated a baseline dataset for further research. To visualize the framework, an analogue planetary environment was set up using Unreal Engine run on SCNN. | OpenCV, SNN, PLanNet v0, and ResNet | A publicly available dataset was used, and a benchmarking custom dataset was generated. | The experimental setup has reduced the search area by 90-95% than by using original methods. | Created a site reference dataset, and generated tool for evaluation and priming their techniques of localization. | The usage of synthetic datasets along with limited planar degrees of freedom on this relatively short study hinders the validation of the results. |
| Alimo et al. (2021) | 2021 | An autonomous navigation algorithm for a robot based on curiosity and Q-learning was implemented on the Sigmoid function where it sways the probability of selection of action in RL way. A case of maze robot was used where the path leading out was the major index defining the quality of the algorithm. | A curiosity function pertaining on a neurological basis was used to design an autonomous navigation model for the maze robot reinforced with Q-learning. These were applied on two different simulations to prove the algorithms' feasibility and validate its' performance. | RL and Q-Learning | No previous dataset used. | Q-learning algorithm with curiosity gave the results for time in seconds as follows: S1-284\| S2-283\| S3-287 | On running the model's simulation, the optimal path was found smoothly, and the agent used the learned strategy to acquire the shortest path from the maze to the exit proving the feasibility of the model. Also, the comparison of the studied model versus the Q-learning algorithm without curiosity further proved the optimum capabilities. | Some lapses of this curiosity driven algorithm that the nodes were required to be placed before along with the fact that the loop problems solving capabilities of the studied model were not good which greatly undermined its true potential. |
| Visca et al. (2022) | 2022 | The effectiveness of three different DL models was compared using the PyGame simulator, which was built with the CNN, RNN, and LSTM algorithms to allow the program to understand to move without colliding with four distinct fixed obstructions. | ANNs were built using Keras, a DL library. Additionally, the LSTM, RNN, and CNN DL algorithms were coded individually. A detailed representation of result was also explained. | Deep-Q Network, RNN, CNN and LSTM | A custom, tailored dataset was generated and implemented. | Algorithm/Reward Mean: CNN/363\| LSTM/21.555\| RNN/3.922 | By computation and analysis of rendering of three diverse DL algorithms, the progressive conclusion of the fact that robust, and secure methods for autonomous vehicles can be obtained by using DRL. | The main issue is dependence of the agent's operation in accordance with the reward. |

*continues on following page*

*Table 1. Continued*

| Reference | Year | Summary | Technique | Algorithm | Dataset | Results | Merits | Demerits |
|---|---|---|---|---|---|---|---|---|
| Piñero-Fuentes et al. (2021) | 2021 | A mobile robotics platform controlled through the robotic operating system (ROS) following the classified output of a modern CNN accelerator AI for FPGAs connected to a neuromorphic dynamic visual sensor was implemented. The results of the validation and test scenarios are tabulated in the study. | The process involves the integration of occurrences into histograms after which they are normalized to create frames implemented to collect data set, and the training of NN. The CNN accelerator allows a viable real-time Edge-AI computing model. | FPGA, and CNN | A publicly available dataset was used, and a benchmarking custom dataset was generated. | Metrics ǁ Loss 0.033ǀ Accuracy 97.62%ǀ Precision 99.32%ǀ Recall 98.31% | The proof of concept that autonomous navigation can be performed on both neuromorphic sensors and edge AI computing devices facilitates future research in this area. | The CNN presented in this study is still conventional as instead of processing events directly, we process histograms built from a set of events. This is a drawback as it consumes more computational resources, and impacts throughput. |
| Staroverov et al.(2020) | 2020 | The landscape was accurately detected based on the quantification of factors relating to the robot-ground encounter. Also, ADNNs provided a precise and trustworthy solution for the issue of classifying different landscape types. | The terrain estimate problem was addressed using sensory data and different DL frameworks. The best possible performance was achieved utilising FFT to generate spectrograms of relevant signals for each topography patch by combining the outcome together using CNN. | RNN, CNN, LSTM, C-LSTM, and SVM | The image dataset was provided by NASA's Planetary Data System and the University of Almería'sFitorobot, while the proprioceptive data was collected from MIT. | The CNN model successfully classified ~92.8% of the observations, 10.5% greater than standard SVM concatenating statistical moments, 16% more than LSTM and 14.2% more than C-LSTM. | In every terrain patch, spectrograms of the available data were computed using FFT, and were then assembled into a multichannel picture within a convolutional network for the best performance. More than 90% of the data were accurately categorised by the CNN model, making it fit for the purpose of study. | The computational load needs to be further examined in order to be optimised for the mounted processor of any robot system to perform a sensory topography categorization system online. |

*continues on following page*

*Table 1. Continued*

| Reference | Year | Summary | Technique | Algorithm | Dataset | Results | Merits | Demerits |
|---|---|---|---|---|---|---|---|---|
| Pflueger et al.(2019) | 2019 | A hybrid recommendation control method was used to resolve the real-time navigational challenge of AGVs in varied obstruction situations. It allocated the time required to solve problem OPC off-line by using GPM while guaranteeing precision in real-time, and robustness due to the trained DNNs. | To build the better correlation between both the ideal state-control pairs; the data feature extraction procedure, network architecture, activation function, and learning rate were carefully chosen. To further evaluate the DNNs training and validation losses under various hyper parameters, cross-validation was used. | DNN and GPM | No previous dataset used. | Number of Obstacles: 6l Position Errors of Final State: 0.0756l Rates of Collision Avoidance: 91%l CPU Time Cost: 0.3283ms | To enhance the autonomy and cognition of AGV guiding control, a combined real-time optimized control methodology based on DNNs was provided. The key benefit is that the calculation process was only carried out offline, which reduced the necessity for repeated AGV preparation due to shifting circumstances. | Applying it to the real world and extending the model for complex coordination procedures will be harder because the model was trained and validated on pre-collected datasets with no real-time performance check. |
| Sun et al. (2020) | 2020 | For the resolution of the perplexity in-house guidance despite the interference of noise in input by RGB-D camera, a HISNav algorithmic model was presented. It was implemented using the sim2real paradigm pre-trained on various modules of the framework. Later, it was loaded into a real robot. | A NN structure for rover co-ordination that is fast and obdurate to probable noise in sensors and actuators via subdivision, preparation, localization, and RL techniques was given. It performed different functions via preparing it on a custom HISNav dataset. | SLAM, HNN, RL, UMDP, and R-CNN | A new HISNav custom dataset was utilized. | The best performance for high noise images was given by the SOLOv2 model with a score of 0, 3527 on mAP (IoU=0.5). | To hasten environmental research and the hunt for a target of interest, the model was pre-programmed by initiation of sub-goals, thereby, saving computation power, and time. Furthermore, the model was noise resistant. | Improving the SPL metric, and further refinement of the framework can be carried out. |

*continues on following page*

*Table 1. Continued*

| Reference | Year | Summary | Technique | Algorithm | Dataset | Results | Merits | Demerits |
|---|---|---|---|---|---|---|---|---|
| Chiodini et al.(2020) | 2020 | A DRL-based navigation strategy for mobile robots, consisting of a point navigation planner was created.It translated dispersed laser range results into moving actions.<br>Using this extremely abstract data as input, the simulation-trained agents may be extended to the real-world setting for practical use, augmented with curiosity to promote traversal capabilities. | An ICM A3C proposal with a standard A3C algorithm as base was developed.It inherited curiosity to handle the problem of sparse rewards in a map-less navigating approach for complex confined spaces. Additionally, incorporation of 2 layers of LSTM acquired a memory function. | DRL, ICM, NN, FFA3C, and LSTM | No previous dataset was used. | The lowest average score for FF A3C was 30.7. Nav A3Cs have an average score of 42.3. Nav A3C + D1D2L and the proposed ICM A3C obviously performed better with average scores of 67.9 and 63.1, respectively. | The challenge of sparse rewards was addressed utilising an approach for a map-less guiding approach for complex interior environments. It used an ICM A3C algorithm.<br>It was achieved with a conventional A3C algorithm as a base via an award signal toincrease inherent stimulus with greater efficiency. | It is limited to the magnitude of traversable distance along with autonomous path selection for optimization of traversal. |
| Wu et al. (2019) | 2019 | An embedded automotive navigation system was based on 3D-RISS and MLEVD was built to work under harsh circumstances such as outer satellite signal interruption and indoor navigation.<br>It was demonstrated by both outdoor and indoor experiments that confirmed the effectiveness of this framework. | An image containing the landmark information was swiftly captured by applying ML, once a landmark with precisely calculated attributes was photographed.<br>After that, 3D-RISS was used by the INS to improve vehicle attributes, and the mistakes were subsequently corrected using the EKF. | CNN, EKF, 3D-RISS, and MLEVD | Custom image dataset used. | Error with landmarks(m)/Error without landmarks(m):<br>0.31/6.21\| 0.45/33.12\|<br>0.41/38.21\| 0.52/37.02\|<br>0.63/20.20 | By incorporating ML, the system can extract key information faster and more accurately, increasing its real-time potential location applications.<br>It is affordable, accurate, suited for non-GPS environments, and uncomplicated to use. | It requires a significant amount of preparation, such as establishing markers and gathering information.<br>Also, the level of light intensity would occasionally impact the perception effect. |

*continues on following page*

*Table 1. Continued*

| Reference | Year | Summary | Technique | Algorithm | Dataset | Results | Merits | Demerits |
|---|---|---|---|---|---|---|---|---|
| Gonzalez et al. (2019) | 2019 | A framework using E-GPR was developed and tested in a series of experiments using a robot.<br>Also, the obstacle is approached from various viewing angles with the model that can approximate an unknown function probabilistically. | The mean TTC error due to sensor fusion was 0.31 seconds, according to a series of experiments with obstacles approaching the robot from different angles.<br>However, the error is effectively decreased when the suggested E-GPR-based predictor was used. | LR, BT, SVM, MLP, and CNN | 900 custom datasets were produced and further utilized in the study. | Average error magnitude for each method across actual TTC intervals was least in interval (4, 5) with E-GPR (0.0189) and maximum in interval (0,1) without using ML (baseline). | Reduction of errors which is vital as an accurate TTC estimation enables a more refined assessment and decision of the risk of a collision. | It is still necessary to execute and perfect an efficient TTC calculation and classification of highly risky obstacles. |
| Jiang et al.(2020) | 2020 | An automated solution assisted real-time operations and reported issues with data transfer, archiving, and manipulation to assist in ground data system analysis to monitor downlink time. | Various supervised ML, and DL pipelines were used to automate and optimize the process of detection of data loss in the system along with quick monitoring of each downlink process was complete or incomplete. | LR, SVM, GNB, FF-DNN, and HPO-DNN. | Custom generated datasets along with a dataset consisting of downlink data from the Curiosity Rover were instated. | HPO-DNN produced the highest F1 score (0.90) whereas the lowest was 0.85 by GNB with implementation on MSL dataset. | Better monitored downlink process with the proposed supervised ML, and DNN pipeline.<br>Additionally, utilization of fast hyper parameter optimization methods made the model robust, and optimized for quick deployment in real use-case scenarios. | The bridging between the frameworks of existing analogue methods for the transition with the automated pipeline assisted methodology remained the only major issue.<br>The cost of implementation of infrastructure and the non-redundancy of inaccuracy of the predicted results is also an issue. |
| Gaudet et al. (2020) | 2020 | A recurrent strategy and value function approximator have been developed as a novel pliant guidance system incorporating meta-learning.<br>The DR/DV guidance law has brought to comparison four separate complexes with unknown but high dynamics to prove the capabilities of a RL meta-perfected strategy to administer the same. | The ML algorithms were used to optimize the result of the occurrence in four different challenging trials in different environments for implementing guidance laws using engineered models. | POMDP, PPO, and Meta-RL | Instead of a dataset, discrete standard values were assumed for study. | Maximum Norm of Terminal Position (m) is by Meta-RL (120 steps) where mu=0.6, sigma=1.6 and max= 139.8 whereas it was minimum for Norm of Terminal Velocity (m/s) with Meta-RL 60 steps as 2.00, 0.92, and 53.73, respectively. | Meta-RL model has clearly shown the significance of the capabilities of the perfecting strategy to implement in concurrency especially the performance results of 60-step Meta-RL trials were beyond encouraging.<br>The significant rise in precision, and incidence handling than DR/DV policies where the former outperformed the latter in many experiments. | It is restricted to landing simulations only which are disadvantageous as the proposed models do show a lot of promise in other aspects as well such as navigation, and autonomy.<br>The parameters that the sensor array has been constrained to can be flexed a bit for even better results. |

*continues on following page*

*Table 1. Continued*

| Reference | Year | Summary | Technique | Algorithm | Dataset | Results | Merits | Demerits |
|---|---|---|---|---|---|---|---|---|
| Daftry et al. (2022) | 2022 | The application of proprioceptive sensors along with a combination supervised/unsupervised ML technique on a Hazard Information System (HIS) which can be utilized by lunar rovers for pre-calculative estimate of rover mobility risks; i.e.; slippage and sink age. | HIS for the rover has been trained on real lunar dataset from the existing lunar rovers along with reinforced training from an analogue lunar habitat to validate its effectiveness and efficiency. | RF, GB, XGB, LGBM, DT, AdaV, SVM, and LR. | The dataset was generated using the data collected experimentally from tele operating the rover on the LunaLab. | The highest percentage of true predictions was given by the RL model (99.81%) whereas the lowest performance was by LR model (92.17%). | It can be utilized for overall better rover navigation, and manoeuvrability by optimized calculations of dormant hazards that may lay in the rover's general mission transition trajectory. | The redundant requirement of human involvement for the actual navigational characteristics was the primary drawback.<br>This exists owing to the fact that the command dispersion time is significantly wasted on transfer of information, and inputs from machine to man, and vice-versa. |
| Zhang et al.(2021) | 2021 | An efficient navigation RL technique combined with the advantages of two variants of approaches into the rule based RL (RuRL) model for reduction in the sample intricacy and temporal expense was developed.<br>It was carried out along with detailed theoretical explanation guaranteeing the optimal path to reduce space. | The SLAM system running on ROS was used to construct a map of the unknown environment via cartographer algorithms, which was then partitioned into hexagonal grids using ML/RL algorithms. | RL, SLAM, and Q-Learning | The maps built using a SLAM mobile robot were used for the study. | Maximal reduction % ǁ Q-learning ("e-greedy) 71.79% ǀQ-learning (SoftMax) 69.93% ǀ SARSA ("e-greedy) 61.69% ǀ SARSA (SoftMax)72.78% | The RuRL algorithm combined with SLAM has shown a significant decrease in sample space complexity, and cost of time along with improved and more success probabilistic path optimization for traversal. | The training of the robot model on real time data can be expensive along with the limitation of the particular training model to work on a particular set of hardware in a single trial. |
| Ziya & KARAKOSE (2020) | 2020 | An approach for generating navigational maps utilizing various aerial imagery sources such as satellites, and LEO vehicles for the implementation of DL methods was provided.<br>This was used to generate a map using two iterations allowing better autonomous navigational prospects for off-road rovers. | The model was designed by implementing U-NET with finite extents of proffered representative-data.<br>It was executed employing Keras and TensorFlow Python libraries which included a rectified linear unit (ReLU) for down sampling of data. | DCNN, U-Net, Keras, and TensorFlow | Navigational maps were generated by aerial robots. | Training for the macro- and micro-feature models was conducted for 250 epochs wherein the macro-feature model clocked an accuracy of 0.9443 on the training dataset.<br>An accuracy of 0.8920 clock was attained on validation dataset. | CNNs, in coordination with aerial data provided measurable improvements to the generation, and the attributes of guidance maps for autonomous outdoor steerage.<br>Thus, future planetary exploration missions can use the described procedures. | More sophisticated expense maps can be generated if the rover processes a cost map developed from a set of input layers generated while flying, and delving further from former deployments which was not undertaken in the study. |

*continues on following page*

*Table 1. Continued*

| Reference | Year | Summary | Technique | Algorithm | Dataset | Results | Merits | Demerits |
|---|---|---|---|---|---|---|---|---|
| Coloma et al.(2022) | 2022 | A detailed comparison of famous ML algorithms for evaluating the discrete slip occurrences in relation with individual wheels in planetary rovers was studied. For analysing the performance with diverse tuning configurations for each algorithm which were then verified by using a vast data set assimilated by a rover having on board kinaesthetic sensors. | A deep characterization of eleven famous ML algorithms estimating discrete slip classes was simulated with individual tyres, speeds, soil contexture, and fixture points for the on board IMU detector. | LR, LDA, SGD, DT, KNN, RNC, RFC, ETC, ADA, MLP, and SVM | Custom datasets with data collected using the LATUV rover was used. | The highest performance accuracy was achieved by DL (84.15%) followed by ADA (83.65) whereas the lowest performance was given by RNN (61.13). | A strong connection between sensor signals and slippage was observed when the rover has off-road tyres. It prompts future tests that consider various types of wheels/tires and suspension mechanisms. | The model is only tested on a single type of wheel design along with a single type of terrain and environmental geometry. |
| Ono et al. (2020) | 2020 | A snapshot of NASA's work included a thorough and advanced enumeration of each proficiency to build science-based scene analysis has been illustrated. It included vision-based traversal gauging; resource-aware, information-theoretic, and integral tactical path planning; and built-in prime resounding settling. | The experiment was carried out by creating a dataset with 100 MSL NAVCAM photos that were retrieved from a multitude of SOLs. Also, the dataset was examined where a smaller shift validated the accuracy of ML algorithm. | ImageNet, CNN, LSTM, SSML, Meta-Learning, MDP, MCS, and Custom Algorithms | NAVCAM image data from the Curiosity rover was used along with several tailored datasets. | 7000 DEM maps covering an area of 30m x 30m and various CFAs (1%, 3%, 5%, 7%, 10%, 12%, 15%, and 20%) were used to generate path times with a mean test accuracy of 92% and success prediction with a mean test accuracy of 81%. | Many scientists who participated in the experiment while refraining from reckoning on automatic captions for future endeavours' coordination, agreed that the method may be valuable for their own study. | Many scientists in the undertaken experiment disputed by refraining for counting on the automated captions for any endeavours' coordination which was partly a disappointing statement. |
| Wu et al.(2019) | 2019 | As a step towards accelerating specificity by minimising search space and time, the presented method for perspective-to-satellite image localization of rover surfaces by ML improved efficiency and accuracy. | Absolute localization accuracy was improved by using ML models to enable robust systematisationand optimized computation. It was used for generating a baseline dataset for further research. | OpenCV, SNN, PLaNNet v0, and ResNet | A publicly available dataset was used, and a benchmarking custom dataset was also generated. | The experimental setup has the capability to reduce the search area by 90-95% than by using original methods. | Experimental results included the creation of a site reference dataset, and generation tool. The community can use the aforementioned for evaluation and priming their techniques of localization. Also, the demonstration of concept for perfect automated localization employing CNN-ResNet was demonstrated. | The usage of synthetic datasets along with limited planar degrees of freedom on this study is too simple for the validation of the results on real-time application of the experimented model. |

*continues on following page*

*Table 1. Continued*

| Reference | Year | Summary | Technique | Algorithm | Dataset | Results | Merits | Demerits |
|---|---|---|---|---|---|---|---|---|
| Vulpi et al.(2021) | 2021 | A ROS-controlled mobile robotics platform adopted the categorized output of a contemporary CNN accelerator AI for FPGAs linked to a neuromorphic dynamic visual sensing element was developed.<br>It was trained on traffic signs dataset wherein the results of the simulations were tabulated in the study. | The process involved the integration of occurrences into histograms. These were normalized to create frames implemented to acquire large datasets and train neural networks.<br>The realistic real-time Edge-AI computing model was enabled by the CNN accelerator. | FPGA and CNN | A traffic signal dataset plus custom data acquired generated datasets were utilized. | Metrics \|\| Loss 0.033\|<br>Accuracy 97.62%\|<br>Precision 99.32%\|<br>Recall 98.31% | The autonomous navigation through the created model can be accomplished on both neuromorphic sensors and edge AI computer devices. | The CNN used in this study is still conventional as instead of processing events directly, the processed histograms are built from a set of events. It consumes more computational resources, and impacts throughput. |
| Ren et al. (2021) | 2021 | The presented GPR architecture was used to create a magnetic map to record the magnetic concentration in the surroundings as part of a guidance system for autonomous mobile robots.<br>By detecting the geomagnetic field's strength, magnetic data was collected and localisation was accomplished by using the disruption of the concentration as a landmark. | Regression may interpolate values with Gaussian distributions and utilise the intensity distribution as a map because the GPR system was used to construct a magnetic footprint. It records the magnetic intensity in the environment.<br>The geomagnetic field's intensity perturbation was used as a marker to help in positioning. | MCL, BF, GPR, SGPR, and SLAM | Custom datasets were generated using SLAM. | SGPR used only the diagonal component for inverse matrix calculation, which was equal to the amount of magnetic data, so the inverse matrix component required for magnetic mapping of HI City was only 2.226 * 10^4. | The geomagnetic field's intensity can be measured, and because it is stable over time and is unaffected by non-magnetic objects like people, such as humans. Thus, localization can be carried out utilizing the perturbation of the field's concentration as a landmark. | Autonomous guidance could malfunction irrespective of the learning methods used for the maps if magnetic objects that were not present during magnetic mapping were incorporated. |
| Shi et al. (2020) | 2020 | A simulation of an IRL framework using SVIN was employed to handle route optimization challenges with planetary rovers.<br>Also, the model considerably approached the effectiveness of the expert controller by employing the method to a fresh set of synthetic pathways on true Mars topography. | An approach for resolving planning challenges was demonstrated based on IRL.<br>It used a set of DCNN and Value Iteration Networks as fundamental internal structures. | MDP, RL, VIN, SVIN, and IRL | A custom grid world dataset as well as a highly realistic synthetic dataset generated from the currently deployed rover were used. | Approximately ~75% accuracy was achieved by the model in terms of overall performance. | The model can significantly approximate the effectiveness of the expert planner used to generate the data. It providesslight indications that it may even be probable to surpass the training data. | The study was performed by training it for short paths on a limited range of small-sized maps which limits the true potential of the model. |

## Path Planning and Navigation

A comprehensive guiding, navigation, and regulation system using RL models can be developed for better execution than current frameworks (Gaudet et al., 2020). Such advancement can also be applicable to resolve problems in orbital statics, and dynamics like refueling, etc. with a combination of models discussed in the research works of Daftry et al. (2022), Abcouwer et al. (2021) and Higa et al. (2019). Energy dynamics of the rover can extend augmentation of diverse models incorporating deviating parameters, variable pace, etc. (Visca et al., 2022). Meta-learning could be implemented to facilitate alternative sequential tasks (Furfaro et al., 2020) to expedite the process, particularly if there is an actuator or sensor failure. The performance can automatically improve in conjugation with the research work of Wu et al. (2019) if the parameters in the composite environment crosses plane restriction. Further, models can be developed to improve obstacle detection and avoidance tactics which are essential for path traversal viability (Arrouch et al., 2022).

## Computational and Energy Requirements

The RL model can be modified to lower the number of loopholes such as advanced input requirement along with restricted ability for solving complex problems (Zhang et al., 2021). The dynamic parameters of the requirement which give maximum throughput enables researchers to use DRL (Mortlock & Kassas, 2021; Ziya & KARAKOSE, 2020). The work on scaling and manageability to disparate robots of various sizes, and propulsion systems can boost the reliability of autonomous decisions (Vulpi et al., 2021). Hyper parameterization of the above algorithms (Ren et al., 2021) may further extend viability of results.

## Mapping and Terrain Hazard

Additional models can be developed to improve and learn through data (Mortlock & Kassas, 2021). The indoor experiments in the works of Shi et al. (2020) and Staroverov et al. (2020) allow low-precision INS visuals performance in harsh surroundings.It could be viable in case of prompt satellite failure or an interior setting failure as discussed by Sun et al.(2020). Different terrain types along with diverse tyre design modification can be used to keep the slip and sink risks in check (Arrouch et al., 2022).

## Communication Relays

Existing ML/DNN models can be improved to monitor the downlink of the data stream (Alimo et al., 2021) and boost efficiency (Gaudet et al., 2020).

## CONCLUSION

The use of AI/ML models in the branch of autonomous navigation, and data processing is tremendous. It is rapidly attracting researchers for further advancement. In this chapter, the existing studies have been analyzed and categorized asdata collection and accounting implementation via LR, SVM, U-Net, along with TensorFlow; hazard mitigation through RF, GB, DT, AdaV, etc.; implementation of ML for navigation systems by using complex algorithms such as EKF, SL, CNN; aerial assistance in dataset collection which can be assimilated using EKF, 3D-RISS, RL; computational and energy resources can be optimized by implementing ImageNet, DeepLabv3+, and Bayes Models. Also, a detailed description of each study has been provided which includes summary, used technique, algorithm, dataset, results, merits and demerits. The open challenges have been identified and delineated for the reference of future researchers working in this area.

Further, the use of ML/DL models for path planning is essential for successful traversal whereas the data accounting algorithms ensure the smooth collection of mission data for the purpose of training future models. Similarly hazard mitigation and aerial assistance are required for ensuring the safety of the rovers' equipment in the environmental dynamics. Additionally, the constrained computational and energy resources require fine-tuned models to work efficiently. Thus, further research may include developing an optimum solution for the true autonomy of navigational tasks.

## REFERENCES

Abcouwer, N., Daftry, S., del Sesto, T., Toupet, O., Ono, M., Venkatraman, S., Lanka, R., Song, J., & Yue, Y. (2021). Machine Learning Based Path Planning for Improved Rover Navigation. *2021 IEEE Aerospace Conference* (50100), (pp. 1–9). IEEE. 10.1109/AERO50100.2021.9438337

Alimo, R., Sam, D., Lakhmiri, D., Kahovec, B., & Divsalar, D. (2021). Automated data accountability for missions in Mars rover data. *2021 IEEE Aerospace Conference* (50100), (pp. 1–8). IEEE. 10.1109/AERO50100.2021.9438220

Alzubi, J., Nayyar, A., & Kumar, A. (2018). Machine Learning from Theory to Algorithms: An Overview. *Journal of Physics: Conference Series*, *1142*(1), 12012. doi:10.1088/1742-6596/1142/1/012012

Arrouch, I., Ahmad, N. S., Goh, P., & Mohamad-Saleh, J. (2022). Close Proximity Time-to-collision Prediction for Autonomous Robot Navigation: An Exponential GPR Approach. *Alexandria Engineering Journal*, *61*(12), 11171–11183. doi:10.1016/j.aej.2022.04.041

Bhushan, M., & Goel, S. (2016). Improving software product line using an ontological approach. *Sadhana*, *41*(12), 1381–1391. doi:10.100712046-016-0571-y

Bhushan, M., Goel, S., & Kaur, K. (2018). Analyzing inconsistencies in software product lines using an ontological rule-based approach. *Journal of Systems and Software*, *137*, 605–617. doi:10.1016/j.jss.2017.06.002

Bhushan, M., Goel, S., & Kumar, A. (2018). Improving quality of software product line by analysing inconsistencies in feature models using an ontological rule-based approach. *Expert Systems: International Journal of Knowledge Engineering and Neural Networks*, *35*(3), e12256. doi:10.1111/exsy.12256

Bhushan, M., Goel, S., Kumar, A., & Negi, A. (2017). Managing software product line using an ontological rule-based framework. *2017 International Conference on Infocom Technologies and Unmanned Systems (Trends and Future Directions) (ICTUS)*, (pp. 376–382). IEEE. 10.1109/ICTUS.2017.8286036

Chiodini, S., Torresin, L., Pertile, M., & Debei, S. (2020). Evaluation of 3D CNN semantic mapping for rover navigation. *2020 IEEE 7th International Workshop on Metrology for AeroSpace (MetroAeroSpace)*, (pp. 32–36). IEEE.

Coloma, S., Martinez, C., Yalçın, B. C., & Olivares-Mendez, M. A. (2022). Enhancing Rover Teleoperation on the Moon With Proprioceptive Sensors and Machine Learning Techniques. *IEEE Robotics and Automation Letters*, *7*(4), 11434–11441. doi:10.1109/LRA.2022.3198794

Daftry, S., Abcouwer, N., Sesto, T. D., Venkatraman, S., Song, J., Igel, L., Byon, A., Rosolia, U., Yue, Y., & Ono, M. (2022). MLNav: Learning to Safely Navigate on Martian Terrains. *IEEE Robotics and Automation Letters*, *7*(2), 5461–5468. doi:10.1109/LRA.2022.3156654

Duo, N., Wang, Q., Lv, Q., Wei, H., & Zhang, P. (2019). A deep reinforcement learning based mapless navigation algorithm using continuous actions. *2019 International Conference on Robots & Intelligent System (ICRIS)*, (pp. 63–68). IEEE. 10.1109/ICRIS.2019.00025

Furfaro, R., Scorsoglio, A., Linares, R., & Massari, M. (2020). Adaptive generalized ZEM-ZEV feedback guidance for planetary landing via a deep reinforcement learning approach. *Acta Astronautica*, *171*, 156–171. doi:10.1016/j.actaastro.2020.02.051

Gaudet, B., Linares, R., & Furfaro, R. (2020). Adaptive guidance and integrated navigation with reinforcement meta-learning. *Acta Astronautica*, *169*, 180–190. doi:10.1016/j.actaastro.2020.01.007

Gonzalez, R., Chandler, S., & Apostolopoulos, D. (2019). Characterization of machine learning algorithms for slippage estimation in planetary exploration rovers. *Journal of Terramechanics*, *82*, 23–34. doi:10.1016/j.jterra.2018.12.001

Higa, S., Iwashita, Y., Otsu, K., Ono, M., Lamarre, O., Didier, A., & Hoffmann, M. (2019). Vision-based estimation of driving energy for planetary rovers using deep learning and terramechanics. *IEEE Robotics and Automation Letters*, *4*(4), 3876–3883. doi:10.1109/LRA.2019.2928765

Jiang, J., Zeng, X., Guzzetti, D., & You, Y. (2020). Path planning for asteroid hopping rovers with pre-trained deep reinforcement learning architectures. *Acta Astronautica*, *171*, 265–279. doi:10.1016/j.actaastro.2020.03.007

Kholiya, P. S., Kapoor, A., Rana, M., & Bhushan, M. (2021). Intelligent process automation: The future of digital transformation. *2021 10th International Conference on System Modeling & Advancement in Research Trends (SMART)*, (pp. 185–190). IEEE.

Matthies, L., Daftry, S., Tepsuporn, S., Cheng, Y., Atha, D., Swan, R. M., Ravichandar, S., & Ono, M. (2022). Lunar rover localization using craters as landmarks. *2022 IEEE Aerospace Conference (AERO)*, (pp. 1–17). 10.1109/AERO53065.2022.9843714

Mortlock, T., & Kassas, Z. M. (2021). Assessing Machine Learning for LEO Satellite Orbit Determination in Simultaneous Tracking and Navigation. 2021 IEEE Aerospace Conference (50100), (pp. 1–8). IEEE. 10.1109/AERO50100.2021.9438144

Negi, A., & Kaur, K. (2017). Method to resolve software product line errors. *Information, Communication and Computing Technology: Second International Conference, ICICCT 2017*, New Delhi, India.

Ono, M., Rothrock, B., Otsu, K., Higa, S., Iwashita, Y., Didier, A., Islam, T., Laporte, C., Sun, V., & Stack, K. (2020). Maars: Machine learning-based analytics for automated rover systems. *2020 IEEE Aerospace Conference*, (pp. 1–17). 10.1109/AERO47225.2020.9172271

Pal, S., Mishra, N., Bhushan, M., Kholiya, P. S., Rana, M., & Negi, A. (2022). Deep Learning Techniques for Prediction and Diagnosis of Diabetes Mellitus. *2022 International Mobile and Embedded Technology Conference (MECON)*, (pp. 588–593). IEEE. 10.1109/MECON53876.2022.9752176

Pflueger, M., Agha, A., & Sukhatme, G. S. (2019). Rover-IRL: Inverse Reinforcement Learning With Soft Value Iteration Networks for Planetary Rover Path Planning. *IEEE Robotics and Automation Letters*, *4*(2), 1387–1394. doi:10.1109/LRA.2019.2895892

Piñero-Fuentes, E., Canas-Moreno, S., Rios-Navarro, A., Delbruck, T., & Linares-Barranco, A. (2021). *Autonomous driving of a rover-like robot using neuromorphic computing*. Advances in Computational Intelligence: 16th International Work-Conference on Artificial Neural Networks, IWANN 2021, Virtual Event, Proceedings, Part II, (pp. 57–68). IEEE.

Prinz, R., Bulbul, R., Scholz, J., Eder, M., & Steinbauer-Wagner, G. (2022). Off-Road Navigation Maps for Robotic Platforms using Convolutional Neural Networks. *AGILE: GIScience Series*, *3*, 55. doi:10.5194/agile-giss-3-55-2022

Rana, M., & Bhushan, M. (2022). Advancements in Healthcare Services using Deep Learning Techniques. *2022 International Mobile and Embedded Technology Conference (MECON)*, (pp. 157–161). IEEE. 10.1109/MECON53876.2022.9752020

Ren, Z., Lai, J., Wu, Z., & Xie, S. (2021). Deep neural networks-based real-time optimal navigation for an automatic guided vehicle with static and dynamic obstacles. *Neurocomputing*, *443*, 329–344. doi:10.1016/j.neucom.2021.02.034

Shi, H., Shi, L., Xu, M., & Hwang, K.-S. (2020). End-to-End Navigation Strategy With Deep Reinforcement Learning for Mobile Robots. *IEEE Transactions on Industrial Informatics*, *16*(4), 2393–2402. doi:10.1109/TII.2019.2936167

Staroverov, A., Yudin, D. A., Belkin, I., Adeshkin, V., Solomentsev, Y. K., & Panov, A. I. (2020). Real-Time Object Navigation With Deep Neural Networks and Hierarchical Reinforcement Learning. *IEEE Access : Practical Innovations, Open Solutions*, *8*, 195608–195621. doi:10.1109/ACCESS.2020.3034524

Sun, Y., Guan, L., Wu, M., Gao, Y., & Chang, Z. (2020). Vehicular navigation based on the fusion of 3D-RISS and machine learning enhanced visual data in challenging environments. *Electronics (Basel)*, *9*(1), 193. doi:10.3390/electronics9010193

Takebayashi, T., Miyagusuku, R., & Ozaki, K. (2021). Development of Magnetic-Based Navigation by Constructing Maps Using Machine Learning for Autonomous Mobile Robots in Real Environments. In Sensors (Vol. 21, Issue 12). doi:10.339021123972

Verma, U., Garg, C., Bhushan, M., Samant, P., Kumar, A., & Negi, A. (2022). Prediction of students' academic performance using Machine Learning Techniques. *2022 International Mobile and Embedded Technology Conference (MECON)*, (pp. 151–156). IEEE. 10.1109/MECON53876.2022.9751956

Visca, M., Powell, R., Gao, Y., & Fallah, S. (2022). Deep meta-learning energy-aware path planner for unmanned ground vehicles in unknown terrains. *IEEE Access : Practical Innovations, Open Solutions, 10*, 30055–30068. doi:10.1109/ACCESS.2022.3155161

Vulpi, F., Milella, A., Marani, R., & Reina, G. (2021). Recurrent and convolutional neural networks for deep terrain classification by autonomous robots. *Journal of Terramechanics, 96*, 119–131. doi:10.1016/j.jterra.2020.12.002

Wang, W., Lin, L., Fan, Z., & Liu, J. (2022). Semi-Supervised Learning for Mars Imagery Classification and Segmentation. *ACM Transactions on Multimedia Computing Communications and Applications.*

Wu, B., WK, P. R., Ludivig, P., Chung, A. S., & Seabrook, T. (2019). Absolute localization through orbital maps and surface perspective imagery: A synthetic lunar dataset and neural network approach. *2019 IEEE/RSJ International Conference on Intelligent Robots and Systems (IROS)*, (pp. 3262–3267). IEEE. 10.1109/IROS40897.2019.8968124

Zhang, X., Liu, Y., Hu, D., & Liu, L. (2021). A Maze Robot Autonomous Navigation Method Based on Curiosity and Reinforcement Learning. *The 7th Int. Workshop on Advanced Computational Intelligence and Intelligent Informatics (IWACIII 2021), Article, M1-6*. IEEE.

Ziya, T. A. N., & KARAKOSE, M. (2020). Comparative study for deep reinforcement learning with CNN, RNN, and LSTM in autonomous navigation. *2020 International Conference on Data Analytics for Business and Industry: Way Towards a Sustainable Economy (ICDABI)*, (pp. 1–5). IEEE.

## ADDITIONAL READING

Barrett, A. M., Balme, M. R., Woods, M., Karachalios, S., Petrocelli, D., Joudrier, L., & Sefton-Nash, E. (2022). NOAH-H, a deep-learning, terrain classification system for Mars: Results for the ExoMars Rover candidate landing sites. *Icarus, 371*, 114701. doi:10.1016/j.icarus.2021.114701

Chen, Y., Marchetti, Y., & Gel, Y. R. (2021, May). Deepening the sense of touch in planetary exploration with geometric and topological deep learning. *Proceedings of the AAAI Conference on Artificial Intelligence*, *35*(17), 15278–15285. doi:10.1609/aaai.v35i17.17793

Chen, Y., Marchetti, Y., Sizikova, E., & Gel, Y. R. (2022, June). TCN: Pioneering Topological-Based Convolutional Networks for Planetary Terrain Learning. *Proceedings of the AAAI Conference on Artificial Intelligence*, *36*(11), 12468–12474. doi:10.1609/aaai.v36i11.21514

Dai, Y., Zheng, T., Xue, C., & Zhou, L. (2022). SegMarsViT: Lightweight Mars Terrain Segmentation Network for Autonomous Driving in Planetary Exploration. *Remote Sensing*, *14*(24), 6297. doi:10.3390/rs14246297

DaPoian, V., Lyness, E., Brinckerhoff, W., Danell, R., Li, X., & Trainer, M. (2021). Science Autonomy and the ExoMars Mission: Machine Learning to Help Find Life on Mars. *Computer*, *54*(10), 69–77. doi:10.1109/MC.2021.3070101

Marchetti, Y., Lightholder, J., Junkins, E., Cross, M., Mandrake, L., & Fraeman, A. (2020, May). Barefoot rover: a sensor-embedded rover wheel demonstrating in-situ engineering and science extractions using machine learning. In *2020 IEEE International Conference on Robotics and Automation (ICRA)* (pp. 6000-6006). IEEE. 10.1109/ICRA40945.2020.9197500

Naziri, S., Cortes, D. D., Ridgeway, C., Ibarra, S., Provenghi, K., & Castelo, J. A. (2021). Bioinspired regolith penetration probe for Lunar exploration. In ASCEND 2021 (p. 4156). doi:10.2514/6.2021-4156

Sanneman, L., & Shah, J. A. (2022). The Situation Awareness Framework for Explainable AI (SAFE-AI) and Human Factors Considerations for XAI Systems. *International Journal of Human-Computer Interaction*, *38*(18-20), 1772–1788. doi:10.1080/10447318.2022.2081282

Shilliday, S. R., McGookin, E. W., & Thomson, D. G. (2022, May). A Comparison of Forward and Inverse Simulation Methods for Fault Detection on a Rover. In *2022 8th International Conference on Control, Decision and Information Technologies (CoDIT)* (*Vol. 1*, pp. 1432-1437). IEEE. 10.1109/CoDIT55151.2022.9804106

Zhu, B., He, J., Xing, Y., & Gao, F. (2021, November). A Plane-based Map for Wheel-legged Rover Efficient Motion Planning during Planetary exploration. In *2021 27th International Conference on Mechatronics and Machine Vision in Practice (M2VIP)* (pp. 90-95). IEEE. 10.1109/M2VIP49856.2021.9665015

## KEY TERMS AND DEFINITIONS

**Activation Function:** It determines whether or not to stimulate a neuron (interconnected nodes), i.e., whether or not the neuron's contribution is significant to the network during prediction using simpler mathematical processes.

**Algorithm:** Itis a logical set of steps outlining how to solve a problem using any type of computational process by which the desired output can be achieved.

**ANN:** It comprises of a computational network based on biological neurons like architecture, which mimics the functions of human-like processing of inputs.

**CNN:** It represents a common ANN for object and image identification with classification.

**Decision Tree:** It is amodel where sampling is divided based on the level of the tree, and the mean as well as standard deviation of the target variable are determined at the leaf nodes.

**Deep Learning:** These algorithms aim to achieve outcomes similar to those produced by the human intellect modelled using Neural Networks.

**Gaussian Process:** It model remains one of the few models of ML that may be solved analytically and yet describe rather complex analyses.

**Karman Filter:** It maintains track of notation by implementing subscripts in the equations which is basically an online learning system that updates its estimation of the weights progressively as new data is received.

**Reinforcement Learning:** It is an algorithm in which learning types are often divided based on goal-oriented ways.

**Regression:** Itmodels the behavior of a quantitative variable (the target variable) based on various predictor variables (components or characteristics) that might be quantitative or qualitative.

**RNN:** It is a form of Neural Networks where one step results act as input to the next step's computations.

**TensorFlow:** It is an open source library designed by the Google Brain Team in 2005 and is distributed under the Apache 2.0 license.

# Chapter 9
# Agricultural Waste Management Systems Using Artificial Intelligence Techniques

**Ashok Kumar Koshariya**
*School of Agriculture, Lovely Professional University, India*

**G. Merlin Suba**
*Panimalar Engineering College, India*

**Sugra Khatoon**
*MJP Rohilkhand University, India*

**Deewakar Baral**
*School of Agriculture, Lovely Professional University, India*

**Asmita Mayuresh Marathe**
*Centre for Energy Studies, Savitribai Phule Pune University, India*

**Sampath Boopathi**
*Muthayammal Engineering College, India*

## ABSTRACT

*In this chapter, the precision agricultural and waste management systems using artificial intelligent techniques are described. The fourth agricultural revolution integrates cross-industry technology to increase precision agriculture output and efficiency. Predictive analytics can be used to incorporate massive amounts of data generated by wireless smart networks and the internet of things. "Smart farming" aims to improve both the quantity and quality of agricultural products. The internet of things (IoT) is being used to improve agro-waste management, as well as crop classification and disease detection. Garbage collection and wireless sensor networks on smart bins are employed in the IOT to improve waste management. Vermicomposting is a process that uses earthworms and other associated bacteria to create incredibly fertile compost. Waste management methods for flower waste, bagasse, banana agro-waste, and agro-industrial wastewater have been depicted. Smart waste management for precision agriculture systems is also illustrated.*

DOI: 10.4018/978-1-6684-8171-4.ch009

## INTRODUCTION: PRECISION AGRICULTURE SYSTEM

The Latin words ager, which means "soil," and "culture," which implies cultivating the land, are combined to form the English word "agricultural." The term "agriculture" today refers to a variety of activities, such as arboriculture, horticulture, and vegeculture. For the 80 percent of the world's poor who depend on farming, agriculture offers the ability to reduce poverty, raise income levels, and improve food security. For the purpose of supplying the growing global population with food, conventional agricultural methods are changing. By utilising irrigation and fertiliser, high-yielding cultivars are being introduced. Global food production has increased enormously as a result of the green revolution, which has also significantly decreased famine, particularly in Asia(Comparetti, 2011). Between 1960 and 2015, agricultural output tripled as a result of Green Revolution technologies. Agriculture currently uses 70% of the world's freshwater and 23% of the planet's liveable land. The various elements of precision agriculture system is shown in Figure 1.

As seeds lack the capacity to effectively fight disease, they are more vulnerable to it. Pesticide use has increased, which has a big impact on environmental externalities. Sustainable agriculture is crucial to preserving the environment and biodiversity. The phrase "transition from sustainability" in agriculture refers to the change from one agricultural framework to another that is based on a more comprehensive theory of sustainable agriculture. Sustainability in food production, consumption, and behaviours entails several things. Precision agriculture (PA) is being introduced specifically to contribute to the potential growth in food demand and to achieve sustainability. For the purpose of supporting farmers in tracking the status of their crops, PA combines various sensing technologies with communication protocols. Agricultural performance assessments assist farmers in determining the conditions required for a healthy crop at a specific point in time. Information about soil nutrients, plant chlorophyll content, and the presence of pest weeds was gathered from several sources. The fourth agricultural revolution integrates cross-industry technology to increase precision agriculture output and efficiency. Technology such as artificial intelligence (AI), blockchain, and unmanned aerial vehicles (UAV) are transforming agricultural activities on digital platforms. Predictive analytics can be used to incorporate massive amounts of data generated by WSN and IoT to understand the behaviour and conditions of the crops. The integration of various networks for detecting, interacting with, and communicating is known as the Internet of Things (IoT). Through internet connectivity, IoT helps establish real-time monitoring for any application. There has been a lot of work with IoT technology in the agricultural sector to enhance smart farming practises. It is currently anticipated that agriculturalists and technologists will use technology to address problems that farmers encounter. These include issues with productivity, cost management, and water shortages. Efforts made on WSN encourage us to acquire and transfer data from sensors to the main servers.

*Figure 1. Elements of precision agriculture system*

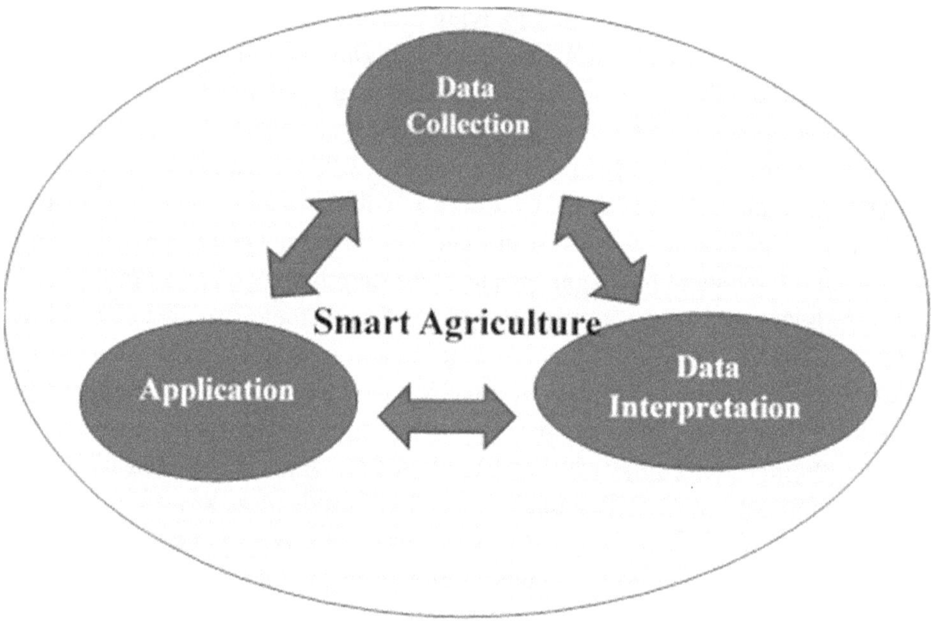


Intelligent irrigation systems built on the Internet of Things and the cloud are designed to collect data on soil moisture, temperature, and soil health in order to use less water. In order to precisely manage the entire gadget, the data collected from sensors generates information about various environmental variables. A further benefit of IoT is the efficient resource scheduling it provides. Using an RF module, IoT and WSN-based agriculture have been set up to monitor the air, temperature, soil moisture, and humidity. With the use of the internet and a gateway, the data is transmitted to a distant server.

The implementation of smart irrigation using Raspberry PI allows for the observation of water level, temperature, and humidity. Arduino and GPRS A UNO-based Internet of Things irrigation system is created to measure soil moisture and send it to a Thing speak server for data trend visualization. To decrease the use of insecticides and fungicides, an IoT-enabled pest prediction and plant disease system is proposed(Obi et al., 2016). The implementation of an IoT and Zigbee protocol-based greenhouse farming system is done to keep track of the weather. For precision agriculture, an IoT and cloud-based architecture is suggested to collect data into a cloud server for analysis. Real-time processing and data analytics are two purposes for which agri-IoT is designed.

The architecture makes it clear that the agricultural field includes integrated sensors for moisture, humidity, acoustics, and sound. For monitoring soil moisture, humidity, and fertility, moisture, humidity level, and a sensor are combined. Bee buzzers and animal presence sensors for use in agricultural fields. These sensors, which are integrated with Arduino and Node MCU, communicate sensor data via the internet to an Android APP or a computer. In the graphical user interface, the user may see the data. The researchers created an architecture for Raspberry Pi-enabled IoT agricultural field monitoring that integrates several sensors for agricultural field monitoring. The Raspberry Pi may transmit environmental data from an agricultural area over the internet with the help of an integrated ESP8266 wi-fi module.

## INTRODUCTION TO SMART AGRICULTURE

IOT sensors have the ability to provide information about agricultural fields and subsequently take action based on user input, making "smart agriculture" an emerging concept. The system can track temperature, moisture levels, water levels, and even movement in the field that can harm crops. The initiative attempts to use smart agriculture using automation and emerging technology, such as IOT. After hardware has been created, it needs to be upgraded based on requirements and technological advancements. To guarantee that changes made to the old version function properly, this new version must be tested. Schematic arrangements of internet of Things in smart agricultural system is illustrated in Figure 2.

A farming management idea called "smart farming" aims to improve both the quantity and quality of agricultural products. Today's farmers have access to technology like GPS, soil scanning, data management, and the Internet of Things. Population growth, climate change, and labour are challenges that smart farming believes must be addressed. In the context of agriculture, "IOT" (Internet of Things) refers to the use of sensors, cameras, and other gadgets to convert every aspect and action related to farming into data. In IOT-based smart agriculture, a system is created for sensor-based crop field monitoring(Boopathi, Lewise, et al., 2021; Boopathi, 2022b; Dhall & Agrawal, 2018).

Every aspect of the life of the average person has been revolutionised by the Internet of Things (IoT). IoT refers to a network of self-configuring objects. IOT-based smart farming gadgets are changing the way agriculture is produced day by day. Extreme weather and climate change are two difficulties that agriculture is facing more and more of. IOT-based smart farming apps will make it possible for the agricultural sector to increase productivity while reducing waste. In IOT-based smart farming, a system is created for sensor-based crop field monitoring (light, humidity, temperature, soil moisture, etc.).

For remote sensing in hydrology and agriculture, soil moisture sensors assess the volumetric water content of the soil. Improved crops are produced with better irrigation management (Kautish et. al 2022), which also uses fewer inputs and boosts profitability(Ahmed et al., 2018). Precision agriculture reduces labour costs, environmental pollution, and input costs while delivering the best input at the best time. The Internet of Things (IoT), IoT-enabled sensor networks, and cloud computing provide real-time data transfer among numerous parties, enabling monitoring, forecasting, and decision-making. The Internet of Things (IoT) is a network of physical things that allows for the collection and exchange of real-time data via sensors, software, connections, and communication services.

*Figure 2. Schematic arrangements of internet of things in smart agricultural system*

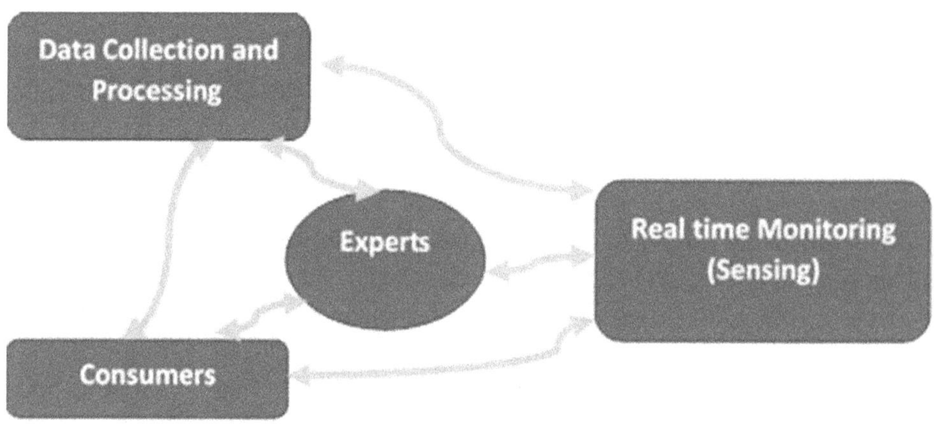


Garbage collection in high-priority regions of smart cities and wireless sensor networks on smart bins are now two examples of how the Internet of Things (IoT) is being used in waste management. IoT and smart sensors have been applied to ecological monitoring and the agricultural sector. Wastes from the food processing industry are produced in vast quantities and include both liquid and solid components. Despite being vitamin and biomass rich, these food wastes are a menace to the ecosystem. Numerous studies have demonstrated that it is possible to transform these organic wastes into valuable goods or even raw materials for other businesses.

Farmers that don't grade and clean their product before selling it at the market create organic waste. After a minimal amount of biological processing, they can also be used as animal feed. The majority of the uncollected trash begins to rot at the market location. As a result, the general public is exposed to threats related to sanitation and hygiene. Ineffective management of organic wastes can cause disease outbreaks and have negative environmental effects. An appealing alternative in environmental management is vermicomposting. In addition, it reduces the negative consequences of improper organic waste management while producing protein for use in animal feed in the form of worm biomass.

A type of organic compost called vermicompost can be used to increase soil fertility and supply nutrients to plants and crops. Earthworms break down organic waste items from food preparation, paper scraps, and animal dung to produce beneficial biofertilizer by-products(Boopathi, Balasubramani, et al., 2021). One of the most effective ways to manage and reduce environmental pollution is through the use of vermicompost. It increases the amount of nutrients in the soil, makes those nutrients more accessible to plants, and enhances the drainage and structure of the soil. Additionally, it prevents insect pest and plant disease attacks (biocontrol).

## MACHINE LEARNING IN PRECISION AGRICULTURE SYSTEM

The agriculture sector is seeing a big impact from machine learning (ML). ML provides unique analytical methods for assessing crop disease, production prediction, and other issues. One of the methods that is frequently used in agriculture to increase the productivity, quality, and output of agricultural goods is machine learning (ML). With the use of cutting-edge ML algorithms, ML is resolving agricultural difficulties like crop selection, the best irrigation water management, disease detection, and pre- and post-soil management of agricultural property. The ML-based predictive model aids the farmer in choosing the appropriate crop based on the weather's unwavering behaviour(Shafi et al., 2019). ML algorithms are being used to optimise water use in relation to the real amount of water needed for food production. Agriculture uses 70% of the water in the world. In relation to India, the use of water for agriculture is particularly important, and problems result from drought and flooding. An ultrasonic sensor measures the water content of the agricultural land and transmits the information to the gateway. Additional data is transmitted to the clod platforms for in-depth study. A productive farm and the development of agriculture have always been closely related. Similar to picture clustering or the classification of good or sick plants, machine learning can be used to identify diseases at an early stage. The algorithm analyses the user-provided real-time data and compares it to the database that already exists and contains the names of all known diseases.

## Artificial Neural Network (ANN)

The central nervous systems of animals serve as inspiration for computational algorithms, which are models of computing. It makes an effort to imitate the behaviour of biological "neurons" in composite form. A multitude of approaches exist for an artificial neural network (ANN) to master machines and comprehend patterns. A machine learning technique called the ANN is used to solve classification, regression, and clustering issues. When the dataset is excessively large and the number of features is excessive, it is mostly used to learn complex non-linear hypotheses. The method is used in PA to predict yield using weather data.

## Support Vector Machine (SVM)

SVMs are supervised learning models that may be applied to classify data and predict future outcomes. They are effective for a variety of real-world applications and can tackle both linear and nonlinear issues. When classifying groups of objects, the algorithm constructs a line dividing them into distinct categories. SVM is used for the identification of infection in PA, plant discrimination, and yield prediction for rice crops. The so-called decision line's objective is to maximise the distance between the spots on either side of it. The model can quickly determine the target classes (labels) for new cases after separation.

## K-Nearest Neighbours (KNN)

A supervised machine learning approach that is frequently used in classification problems is the K nearest neighbours algorithm. The fundamental tenet of it is that "the fruit does not fall far from the tree," which implies that related things are still close by. This algorithm classifies data points based on the classification of their neighbours. KNN is a computer algorithm that classifies objects by looking at k of its previously classified nearest neighbour items and calculating the maximum votes from nearest neighbours with the same properties as those chosen to identify the items. In PA, KNN is used for crop yielding, yield categorization, and disease detection(Chaurasiya, 2020; Yupapin et al., 2022).

## Convolutional Neural Network (CNN)

Neural networks are effective processing tools for data that has the shape of a 2D matrix, like photographs. CNNs are often employed for picture classification and recognition. In order to identify or characterise a picture, CNN uses 2D pixel matrices that represent the image. In image classification, we start with an image as our input, assign significance to each of its different elements/features, and then separate them out. Other classification algorithms require significantly more pre-processing than CNN, which requires a lot less. It is used in PA to detect diseases and determine nutritional deficiencies.

## LONG RANGE (LoRa) IN PRECISION AGRICULTURE SYSTEM

LoRa is a reliable method of long-distance wireless channel communication. LoRa operates on ISM bands, which include 868 MHz in Europe, 995 MHz in North America, and 433 MHz in Asia, depending on the country. End node devices, a network server, a gateway, and an application layer make up the LoRa-WAN architecture(Heideker et al., 2020). Lora is appropriate for applications requiring long-distance transmission at a modest data rate. When compared to Wi-Fi, LoRa has a lower bandwidth of 20 to 30 kbps. Schematic view of Long Range (LoRa) in precision agriculture system is depicted in Figure 3.

**End node devices:** The physical characteristics of an agricultural field are sensed by end node devices, which include pH sensors, temperature sensors, humidity sensors, and more.

**Gateway:** The gateway serves as a bridge to connect the cloud server and end node devices. The integration of the gateway with various communication systems enables communication between the cloud server and end nodes.

**Network Server:** The network server has the ability to log end node authentication information. Through Wi-Fi and Ethernet communication, the network server and gateway are connected.

**Application Layer:** All operations at sensor nodes, etc., are under the control of the device interface. Any interface is acceptable, including MQTT and the Graphical User Interface (GUI) of Blynk. This layer has direct communication with the user. Since LoRa uses an unlicensed frequency, transmission security is a risk. Three keys are used by LoRa-WAN during transmission: a specific key for the end computer, a unique network key, and a unique application key. It is the responsibility of the Unique Network Key (EUI64) to keep data safe and secure as it moves between various networks. Data protection is made possible during transmission from one node to another by the unique application key. Each end device specific key (EUI128) is a device-specific key that enables wireless device authentication for LoRa network-connected devices. Based on the parameters of the spreading factor (SF), link budget, signal-to-noise ratio (SNR), bandwidth (BW), and bit error rate (BER), LoRa communication in various applications can be analysed(Boopathi, Haribalaji, et al., 2022; Mekonnen et al., 2020).

*Figure 3. Schematic view of long range (LoRa) in precision agriculture system*

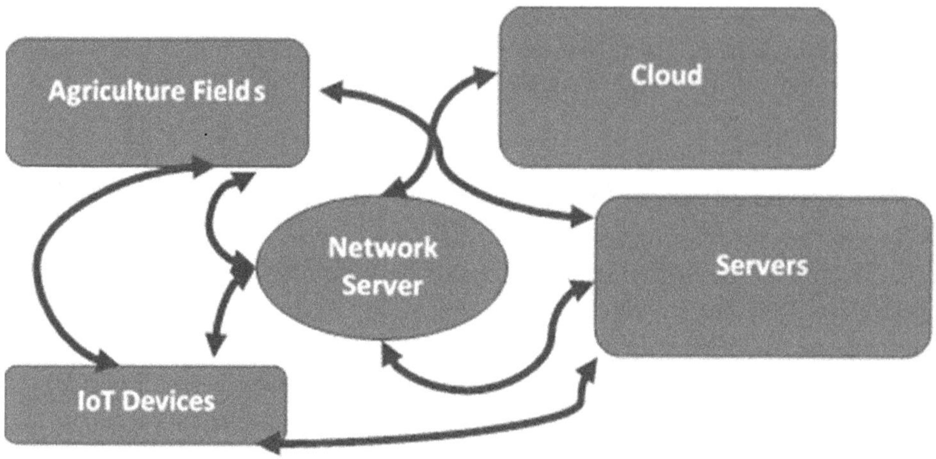


**Spreading Factor:** It is possible to identify a chirp's beginning frequency as a symbol. The spreading factor is a special parameter that determines how many bits are encoded in a symbol (SF). This implies that one symbol is described by numerous chips that are spread spectrum code pulses since each symbol represents two SF bits.

**Signal-to-Noise Ratio (SNR):** SNR, or signal-to-noise ratio, measures the proportion of transmitted signal power to noise power. Lora uses spreading factors and forward error correction (FEC) methods to improve LoRa performance and has an SNR range of -7.5 to -20 dB(Boopathi, 2022a, 2022e, 2022c; Boopathi et al., 2023; Boopathi, Jeyakumar, et al., 2022; Boopathi, Lewise, et al., 2021; Boopathi & Sivakumar, 2013; Haribalaji, Boopathi, & Asif, 2021a; Haribalaji, Boopathi, Asif, et al., 2021; Jeyakymar et al., 2022; Myilsamy & Sampath, 2021; Trojovský et al., 2023).

**Sensitivity:** The spreading factor, noise, and bandwidth are taken into consideration in LoRa as inputs for a radio receiver's sensitivity output. Sensitivity is the capability of the receiver in a LoRa system to amplify the weak signals that are collected by the receiver.

**Bit Rate/data rate:** There is a transfer of bits from one place to another. The LoRa bit rate (Rbit) is a numerical value.

**Bit Error Rate (BER):** The BER of the transmitted signal is defined as the rate of error bits divided by the total number of bits added together. If the BER is 10-4, then 10,000 bits have been sent with one bit being incorrect. A higher BER reveals subpar network performance.

**Packet Error Rate (PER):** LoRa technology has the potential to improve connectivity and communication in WSN-based PA. The PA can benefit from LoRa's long-range transmission, low data rate, and low power consumption capabilities. A packet is a data unit used in a radio transmission that is subject to FEC.

## CASE STUDY ON PRECISION AGRICULTURE SYSTEM

The hardware design and system description for achieving the goals of a precision agriculture system consists of three parts: a sensor module, a gateway, and a portable IoT device. This chapter includes the description of the system and the final prototype.

### System Description

The implementation of real-time monitoring of agricultural fields using LoRa and WiFi modules is offered as an architecture. The sensor module, LoRa gateway, and IoT-based handheld device are the three key parts of the suggested architecture. The data flow between the sensor nodes, the gateway, and the handheld devices is shown in this design. This architecture makes it easier to construct an IoT-based agriculture monitoring system in locations with poor internet connectivity. The main phase for developing and customising is the hardware design of the sensor module and gateway. The prototype is typically modified to provide a system that is dependable and simple while also lowering the cost(Boopathi, 2021a; Singh et al., 2021).

The sensor mote is a piece of equipment used in agriculture that communicates data to an external power source. Researchers at Texas A&M University in the United States claim that choosing wireless communication plays a big influence in preventing power loss during data transmission. For the development of an energy-efficient prototype, it is critical to have an energy-efficient prototype. Therefore, a wireless communication module with low power consumption is selected to send data over a long distance. The microcontroller unit is another vital part for customization. The ATMEGA328P microcontroller is regarded as having minimal data complexity and low power requirements. The +5 V and +3.3 V voltages are included in the sensor gateway to convert and supply the correct voltage for tracking the complexity of the sensors' data.

### Sensor Mote

The main element for detecting and communicating the various characteristics of an agricultural field is a sensor mote. The sensor mote's parts and how they work together to give a comprehensive picture of the land environment are explored.

Numerous sensor modules are included with the sensor mote in order to perceive the current conditions in the agricultural field, such as temperature, humidity, pH, etc. The controller unit is interfaced with all of these sensors. When the controller unit receives data from the sensors, it starts to transmit that data. As they do not rely on the same source of electricity as the human eye, the components incorporated in the sensors need a distinct power supply in order to operate. When it comes to the essential power for the components, a power supply is employed to maintain the power supply. The controller unit and display unit are connected via a digital interface that shows sensor data as well as other settings(Aguida et al., 2021; Myilsamy et al., 2021).

The ATMEGA328 is an 8-bit microcontroller that uses RISC architecture, or Reduced Instruction Set Computer. Low-cost IoT and WSN applications benefit from RISC architecture. SEMTECH created the transceiver radio frequency (RF) module known as the SX1278 LoRa module. The controller's operational voltage varies from 1.8 to 5 Volts. The sensor mote's power source is created by fusing a bridge rectifier and an AC step-down transformer. The LoRa module's specifications are as follows. 20dBm - 100mW constant RF output vs. V supply, +14dBm high-efficiency PA, and 168dB link budget in LoRa Noise immunity, 200nA retention in the register, and low current of 9.9mA. Communication with sensors such as soil pH sensors, DHT11, etc. is accomplished using these pins. Digital and analogue sensors, such as soil humidity and temperature sensors, are coupled to digital and analogue pins.

A customised sensor mote that is constructed on a PCB layout with essential components houses the ATMEGA 328P controller and LoRa modem. The sensor mote circuit's connections are as follows: To provide power to sensors and actuators, the board has power rail pins with voltages of +12V, 5V, and 3.3V. The +5V, GND, and A0 pins of the modified Atmega328 board are used to connect the soil moisture sensor. The atmega328 microcontroller's +Vcc and GND pins are coupled to an internal +12V to +5V converter. The LoRA modem's MOSI, MISO, SCK, GND, Vcc, and RESET pins are connected to a specially designed Atmega328 board.

## Gateway With NUTTYFI/Node MCU

The transceiver unit is the LoRa module, which can receive data from the sensor mote and, with the help of the NUTTYFI Wi-Fi module, send it to the cloud server via internet access. It includes the parts that are integrated into the gateway. The 2.4 GHz frequency range is used by the Wi-Fi module, which is based on the 802.11b wireless ethernet standard, to link the gateway to the internet. The gateway also has a power converter accessible to provide the proper power to the components housed inside the enclosure of the device. The ATMEGA 328 P microcontroller and a comparable SX 1278 LoRa modem are incorporated in the NUTTYFI gateway, as illustrated in the image below. In the table below, the circuit architecture of the gateway using the NUTTYFI Wi-Fi module is presented.

The ESP8266 12e series is the foundation of the NUTTYFI cloud device. This gadget has nine digital pins, one analogue pin, 128 kilobytes of memory, and four megabytes of storage. The LoRa modem functions as a receiver in the gateway, helping it receive data. The gateway includes an LM7812 regulator that regulates the output voltage at 12 volts. It is a 32-bit single-board microprocessor with 3V and 5V power supplies. The following benefits are used to select the NUTTYFI cloud device: reduced board size, integrated WiFi network connectivity, and an inexpensive price.

## Gateway With Raspberry Pi

In our system, the raspberry pi may connect to the cloud server's internet network thanks to the gateway. It presents the parts that make up the LoRa-based gateway. The gateway logs the relevant data using a WiFi module that is included inside the Raspberry Pi. In our system, the raspberry pi may connect to the cloud server's internet network thanks to the gateway. The gateway logs the relevant data using a WiFi module that is included inside the Raspberry Pi. The gateway is able to establish communication between the gateway and cloud server thanks to the Raspberry Pi's interface with an ATMEGA 328P microcontroller. The gateway contains an embedded SX 1278 LoRa modem of a similar design, which serves as a receiver for the data coming from sensor nodes.

The -5V board's +5V pin is connected to the 1, 16, 2, and 15 LCD pins. The LCD contrast is adjusted using a 10K POT variable pin. LoRA modem pins are connected to customised Atmega328 boards' pins 11, 12, 13, GND, +5V, and RESET. The LCD's D4, D5, D6, and D7 pins are linked to the LoRA modems' 6, 5, 4, and 3 pins. The atmega328 microcontroller's +Vcc and GND pins interact with on-board +12V to +5V converters.

## IoT Based Hand-Held Device

Due to the integration of LoRa communication and Wi-Fi, farmers may now monitor the agricultural field from any location. The IoT-based handheld device is a user-friendly portable gadget targeted particularly at rural farmers in the target market. They can view the data from the agricultural fields using the device's embedded graphical display. A designed and constructed IoT-based handheld gadget is used to deliver a portable device that will help the farmer monitor and oversee the agricultural area. The Arduino Mega is based on the ATmega1280 or 2560.With its 16 analogue input pins, every type of peripheral device can be read or sensed.

The handheld device includes an SX 1278 LoRa modem to receive data from the sensor module.For the purpose of viewing the data received from the cloud server and gateway, a liquid crystal display (LCD) is incorporated. The handheld device includes one analogue pin and an LCD with eight digital pins.An Arduino microcontroller with built-in LCDs is called the Mega 2560. The +5V and -5V pins of the Arduino board are wired to the LCD's pins. The display's colour can be altered using the 10K POT variable pin. Hardware design and system description of a LoRa-based architecture for monitoring farmland. Here, a block diagram is used to explain the importance of each element in the architecture. In-depth discussions are held regarding the customizability of sensor mote, gateways with NUTTYFI, and handheld devices with IoT capabilities. This chapter also covers the integration of the ATMEGA 328P controller and SX 1278 LoRa module for sensor nodes and gateways. The gateway includes controller switches that let the farmer use the LoRa network to control the actuators in the agricultural field.

## AGRICULTURE WASTE MANAGEMENT SYSTEM

### Vermicomposting Method

Vermicomposting primarily uses earthworms' biodegradation activity to keep nutrients moving from one system to another. Both micro- and macro-organisms in the vast microflora have the capacity to transform organic wastes into useful resources. Earthworms contribute to the breakdown of cellulose, soil development, and humus accumulation.

The biological, chemical, and physical characteristics of soil are negatively impacted by earthworms' biological activities. Earthworms are remarkable in that they feed on organic wastes and only use a small amount of them for growth, excreting the majority of them in a partially digested state. This is due to the variety of microorganisms, hydrolytic enzymes, and hormones found in earthworm intestines, which promote the quick breakdown of partially eaten food. The porosity improves root penetration, aeration, water absorption, and drainage. The soil aggregates created in this way by earthworms and related bacteria contribute to the maintenance of the soil ecology(Bagha et al., 2022; Kannan et al., 2022). Vermicomposting is a non-thermophilic, bio-oxidative process that uses soil-dwelling earthworms and other associated bacteria to create incredibly fertile compost. By reestablishing beneficial microorganisms, earthworms hasten soil restoration and increase soil productivity. Vermicompost has finely divided, peat-like, porous texture, good aeration and drainage, high water holding capacity, and increased microbial activity and buffering capacity as some of its biophysico-chemical characteristics.

## Management of Flower Waste

A research study in Indore, India, looked at how floral debris was handled in temples. Typically, these wastes are biodegradable and easily accessible for microbial development. In order to control temple garbage, vermicomposting technology has been used. Cow manure and waste materials are combined in the process, and it is left to break down for 45 days at 30 °C. Temperature (25 °C), pH 8.0, electric conductivity (200 micro Siemens/cm), and C/N ratio (12.3) after 45 days are the ideal conditions for vermicomposting floral waste. This method's costs were calculated, and its applicability to the city of Indore was also supported.

## Management of Bagasse

Earthworms are introduced during the vermicomposting process, and they break down organic materials and produce castings that are useful fertiliser. With a temperature of 26 °C and 62–82 percent moisture content, the digestive enzymes of Eisenia foetida break down organic matter(Bhat et al., 2019; Haribalaji, Boopathi, & Asif, 2021b). In comparison to compost and initial agro-industrial waste, vermicomposting reduces carbon emissions while increasing the amount of nitrous oxide in the soil (bagasse). Additionally, VM Costing enhances the biological, chemical, and physical qualities of soil.

## Management of Banana Agro-Waste

As a significant crash crop in India, bananas produce a significant amount of garbage. Earthworms, Eudrilus eugeniae, were used to combine organic waste with cow dung. It was discovered that a mixture of 200 g banana waste and 800 g cow manure produced the best growth and reproduction. The earthworm did not survive treatments of 1000 g banana waste alone or 800 g banana and 200 g cow dung combined.Additionally, the combination of 200 grammes of banana waste and 600 grammes of cow dung mixture led to favourable development and reproduction. Vermicomposting is a simple way to handle banana waste if cow dung (CD) is added in the right amounts. According to research, using more banana waste in feed combinations is harmful(Boopathi, 2019, 2021b; Boopathi et al., 2012; Kalidas et al., 2012; Sampath et al., 2021; Saravanan et al., 2022).

## Management of Agro-Industrial Wastewater

Vermicomposting can lower the C/N ratio of agro-industrial waste water by 0.69 to 79 percent and the soluble chemical oxygen requirement by 20 to 88 percent.

According to a study, vermicompost was produced when highly polluted effluent from a palm oil plant was cleaned with earthworms(Boopathi, 2021a, 2022d; Kannan et al., 2022; Sampath, Pandian, et al., 2022). Due to its higher nutrient content, greater reduction in soluble chemical oxygen demand (COD), and lower C/N ratio, rice straw was proven to be a superior amendment and absorbent than soil. The pH, electrical conductivity, and nutritional content all significantly increased during the vermicomposting process(Boopathi, 2021c; S. et al., 2022; Sampath, C., et al., 2022). A study indicated that the best grade vermicompost with a good nutritional status was created using rice straw and palm oil mill effluent. Earthworm growth was similarly slowed in every compost pile used for the study. According to a study, filter cake and orange peel can be combined with cow dung to create a co-vermicompost that can be used in sustainable agriculture. After composting the heaps for the first 60 days, earthworms were added to start the vermicomposting process.

*Figure 4. Schematic arrangements for Agro-waste management system*

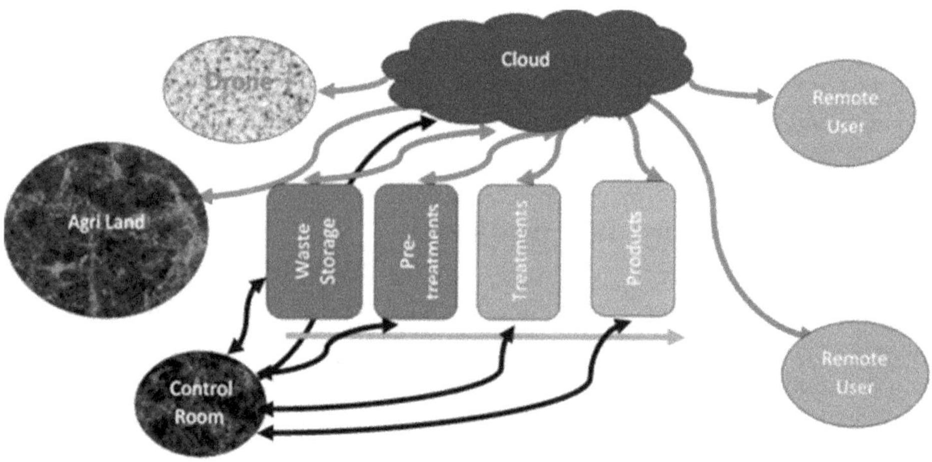


## SMART WASTE MANAGEMENT FOR PRECISION AGRICULTURE

The report was presented in two parts, with the second section focusing on the possible components for designing the architecture layer based on the IoT platform. The schematic arrangements for the agro-waste management system are illustrated in Figure 4.

**Framework:** "Smart" bins with RFID tags and global positioning system (GPS) modules are placed. When the bins are a quarter full, signals will be sent to "smart" trucks. Treatment units for the AGW can be classified as waste-to-energy generation or product generation.

**Possible components for the architecture layer:** The key components of such a framework include the agricultural stage, collection and storage, pre-treatment stage, treatment stage, and products and delivery stage. The communication and service layer represent the type of communication services used for information acquisition, transfer, storage, analysis, and access. layer proposes the infrastructure where the "smart" element can be applied and obtain the information that serves as the input for the management layer. RFID-tagged "smart" bins, dynamic weight sensors, and an integrated circuit card reader for the collection truck are for restaurant food waste management. a 20.5% increase in waste collection in a case study in Suzhou, China, that practised this in their case study(Boopathi, Thillaivanan, et al., 2021; Boopathi & Myilsamy, 2021; Heideker et al., 2020; Shafi et al., 2019). The quantification of waste available to be collected is important in initiating an effective management loop. Sensors and the Internet of Things have offered real-time critical data to facilitate optimal design and a feedback loop for waste management systems. The allocation of RFID-tagged smart bins and automatic weight sensors at all stages can reduce the illegal transportation of waste outside the control of pre-treatment or treatment units(Babu et al., 2022; Boopathi, Arigela, et al., 2022; Jeevanantham et al., 2022). Energy crops and plant residues are classified as lignocellulosic biomass. Pre-treatment is needed to break down its physical structure and enhance its biodegradability and solubilization. Mechanical, alkali, and fungal treatment, as well as enzymatic digestion and ensiling, are all part of the pre-treatment. Pre-treatment can be carried out by using a mechanical or biological pre-treatment system, with the latter having the advantage of low energy demand and scalability but being associated with cost and substrate specificity. Each treatment unit has its own requirements for the input feedstock and different products, and each treatment unit is equipped with its own special treatment unit. Monitoring of the AD is complex as it is a multi-variate biological process with few reliable online sensors for measuring important parameters. The process of decomposing organic matter (OM) to produce biochar, bio-oil, and gases is complex and involves a number of different technologies. Biogas may require post-treatment such as $NH_3$ stripping, upgrading for methanation, or enrichment. Liquid digestate will need a curing area for stabilisation and pathogen deactivation.

## SUMMARY

The fourth agricultural revolution integrates cross-industry technology to increase precision agriculture output and efficiency. Technology such as artificial intelligence (AI), blockchain, and unmanned aerial vehicles are transforming agricultural activities on digital platforms. Predictive analytics can be used to incorporate massive amounts of data generated by WSN and IoT. "Smart farming" aims to improve both the quantity and quality of agricultural products. IOT sensors have the ability to provide information about agricultural fields and take action based on user input.

Garbage collection in smart cities and wireless sensor networks on smart bins are examples of how the Internet of Things are being used in waste management. Algorithms are being used to optimise water use in relation to the real amount of water needed for food production. supervised machine learning (SVM) is used for crop classification and disease detection. Convolutional neural networks (CNN) are effective processing tools for data that has the shape of a 2D matrix like photographs.

LoRa is a reliable method of long-distance wireless channel communication. It operates on ISM bands, which include 868 MHz in Europe and 995 MHz in North America. LoRa technology has the potential to improve connectivity and communication in WSN-based PA. The NUTTYFI gateway is based on the ESP8266 12e series of microcontrollers. This device has nine digital pins, one analogue pin, 128 kilobytes of memory, and four megabytes of storage. The gateway contains an embedded SX 1278 LoRa modem of a similar design.

Vermicomposting is a non-thermophilic, bio-oxidative process that uses earthworms and other associated bacteria to create incredibly fertile compost. Earthworms are remarkable in that they feed on organic wastes and only use a small amount of them for growth.

## Abbreviations

AI - Artificial Intelligence
AI - Artificial Intelligence
ANN -Artificial Neural Network
BER -Bit Error Rate
CD - Cow Dung
CNN -Convolutional Neural Network
CNN -Convolutional Neural Networks
COD - Chemical Oxygen Demand
GPRS - General Packet Radio Services
GUI -Graphical User Interface
IoT – Internet of Things

KNN -K-Nearest Neighbours
LCD Liquid Crystal Display
LoRa – Long Range
ML- MACHINE LEARNING
NUTTYFI - Self-contained WiFi Networking solution hardware platform
OM - Organic Matter
PA -Precision Agriculture
PER Packet Error Rate
RFID - Radio-Frequency Identification
SML - Supervised Machine Learning
SNR -Signal-to-Noise Ratio
SVM -Support Vector Machine
UAV - Unmanned Aerial Vehicles
WSN - Wireless Sensor Network

## REFERENCES


Aguida, M. A., Ouchani, S., & Benmalek, M. (2021). An IoT-based Framework for an Optimal Monitoring and Control of Cyber-Physical Systems: Application on Biogas Production System. In *ACM International Conference Proceeding Series* (Vol. 1, Issue 1). Association for Computing Machinery. 10.1145/3494322.3494341

Ahmed, N., De, D., & Hussain, I. (2018). Internet of Things (IoT) for Smart Precision Agriculture and Farming in Rural Areas. *IEEE Internet of Things Journal*, *5*(6), 4890–4899. doi:10.1109/JIOT.2018.2879579

Babu, B. S., Kamalakannan, J., Meenatchi, N., Karthik, S., & Boopathi, S. (2022). Economic impacts and reliability evaluation of battery by adopting Electric Vehicle. *IEEE Explore*, (pp. 1–6). IEEE.

Bagha, H., Yavari, A., & Georgakopoulos, D. (2022). Hybrid Sensing Platform for IoT-Based Precision Agriculture. *Future Internet*, *14*(8), 233. doi:10.3390/fi14080233

Bhat, M. A., Adil, A. W., Sikander, B. M., Lone, Y., & Malik, J. A. (2019). Waste Management Technology for Sustainable Agriculture. In *Waste Management Technology for Sustainable Agriculture* (pp. 156–176). Issue September. doi:10.4018/978-1-7998-0031-6.ch009

Boopathi, S. (2019). Experimental investigation and parameter analysis of LPG refrigeration system using Taguchi method. *SN Applied Sciences*, *1*(8), 892. doi:10.100742452-019-0925-2

Boopathi, S. (2021a). An investigation on gas emission concentration and relative emission rate of the near-dry wire-cut electrical discharge machining process. *Environmental Science and Pollution Research International*, *29*(57), 86237–86246. doi:10.100711356-021-17658-1 PMID:34837614

Boopathi, S. (2021b). Improving of Green Sand-Mould Quality using Taguchi Technique. *Journal of Engineering Research*, 1–12. doi:10.36909/jer.14079

Boopathi, S. (2021c). Improving of Green Sand-Mould Quality using Taguchi Technique. *Journal of Engineering Research*. doi:10.36909/jer.14079

Boopathi, S. (2022a). An experimental investigation of Quench Polish Quench (QPQ) coating on AISI 4150 steel. *Engineering Research Express*, *4*(4), 045009. doi:10.1088/2631-8695/ac9ddd

Boopathi, S. (2022b). An Extensive Review on Sustainable Developments of Dry and Near-Dry Electrical Discharge Machining Processes. *Journal of Manufacturing Science and Engineering*, *144*(5), 1–37. doi:10.1115/1.4052527

Boopathi, S. (2022c). An investigation on gas emission concentration and relative emission rate of the near-dry wire-cut electrical discharge machining process. *Environmental Science and Pollution Research International*, *29*(57), 86237–86246. doi:10.100711356-021-17658-1 PMID:34837614

Boopathi, S. (2022d). (in press). Cryogenically treated and untreated stainless steel grade 317 in sustainable wire electrical discharge machining process: A comparative study. *Environmental Science and Pollution Research International*, 1–10. doi:10.100711356-022-22843-x PMID:36057706

Boopathi, S. (2022e). Performance Improvement of Eco-Friendly Near-Dry Wire-Cut Electrical Discharge Machining Process Using Coconut Oil-Mist Dielectric Fluid. *Journal of Advanced Manufacturing Systems*, 1–20. doi:10.1142/S0219686723500178

Boopathi, S., Arigela, S. H., Raman, R., Indhumathi, C., Kavitha, V., & Bhatt, B. C. (2022). Prominent Rule Control-based Internet of Things: Poultry Farm Management System. *IEEE Explore*, (pp. 1–6).

Boopathi, S., Balasubramani, V., Kumar, R. S., & Singh, G. R. (2021). The influence of human hair on kenaf and Grewia fiber-based hybrid natural composite material: An experimental study. *Functional Composites and Structures*, *3*(4), 045011–045021. doi:10.1088/2631-6331/ac3afc

Boopathi, S., Haribalaji, V., Mageswari, M., & Asif, M. M. (2022). Influences of Boron Carbide Particles on the Wear Rate and Tensile Strength of Aa2014 Surface Composite Fabricated By Friction-Stir Processing. *Materiali in Tehnologije*, *56*(3), 263–270. doi:10.17222/mit.2022.409

Boopathi, S., Jeyakumar, M., Singh, G. R., King, F. L., Pandian, M., Subbiah, R., & Haribalaji, V. (2022). An experimental study on friction stir processing of aluminium alloy (AA-2024) and boron nitride (BNp) surface composite. *Materials Today: Proceedings*, *59*, 1094–1099. doi:10.1016/j.matpr.2022.02.435

Boopathi, S., Lewise, K. A. S., Subbiah, R., & Sivaraman, G. (2021). Near-dry wire-cut electrical discharge machining process using water-air-mist dielectric fluid: An experimental study. *Materials Today: Proceedings*, *49*(5), 1885–1890. doi:10.1016/j.matpr.2021.08.077

Boopathi, S., & Myilsamy, S. (2021). Material removal rate and surface roughness study on Near-dry wire electrical discharge Machining process. *Materials Today: Proceedings, 45*, 8149–8156. doi:10.1016/j.matpr.2021.02.267

Boopathi, S., & Sivakumar, K. (2013). Experimental investigation and parameter optimization of near-dry wire-cut electrical discharge machining using multi-objective evolutionary algorithm. *International Journal of Advanced Manufacturing Technology*, *67*(9–12), 2639–2655. doi:10.100700170-012-4680-4

Boopathi, S., Sivakumar, K., & Kalidas, R. (2012). Parametric Study of Dry WEDM Using Taguchi Method. *International Journal of Engineering Research and Development*, *2*(4), 1–6.

Boopathi, S., Thillaivanan, A., Pandian, M., Subbiah, R., & Shanmugam, P. (2021). Friction stir processing of boron carbide reinforced aluminium surface (Al-B4C) composite: Mechanical characteristics analysis. *Materials Today: Proceedings*, *50*(5), 2430–2435. doi:10.1016/j.matpr.2021.10.261

Boopathi, S., Venkatesan, G., & Anton Savio Lewise, K. (2023). Mechanical Properties Analysis of Kenaf–Grewia–Hair Fiber-Reinforced Composite. In *Lecture Notes in Mechanical Engineering* (pp. 101–110). Springer. doi:10.1007/978-981-16-9057-0_11

Chaurasiya, P. K. (2020). A Detailed Study of Cattle Dung based Biogas Plant for Cooking Fuel and Organic Fertilizer. *September*. doi:10.13140/RG.2.2.18496.97282

Comparetti, A. (2011). Precision Agriculture: Past, Present and Future. *Agricultural Engineering and Environment*, *16*, 17.

Dhall, R., & Agrawal, H. (2018). An improved energy efficient duty cycling algorithm for IoT based precision agriculture. *Procedia Computer Science*, *141*, 135–142. doi:10.1016/j.procs.2018.10.159

Haribalaji, V., Boopathi, S., & Asif, M. M. (2021a). Optimization of friction stir welding process to join dissimilar AA2014 and AA7075 aluminum alloys. *Materials Today: Proceedings*, *50*, 2227–2234. doi:10.1016/j.matpr.2021.09.499

Haribalaji, V., Boopathi, S., & Asif, M. M. (2021b). Optimization of friction stir welding process to join dissimilar AA2014 and AA7075 aluminum alloys. *Materials Today, 50*(xxxx), 2227–2234. doi:10.1016/j.matpr.2021.09.499

Haribalaji, V., Boopathi, S., Asif, M. M., Jeyakumar, M., Subbiah, R., & Lewise, K. A. S. (2021). Influences of Friction stir tool parameters for joining two similar AZ61A alloy plates. *Materials Today: Proceedings*, *50*(5), 2547–2553. doi:10.1016/j.matpr.2021.12.074

Heideker, A., Ottolini, D., Zyrianoff, I., Neto, A. T., Salmon Cinotti, T., & Kamienski, C. (2020). IoT-based Measurement for Smart Agriculture. *2020 IEEE International Workshop on Metrology for Agriculture and Forestry, MetroAgriFor 2020 - Proceedings*, (pp. 68–72.). IEEE doi:10.1109/MetroAgriFor50201.2020.927754610.1109/MetroAgriFor50201.2020.9277546

Jeevanantham, Y. A., Saravanan, A., Vanitha, V., Boopathi, S., & Kumar, D. P. (2022). Implementation of Internet-of Things (IoT) in Soil Irrigation System. *IEEE Explore*, (pp. 1–5). IEEE.

Jeyakymar, M., Munusami, V., Shanmugam, P., Boopalan, N., Boopathi, S., & Sureshkumar, M. (2022). An investigation on wear loss and hardness of nitro-carburizing coated stainless-steel grade-316. *Materials Today: Proceedings*, *66*, 1398–1404. doi:10.1016/j.matpr.2022.05.223

Kalidas, R., Boopathi, S., Sivakumar, K., & Mohankumar, P. (2012). Optimization of Machining Parameters of WEDM Process Based On the Taguchi Method. *IJEST, 6*(1\&2), 91.

Kannan, E., Trabelsi, Y., Boopathi, S., & Alagesan, S. (2022). Influences of cryogenically treated work material on near-dry wire-cut electrical discharge machining process. *Surface Topography : Metrology and Properties*, *10*(1), 015027. doi:10.1088/2051-672X/ac53e1

Kaur, R., & Kautish, S. (2022). Multimodal sentiment analysis: A survey and comparison. *Research Anthology on Implementing Sentiment Analysis Across Multiple Disciplines*, 1846-1870.

Kautish, S., Peng, S. L., & Obaid, A. J. (Eds.). (2021). *Computational intelligence techniques for combating COVID-19*. Springer International Publishing. doi:10.1007/978-3-030-68936-0

Kautish, S., Reyana, A., & Vidyarthi, A. (2022). SDMTA: Attack detection and mitigation mechanism for DDoS vulnerabilities in hybrid cloud environment. *IEEE Transactions on Industrial Informatics*, *18*(9), 6455–6463. doi:10.1109/TII.2022.3146290

Madhu, G., Govardhan, A., Ravi, V., Kautish, S., Srinivas, B. S., Chaudhary, T., & Kumar, M. (2022). DSCN-net: A deep Siamese capsule neural network model for automatic diagnosis of malaria parasites detection. *Multimedia Tools and Applications*, *81*(23), 34105–34127. doi:10.100711042-022-13008-6

Mekonnen, Y., Namuduri, S., Burton, L., Sarwat, A., & Bhansali, S. (2020). Review—Machine Learning Techniques in Wireless Sensor Network Based Precision Agriculture. *Journal of the Electrochemical Society*, *167*(3), 037522. doi:10.1149/2.0222003JES

Myilsamy, S., Boopathi, S., & Yuvaraj, D. (2021). A study on cryogenically treated molybdenum wire electrode. *Materials Today: Proceedings*, *45*(9), 8130–8135. doi:10.1016/j.matpr.2021.02.049

Myilsamy, S., & Sampath, B. (2021). Experimental comparison of near-dry and cryogenically cooled near-dry machining in wire-cut electrical discharge machining processes. *Surface Topography : Metrology and Properties*, *9*(3), 035015. doi:10.1088/2051-672X/ac15e0

Obi, F., Ugwuishiwu, B., & Nwakaire, J. (2016). Agricultural Waste Concept, Generation, Utilization and Management. *Nigerian Journal of Technology*, *35*(4), 957. doi:10.4314/njt.v35i4.34

S., P. K., Sampath, B., R., S. K., Babu, B. H., & N., A. (2022). Hydroponics, Aeroponics, and Aquaponics Technologies in Modern Agricultural Cultivation. In *Trends, Paradigms, and Advances in Mechatronics Engineering* (pp. 223–241). IGI Global. doi:10.4018/978-1-6684-5887-7.ch012

Sampath, B., Naveenkumar, N., Sampathkumar, P., Silambarasan, P., Venkadesh, A., & Sakthivel, M. (2021). Experimental comparative study of banana fiber composite with glass fiber composite material using Taguchi method. *Materials Today: Proceedings*, *49*(5), 1475–1480. doi:10.1016/j.matpr.2021.07.232

Sampath, B., Pandian, M., Deepa, D., & Subbiah, R. (2022). Operating parameters prediction of liquefied petroleum gas refrigerator using simulated annealing algorithm. *AIP Conference Proceedings*, *2460*(1), 070003. doi:10.1063/5.0095601

Sampath, B. C. S., & Myilsamy, S. (2022). Application of TOPSIS Optimization Technique in the Micro-Machining Process. In Trends, Paradigms, and Advances in Mechatronics Engineering (pp. 162–187). IGI Global. doi:10.4018/978-1-6684-5887-7.ch009

Sampathkumar, A., Rastogi, R., Arukonda, S., Shankar, A., Kautish, S., & Sivaram, M. (2020). An efficient hybrid methodology for detection of cancer-causing gene using CSC for micro array data. *Journal of Ambient Intelligence and Humanized Computing*, *11*(11), 4743–4751. doi:10.100712652-020-01731-7

Saravanan, M., Vasanth, M., Boopathi, S., Sureshkumar, M., & Haribalaji, V. (2022). Optimization of Quench Polish Quench (QPQ) Coating Process Using Taguchi Method. *Key Engineering Materials*, *935*, 83–91. doi:10.4028/p-z569vy

Shafi, U., Mumtaz, R., García-Nieto, J., Hassan, S. A., Zaidi, S. A. R., & Iqbal, N. (2019). Precision agriculture techniques and practices: From considerations to applications. *Sensors (Switzerland)*, *19*(17), 1–25. doi:10.339019173796 PMID:31480709

Singh, R. K., Berkvens, R., & Weyn, M. (2021). AgriFusion: An Architecture for IoT and Emerging Technologies Based on a Precision Agriculture Survey. *IEEE Access : Practical Innovations, Open Solutions*, *9*, 136253–136283. doi:10.1109/ACCESS.2021.3116814

Trojovský, P., Dhasarathan, V., & Boopathi, S. (2023). Experimental investigations on cryogenic friction-stir welding of similar ZE42 magnesium alloys. *Alexandria Engineering Journal*, *66*(1), 1–14. doi:10.1016/j.aej.2022.12.007

Yupapin, P., Trabelsi, Y., Nattappan, A., & Boopathi, S. (2022). Performance Improvement of Wire-Cut Electrical Discharge Machining Process Using Cryogenically Treated Super-Conductive State of Monel-K500 Alloy. *Iranian Journal of Science and Technology. Transaction of Mechanical Engineering*, *0123456789*. doi:10.100740997-022-00513-0

# Chapter 10
# Additive Manufacturing Using Robotic Programming

**Akash Mohanty**
*Vellore Institute of Technology, India*

**Ranjit P. S.**
*Aditya Engineering College, India*

**B. Jothi**
*SRM Institute of Science and Technology, India*

**J. Samson Isaac**
*Karunya Institute of Technology and Sciences, India*

**J. Jeyasudha**
*SRM Institute of Science and Technology, India*

**Sampath Boopathi**
*Muthayammal Engineering College, India*

## ABSTRACT

*Due to the development of industrial 5.0 concepts, the integration of multi-disciplinary concepts is evolving in the manufacturing industries. In this chapter, the integration of robot technology with additive manufacturing processes is discussed to improve the processes for making complex objects. A case study has been used to explain two-phase and multi-plane three-dimensional printing using industrial robots. Using previous research activities, a brief literature on 3D printing technology, parameters, and applications has been developed. The case study on the aerofoil wing made by industrial robotics programming has been explained. The flow chart that can exhibit both three phase and multi-plane processes has been illustrated.*

DOI: 10.4018/978-1-6684-8171-4.ch010

## INTRODUCTION

Robotics is an interdisciplinary discipline of technology that combines mechanical, electronic, and computer science to create and build intelligent devices that can function in a variety of settings. Robots are intelligent devices with individual capabilities and traits that can help people with repetitive chores. Robotic manipulators are used in production to lower costs and improve quality. Industrial robots are flexible, adaptable, high-precision machines with several degrees of freedom that can operate in a variety of environments. Fixed and mobile industrial robots are the two main categories. Robot motions are regulated by control systems, enabling them to carry out tasks in a particular order. playback control systems, intelligent control systems, and limited sequence control systems. Robots can do jobs repeatedly by being taught positions that have been recorded in memory using playback control systems. With a predetermined path stored in memory, the PTP control system enables the robot to advance progressively from one position to another. A high-level microcontroller is used in playback with continuous path control to specify the intermediate path between desired spots. Robots with intelligent control systems are programmed using a microcomputer and a high-level programming language (Hua et al., 2021; Top et al., 2021).

Robot Programming: Lead via programming, text programming, simulation and offline programming are the three techniques used in robot programming. The robot is moved through a motion sequence using a teach pendant or flex pendant, and commands or instructions are typed into the controller. The created control algorithm is tested using simulation or off-line programming before being inserted into the robot controller. By mimicking the motion cycle in a virtual setting, this technique decreases downtime and enables the programmer to create novel time-saving strategies. Engineering applications for industrial robots include material handling, machine loading and unloading, part transfer, picking, and placing, assembly, disassembly, adhesive sealing and glueing, spray coating and painting, mechanical cutting, grinding, deburring, and polishing, water jet and laser cutting, product inspection and testing, packaging, and pelletizing. For additive manufacturing, fusing the fused deposition modelling method with a 6-axis industrial robotic arm is helpful since it gives the extruder horizontal movement and more freedom for the end-effector. Ishak et al. created a 3D printer by mounting the head to the flange of a Motoman SV3X industrial robot. Printing components with precise geometry required the development of nonplanar toolpaths.

Improved surface quality and mechanical qualities were achieved by using new software to create a toolpath for a robotic printing system to deposit material on curved surfaces. It is intended and built to build 3D things utilising an FDM extruder and a FANUC LRMATE 200iB robotic arm. To translate G-code into robotic commands, a translator is created. To increase mechanical strength, a robotic additive manufacturing process method is created to deposit material on nonplanar layers. The end-effector is manipulated throughout the entire workplace by a six-axis industrial robot. In this study, a fresh method for mass 3D printing lightweight mesh structures using an industrial robot was devised. Polyurethane foam is used to print a 2 m tall statue, and a robotic system was created to print massive sections made of thermoplastic material that weigh 25 kg(Lawlor et al., 2021). To get over the limitations of traditional FDM while printing overhang structures, various strategies were devised. The requirement for support material for overhanging structures is decreased by a novel technology that prints items utilising an industrial robot with multi-axis mobility. In order to speed up printing and cut down on waste, the "Freeform Printing" technique, which uses a 6-axis robot to print without any additional supports, focuses on biologically inspired design. Using multi-plane FDM and an industrial robot arm, it is possible to print layers in several directions. To save printing time and eliminate waste, the "Freeform Printing" method uses a 6-axis robot to print without the usage of any additional support structures. Contrary to conventional FDM, which is limited to single plane layering, multi-plane FDM using an industrial robot arm allows the printing of layers in multiple planes.

*Figure 1. FDM: infilling patterns*

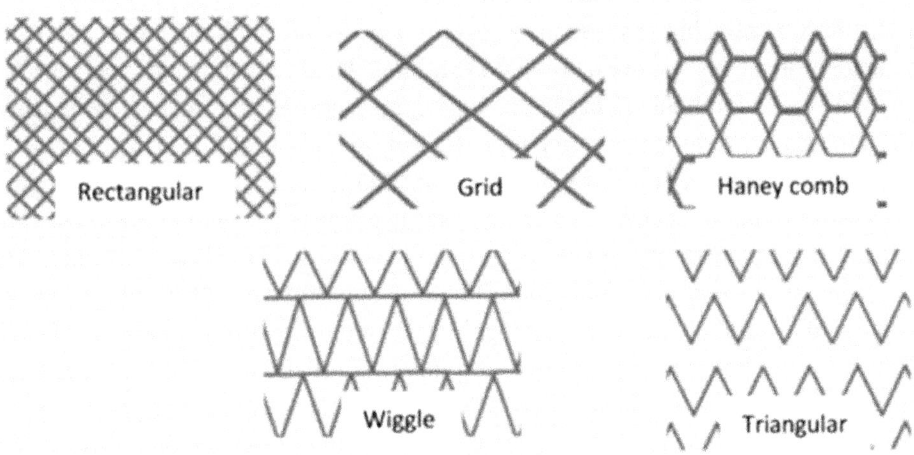


A novel algorithm makes it possible to print lattice structures with an industrial robot without a support framework and in multiple directions. In order to reduce material waste and print material on already existing components, path creation algorithms for the FDM process use a robotic arm to deposit material on curved surfaces. The industrial robot's integrated FDM allows for the creation of nonlinear toolpath printing on curved planes. Depending on the geometrical characteristics of the object, the orientation of the FDM extruder can be altered by attaching it to the robotic arm as an end effector. We reduce the number of layers and remove support structures using two nonplanar slicing methods. Selective volumes yield curved slices, and the curved section is printed using nonplanar toolpaths. An automated fused deposition modelling method is utilised to produce these curved slices. It was suggested to use a collision-free trajectory planning method to print 3D models with many curves. It showed that by carefully regulating the trajectory parameters with respect to surface curvature, collision with the printing surface was avoided. Using the suggested trajectory, complex 3D structures of diverse curvatures were successfully created, and a suitable level of surface polish was attained. For these robot manipulators, collision-free trajectories were constructed using algorithms for breaking down parts into multi-resolution layers made up of both planar and non-planar layers. In order to prevent the extruder from colliding with previously deposited layers, trajectory planning algorithms for a 6 DOF industrial robot were created(Mahmood et al., 2021; Özcan et al., 2021). By avoiding sharp corners, a B-spline-based smoothing technique was employed to provide a smooth printing route. The Scanline polygon filling technique uses horizontal lines, or scanlines, to fill inner pixels of a polygon given only its vertices. In order to fill the polygon between pairs of intersections, the scanline is intersected with its edges. The operation of this algorithm is shown in the subsequent phases. Due to its simplicity and capacity to create complicated structures, the FDM technique has a wide range of applications for lightweight components in the aerospace, automotive, and medical industries. Although it has three degrees of freedom, it can only print on a flat, horizontal surface.

The FDM technology prints items by slicing them parallel to the horizontal plane, which leaves curved surfaces with a subpar surface polish. Normally, two nozzles are used to extrude construction material and support material. However, if the intended object does not have any overhang characteristics, the second nozzle sits idle the entire duration of printing, increasing printing time and requiring printing in many planes. Manufacturing will benefit from the integration of robotic and FDM technologies since the robot arm can move freely around a sizable workspace. If the extruder functions as an end effector, it is feasible to print with it in several orientations and on several planes. By shifting the extruder's orientation, the staircase effect can be avoided. Since the six-axis industrial robot has more degrees of freedom (6) than the traditional FDM machine, a novel application for it is being investigated. The FDM extruder can be used as the industrial robot's end effector to create new methods of material deposition. This enables the material to be deposited over nonplanar layers of curved structures as well as simultaneous printing over several planes and directions. By simultaneously extruding the same material from both nozzles while using a new toolpath pattern, a layer's printing time can be shortened (Figure 1).

## LITERATURE SURVEY

A revolutionary method of industrial production, additive manufacturing (AM) enables the development of components and systems that are both lighter and more robust. It deposits material in exact geometric shapes layer by layer using data from CAD software or 3D object scanners. One of the most widely used additive manufacturing technologies is FDM. The effectiveness of items printed using the FDM method is studied, and the results are reported. Process factors studied include layer thickness, raster angle, raster width, deposition speed, deposition rate, nozzle diameter, nozzle temperature, and bed temperature. The use of an industrial robot to create items in various directions using various slicing techniques is also mentioned. The mechanical characteristics, quality, and printing time of the object are all impacted by a number of parameters in the FDM process. Each layer's outside edge is known as a "shell," sometimes known as a "contour." Infill is a repeated construction used to fill empty space inside the shell(Badekila et al., 2021; Magalhães et al., 2020). The operating parameters inside a layer are displayed in Figure 2.1. The mass and general strength of the printed object are greatly influenced by the infill density, which establishes the quantity of material used to print the infill. The choice of infill density depends on the intended application, such as the finished product's visual appeal. A higher infill density results in a longer print run. The road is made of a thin thermoplastic strip that is extruded through the extruder's nozzle. The nozzle's diameter determines its width, while the raster's direction and the X-axis determine its angle. The space in a layer between two neighbouring highways is known as the "air gap." The thickness of the deposited layer, which has an impact on the overall quality of the printed object, is known as the "layer thickness." By choosing the smallest layer thickness, it can be reduced, but the printing process takes longer. It also has an impact on the printed part's mechanical strength and the requirement for support material. The printing of extra support material will incur extra time and material costs. Build orientation is also an important factor. Deposition rate is the rate at which material is extruded through the nozzle, whereas printing speed is the speed at which the extruder moves while depositing material in the XY plane. Additionally, nozzle temperature is managed to maintain a constant raster width.

This article outlines a process for producing intelligent quick prototypes that can achieve the appropriate level of precision, produce items with enhanced mechanical qualities, and shorten the time and cost of construction. The FDM 1600 machine was used to print samples using ABS. This study used the traditional laminate theory to analyse the effects of build orientation and toolpath on the mechanical strength of printed components created using the FDM technique. The findings demonstrated that specimens with contour construction have excellent tensile strength. For tensile testing, specimens were created in accordance with ASTM D 5937-96 criteria(Boopathi, Balasubramani, et al., 2023; Boopathi, Siva Kumar, et al., 2023; Kumara et al., 2023; S. et al., 2022; Sampath et al., 2022). This study looked at how different deposition routes affected the mechanical strength of FDM products. The thermal expansion coefficient and tensile characteristics (Young's modulus) of the printed specimens were found to be mostly influenced by the build orientation. A stiffness matrix was created for each of the two new deposition pathways (spiral and modified raster). Testing for tensile strength was done to assess mechanical strength. Because the theoretical and experimental results were nearly identical, it was determined that either method could be used to compute the stiffness qualities(Saravanan et al., 2022; Trojovský et al., 2023).

To print functioning items, this system has the capacity to deposit four different types of material within a layer. Software created in Visual Basic separately controls the positioning and depositing operations. Each layer is deposited, and then its image is taken and checked for flaws. To satisfy the functional requirements, tight tolerances are required. Five-layer orientations were used to examine the impact of layer orientation (raster angle) on the mechanical characteristics of an ABS-printed FDM item. The specimen with a raster angle had a higher modulus of rapture value than the others, according to the results. The Taguchi technique was used to assess the effect of process variables such as layer thickness, raster width, and deposition speed on the calibre of FDM-printed specimens. Using a Surtronic roughness tester to quantify the roughness of the surface, an ANOVA analysis revealed that the layer thickness had the greatest impact.

The elastic modulus was reduced by up to 10 percent after printing, and the tensile strength was decreased by between 20 and 50 percent, according to a comparison of elastic moduli and strengths for ABS filament and FDM produced items. This was brought on by the existence of voids in the printed component and a change in molecular orientation following the deposition of ABS material. The temperature, airgap, raster orientation, and raster width were used to calculate the compressive and tensile strengths of ABS specimens produced on FDM 1650 machines. Results showed that the compressive strength of the ABS material was 82–94%, while the tensile strength was 67–78% of that of the injection-molded ABS material.

The socket mould is scanned with a laser to get 3D data, which is then used to produce a CAD model and be printed using an FDM 1600 printer with nylon P301 as the printing medium. Clinical testing revealed that after several months of use, the patient finds the manufactured prosthetic socket to be comfortable. The new metal-polymer composite, which may be utilised to create injection moulding tool inserts with appropriate mechanical qualities and cost and time savings, is made up of iron particles and a nylon polymer matrix. According to this experiment, the temperature of the bed or envelope significantly affects dimensional accuracy. The process variables that determine how elastic the ABS-made prototype performs include layer thickness, raster width, raster angle, and air gap. ANOVA is used to research how to get the best elastic performance(Boopathi, Thillaivanan, et al., 2022)(Boopathi, 2023; Boopathi & Sivakumar, 2014; Sampath et al., 2021; Sampath & Myilsamy, 2021). The mathematical model was created to investigate the effects of the number of layers, length of prototype stacking, chamber temperature, and material shrinkage rate(Boopathi, 2022; Boopathi, Haribalaji, et al., 2022; Kannan et al., 2022; Yupapin et al., 2022). The findings demonstrated that deformation decreases as layer number rises, however, wrap deformation increases when chamber temperature rises. Thermal analysis revealed that production method and bed temperature have an impact on the bonding strength and mesostructured of deposited filaments, whereas sintering and creep deformation have a substantial impact on bonding quality. The residual stresses in large prototypes were investigated in this work in relation to the FDM process variables. The findings demonstrated that printing speed is more important than layer thickness. With increasing layer thickness, more residual stress is being retained (Kautish et. al 2022).

This study investigated the effects of process variables on the dimensional accuracy of an FDM part, including raster angle, raster width, air gap, orientation, and layer thickness. The dimension along the thickness is always greater than the required value, and there is significant shrinking along the length and width. The bending of the FDM part caused by thermal stress can be avoided by selecting the best parameters, according to the results of a FEM model that was created to validate the derived numerical model. This study recommended employing water-soluble supports for simple removal following part production. Additionally, it advised against using narrow raster angles while using FDM. Test specimens were printed, and central composite design was employed to reduce the number of experiments. The strength increases with layer count; however, distortion is to blame for the printed items' poor strength(Datta et al., 2020).

This method acquires real-time depth data using a cheap Kinect sensor. Within its field of view, it is accurate in tracking and mapping scenes, and it operates at a 30 Hz frame rate. For bending, impact, and deflection tests, experimental results demonstrated that specimens with a raster orientation have higher tensile strength than specimens without a raster orientation. The tensile strength of the FDM specimen is essentially identical to the specimen made with injection moulding, demonstrating the influence of part orientation (build directions X, Y, and Z) and toolpath approach on the mechanical properties of FDM parts. Due to its internal structure, the FDM specimen displayed less elongation than the injection-moulded specimen.

According to this research, anisotropy and distortion are to blame for the poor compressive strength of the FDM part, and there is a complex non-linear relationship between the process parameters and compressive strength. A neural network was used to create this association. An analytical model was developed to determine the tensile strength and stiffness of printed specimens using ABS material, building direction, and shapes. Results indicated that there was a tiny margin of error and that the results were nearly identical to the experimental results. Specimen residual stresses were measured using an optical sensor with a short fibre Bragg grating employing layer thickness and raster angle. To collect depth and colour information, Kinect is mounted on a swivelling chair and kept still. To cut down on time and expense, models with hollow shells were created. For validation, a significant number of scans and prints were made. On FDM specimens printed at various raster angles and build orientations, mechanical characteristics were studied. Tensile and bending tests on the ABS material that was used for printing were also performed. It was discovered that surface roughness was more significant than raster angles. Tensile, bending, and impact tests on ABS material revealed that the price and printing time were negatively correlated with the infill pattern's strength, showing that specimens with higher densities demonstrate superior mechanical qualities. Tensile testing revealed that at around 250 orientations, the behaviour switches from ductile to brittle(Li et al., 2020). Colored PLA has the highest crystallinity percentage, whereas natural PLA has the lowest, according to X-ray diffraction and tensile testing. This shows that crystallinity, deposition temperature, and tensile strength are strongly correlated. With two subsystems for positioning and depositing material, this system has the ability to print four different types of materials inside the shell of a layer. A camera is installed to record the printing process in real time, and software is built to construct toolpaths. It is able to create functioning ceramic actuators with a variety of compositions. The created extruder has four nozzles that may simultaneously deposit biopolymers to manufacture tissue scaffolds. During this deposition, which is intended to take place below room temperature, a multi-nozzle extruder is used to extrude various polymers. This creates a material gradient. The extruder-printed multilateral scaffolds are demonstrated to have adequate strength and a better hydrophilic nature(Oladapo et al., 2019).

Using several nozzles, a multi-material 3D printing system may print various materials in independent layers. A sandwich construction was made to show off the capabilities, and a programme was written to carry out the printing process. In order to employ five distinct filament types simultaneously and speed up printing, this paper suggests an extrusion model with five nozzles. This platform can print items with various material proportions and colours since it uses a single printing head with several capillaries to deposit bio inks. It is able to produce objects with exterior geometries and interior gradient structures. The created system is more affordable and effective for multi-material additive manufacturing. By adjusting the ratios of each material within the layer, specimens were printed, and voids developed between the sections of two materials. The findings of the tensile test demonstrated that the structural configuration had an impact on the tensile property. By preventing collisions, an algorithm is created to build toolpaths for individual print heads. Large parts may be built faster using this technology, which can also print numerous materials precisely and in voxel-by-voxel form. The fact that the system was used to print two separate objects shows how quickly things may be created. Each layer is divided into a variety of contours by the toolpath planning algorithm, and robotic actuators are then employed to print material simultaneously inside each contour. According to simulation studies, it is effective at producing single and multi-material components quickly.

## ROBOTIC APPLICATIONS

Robots are necessary for space exploration because they can withstand dangerous environments and run on solar power. Military: Human soldiers can be replaced by robots in combat operations, lowering risk. Medical: Reducing danger, robots can operate either with or without a human surgeon. Industrial Robots can produce high-quality goods with little waste, which can save time and money. Agriculture: Using robots in agriculture can increase output. Rescue robots, which can go on land, water, and the air, are employed for search and rescue during natural catastrophes. Domestic: Domestic or personal robots can be employed for security, service, education, and amusement(Yan et al., 2018). It can be integrated with internet of things to improve the efficiencies(Babu et al., 2022; Boopathi, Arigela, et al., 2022; Boopathi, Khare, et al., 2023; Jeevanantham et al., 2022; Palaniappan et al., 2023; Senthil et al., 2023).

## PARAMETERS OF ROBOT

With three translations, three rotations, and six degrees of freedom, a robot's configuration is defined by its degrees of freedom (DOF). Number of axes: A robot has an equal number of axes as degrees of freedom. A work area is a solid area that indicates the positions that a robot is currently holding. The weight of a robot's manipulability end-effector and product is known as its payload. Industrial robots can move at two different speeds: angular and linear.

Acceleration: When every joint is moving simultaneously, compound acceleration is calculated. The smallest movement increment a robot can make within its work volume is called spatial resolution. Robot accuracy is the difference between the task that is desired and the task that is actually completed. Robots can go back to places they've already visited by again doing the same tasks. Compliance: The displacement of the robot axis in response to a specific torque or force is known as compliance(Jiao et al., 2018). The parameter optimization can be done based on Taguchi method and it can be optimized by multi-criteria concepts(Boopathi, 2019, 2021; Sampath et al., 2021; Saravanan et al., 2022).

## DUAL NOZZLE – FDM PROCESS

In dual nozzle FDM machines, build material and support material are deposited by two nozzles that are spaced apart uniformly. To speed up printing, the extruder's orientation must be altered so that the direction of the infill for each layer is perpendicular to that of the preceding layer. The orientation of the extruder cannot be changed in conventional FDM. A new methodology is created for twin nozzle material deposition using a 6-axis industrial robot (Figure 2).

### Dual Nozzle Extrusion

A CAD model is created by the industrial robot with dual nozzle extrusion, which then converts it to STL format, slices it into horizontal layers, and extracts the boundary of each layer. To create the toolpath for single nozzle extrusion, the form of each layer's border is taken into consideration. Create a twin nozzle extrusion toolpath for infill by grouping all the squares. Both single nozzle and twin nozzle extrusion toolpaths are produced. To achieve strong bonding between layers, the infill's toolpath is positioned perpendicular to the direction of the preceding layers. All layer toolpaths are connected from top to bottom. To control robot movement and material deposition, MATLAB software translates a toolpath into a fast programme. This quick software is checked using simulations before being uploaded to the controller and printed by the robot(Klein et al., 2013).

### Toolpath Generation

In order to decrease overhangs, dual nozzle extruders use two nozzles to deposit building material and support material. The intersection points of the STL format and the slicing horizontal plane are joined to build the toolpath for a layer of any shape. To create a twin nozzle extrusion toolpath, a polygon's area is divided into squares. A "2x" distance separates the two perpendicular sets of straight lines that make up the reference grid. A "2x" square serves as the grid element, while rectangles are created by grouping squares vertically. Rectangles have two dimensions: width and length. A, B, C, D, E, and F are used to denote the vertical groups of rectangles that are located between two grid lines. The black and yellow halves of rectangles are separated by other colours. The first nozzle deposits the material for the black portion, while the second nozzle deposits the material for the yellow portion. By simultaneously depositing material through both nozzles, two nozzles are employed to speed up the printing of freshly produced rectangles. The two nozzles' toolpaths are produced inside the polygon's black and yellow regions, enabling material deposition. The printing time can be cut in half compared to single-nozzle extrusion with dual nozzles. When compared to single-nozzle extrusion, dual nozzle extrusion can shorten the time required to print an infill layer. Each layer's infill toolpath is oriented to improve the overall mechanical qualities of the item by fostering strong bonds between adjacent layers. Squares with full or partial enclosures are arranged in horizontal groups throughout the layer's interior. The direction of the layer's infill toolpath will be made easier by the newly generated rectangles. With dual nozzle extrusion, two toolpath patterns are employed to deposit material for the layer's infill. The dual nozzle extruder is moved and oriented by the robot following a toolpath pattern to transfer the tips of nozzles 1 and 2 to points 1 and 3, respectively. The direction of the extruder is maintained constant while Nozzle 1 follows the black toolpath and Nozzle 2 follows the yellow toolpath. For dual nozzle extrusion, two nozzles move in accordance with a toolpath pattern that is highlighted in black and yellow. The dual nozzle extruder is moved and positioned by the robot in a manner similar to toolpath pattern 1. Combining toolpaths for single and multiple nozzle extrusion can be used to create rectangles, which can shorten printing time. For combined toolpaths with rectangles measuring 30 mm by 25 mm, the extruder orientation changes. In sub-areas A1 and A2, the toolpath is generated using toolpath pattern 1, and dual nozzle extrusion is employed in B1 and B2. By merging the toolpaths for single and dual nozzle extrusion, the toolpath for sub-area C can be decreased to "0.6t."

## Simulation for Printing a Cube

A cube with a 41-mm edge length is used to demonstrate the developed methodology. To create a set of layers, the cube is converted into STL format and sliced from bottom to top with a 0.5 mm increment. Each layer has a square shape with a side of 41 mm. To produce a toolpath for single nozzle extrusion along the shells of each layer, MATLAB software is created. The extruder orientation is modified to be about the Z-axis, and the toolpath direction for the infill is set to be perpendicular to that of the base layer. To create toolpaths, the infill direction of each layer is configured to be perpendicular to the infill direction of the layers above it. To control robot movement and material deposition, MATLAB software translates a toolpath into a fast programme. The imported robot has a twin nozzle extruder model with TCP defined at the first nozzle tip. To distinguish between nozzles 1 and 2, Robot Studio creates a virtual controller and shows a toolpath. The robot positions and manoeuvres the dual nozzle extruder model along the Robot Studio software-generated toolpath. The robot switches the extruder's orientation to test the quick programme(Derby, 2012).

## Printing of the Cube

The ABB IRB 1600 industrial robot's dual nozzle extruder is connected to it, and it is driven by two stepper motors with two distinct digital output signals. Material for the shell is deposited by nozzles 1 and 2. The orientation of the extruder is altered to make any two successive layers perpendicular. The orientation of the extruder is altered to make any two successive layers perpendicular. Setting the temperature to 195 degrees and printing the cube with black PLA material, the industrial robot drove the dual nozzle extruder along a toolpath and deposited material at appropriate areas. The created process cuts each layer's infill printing time in half, yielding a total printing time of 61.07 minutes. 50 percent less time is spent printing when using dual nozzle extrusion. The twin-nozzle extrusion technique allows for faster large-scale printing than a traditional FDM machine. In the dual nozzle FDM method, two toolpath patterns can be produced for rhombus cross sections. Printing time can be cut in half by using a dual nozzle material deposition technology instead of an FDM procedure(Van Noort, 2012; Xu et al., 2011).

# MULTI-PLANE FDM PROCESS

Planar and nonplanar layers are used in two different ways to print items on curved surfaces and numerous planes. Cutting objects in parallel on two separate planes is used for printing in multiple planes (Figure 2).

## Modelling and Toolpath Generation

Using an industrial robot, the fused deposition modelling process prints items simultaneously in many planes. Using a revolutionary slicing technique, the industrial robot may be used to print items over several planes. The simultaneous printing over many planes made possible by the integration of an industrial robot and FDM allows for the creation of an object as a whole. The strength of overhangs is decreased when

*Figure 2. Flow chart for the two-phase and multi plane FDM processes*

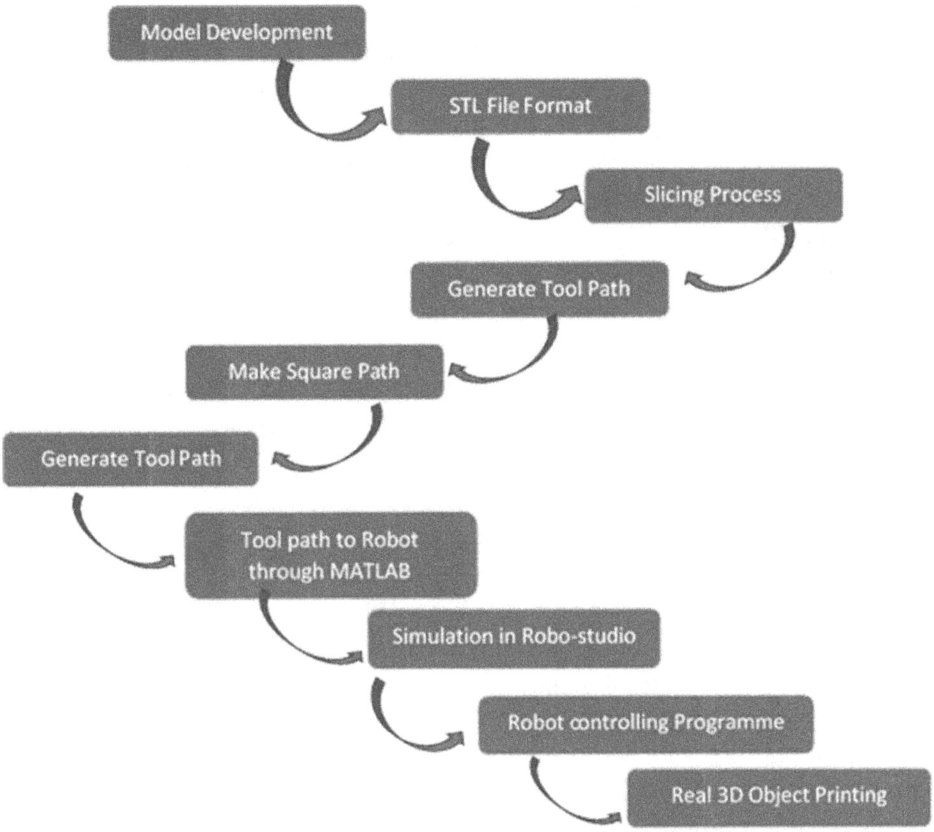


printed individually using algorithms based on the object's structure, eliminating the requirement for multiple plane printing. FDM and an industrial robot are combined in a novel way for multiple plane slicing and printing to boost strength. In order to achieve uniform deposition in the intersection zone, a new printing technique is created that continuously changes the extruder's direction.

**Methods:** The CAD model is converted to an STL file and split into layers with a fixed layer thickness for inclined plane printing. The scanline polygon filling procedure is used to build the toolpath, which MATLAB then converts into quick commands. Before being sent to the robot controller and printed, the extruder is simulated and confirmed.

**Slicing:** To demonstrate the capabilities of an industrial robot, a plate with an angle of 300 and a thickness of 7.5 mm is employed. To print without any support structure, a unique, customizable setup with glass plates is created. The robot must move the extruder perpendicular to the horizontal plane, perpendicular to the inclined plane, and perpendicular to the third plane in order to print the stiffener because the support frame is insufficient for items with an inclination angle of 90° or more. The horizontal and inclined plate serve as one item, and the stiffener serves as the second, and are cut into layers with a 0.5 nozzle diameter by an extruder. A front view of the model is produced by aligning two planes parallel to one another and using a layer thickness of 0.5 mm. Points outside of the intersection line are not included in the layer, which is composed of two intersecting planes. The excess section is cut off by the MATLAB software when it determines the line segment between two intersecting planes. The stiffener is oriented at 15° degrees to the horizontal plane, and the object is cut horizontally and inclined. To illustrate the capabilities, the stiffener is printed as a number of layers with size gradations. The excess section is cut off by the MATLAB software when it determines the line segment between two intersecting planes. The stiffener is oriented at 15° degrees to the horizontal plane, and the object is cut horizontally and inclined(Moon et al., 2010). To show off the potential of multiple plane slicing and printing using an industrial robot, the stiffener is produced as a number of layers with gradations in size.

**Toolpath Generation:** The angle plate was produced using two separate toolpath patterns printed simultaneously over two planes. When two parallel planes intersect, the tool proceeds along a route with distinct orientations for each layer, producing a line segment. Two examples show how the tool's orientation changes as it transitions from a horizontal to an inclined plane. To ensure consistent deposition, the nozzle's direction is changed as it moves into the intersection zone. Up until point 8, the tool is positioned perpendicular to the inclined plane. Similar to toolpath pattern 1, toolpath pattern 1 is used to print material on an inclined plane. To boost strength, toolpath pattern 2 is created for FDM layers that follow one another. The toolpath for the shell is created using toolpath pattern 1. Similar to the traditional FDM method, the stiffener's toolpath is created, but the tool's orientation is kept perpendicular to the inclined plane. Instead of G and M codes, robot motion commands are used to construct the toolpath.

**Simulation of 3D Printing:** Robot Studio generates toolpaths in quick commands to replicate extruder movement using an ABB IRB 1600 robot and CATIA software. The extruder's nozzle tip followed the generated toolpath and orientation, while the tool speed was kept at 50 mm/s.

**Simultaneous of 3D Printing of Angle Plate:** When printing on the horizontal and slanted planes, the extruder rotates. The support structure is necessary for the traditional FDM procedure to print the angle plate. The contact surface area between the layers varies, which causes variations in the object's strength. The relative benefits of the robotic FDM technique's slicing methodology over the traditional FDM process. To prevent the deposition of excess material, the object is printed using an extruder at a consistent velocity (50 mm/sec) and constant feed rate (15 mm/min). To prevent additional material from being deposited where two planes of the gable roof overlap, the direction of the extruder might be altered. The extruder's orientation is changed as it moves closer to the corner, thus minimising the amount of material that accumulates there and boosting robot speed. Strong adhesion is produced by slicing parallel to both planes, which uniformly and maximally increases the contact area between the layers. The ILP process prints overhang without a support framework using sliced layers, which causes variations in bonding strength. The maximum tilting angle of an extruder is determined by the half taper angle of the nozzle cone and the extruder assembly width. The results of the experiments show that printing on an inclined plane with a sharp corner is possible by rotating the extruder. The suggested approach may produce inclined planes with a maximum inclination of 00 to I in both clockwise and counter clockwise directions. Curved corners can be created to prevent interference between the nozzle and the printing surface, but due to the gravitational pull, the proposed method cannot be used to print the object if the angle of inclination is too great. The anticlockwise inclination of the horizontal axis with respect to the inclined plane is denoted by the symbol. The suggested slicing methodology allows for consistent deposition at the junction zone while printing objects with various combinations of curved and flat surfaces. Objects with overhang structures are printed using an industrial robot equipped with an FDM extruder at a constant robot velocity and consistent deposition rate to cut down on printing time and material waste.

## CASE STUDY: FDM FOR AIRCRAFT WING PRINTING PROCESS

The staircase effect, which produces a rough surface finish along an object's curved surface, is brought on by the layered deposition of material. Although adaptive slicing can lessen the staircase effect, it can also result in a subpar surface finish. Due to the finite degrees of freedom, typical FDM machines cannot produce nonplanar layers. A dual nozzle extruder is used by a 6-DOF industrial robot to create nonplanar layers along curved surfaces, which eliminates the staircase effect. A dual-nozzle extruder is controlled by a robot to follow non-linear toolpaths and deposit material along curved surfaces. To provide a uniform deposition of material, the extruder is kept perpendicular to the curvature of the nonplanar layers. Robot Studio simulation software is used to evaluate the extruder orientation after it has been determined using a MATLAB programme. The staircase effect was successfully eliminated when a NACA 0015 air foil wing was produced using the ABB IRB 1600 industrial robot.

**Wing Design:** A wing is a fin with two curved surfaces that generates aerodynamic force. By generating a pressure difference between two surfaces, an air foil provides lift force. Wing nomenclature is dependent on the kind of air foil, with the leading edge having the most curvature. The suggested methodology is assessed using an air foil wing's symmetrical angle of attack.

**Select Orientation of the Wing:** The NACA 0015 air foil-shaped wing segment has dimensions of 80 mm and 20 mm, however, only three printing orientations allow for printing with the bare minimum of support. A horizontal plane parallel to the heat bed is used for slicing. Figures display the created set of layers for each printing orientation. The wing must have support materials deposited since cutting it from bottom to top results in a poor surface finish and lack of uniform strength. For the second printing orientation, the wing is cut parallel to virtual plane "P" and perpendicular to "Q," from bottom to top. When flying, the wing's strength is uniform, yet its tensile strength is weak in the printing direction. A FDM-fabricated part fails if a compressive load acts in the opposite direction to the printing direction, causing delamination and bending out of plane. In order to print wing segments without experiencing the staircase effect, a new slicing technique is suggested. While support layers are formed up to the height of the chord line, nonplanar layers are generated in parallel to one another and in a circle that is centred on the curved surface of the wing. To provide a smooth surface finish on the lower surface, classic horizontal slicing is used to make planar and nonplanar layers. Because interference with previously deposited material is avoided by maintaining the extruder's orientation perpendicular to the nonplanar printing surface, there is greater bonding strength and improved mechanical performance. The orientation of the extruder can be changed by a 6-DOF industrial robot as it deposits material along nonplanar layers, creating a nonlinear toolpath.

## Printing Process

The wing segment is printed in four stages: generation, toolpath generation, simulation, and printing itself.

**Layers Slicing:** The wing segment and its support structure are printed in layers to produce a smooth lower surface.

- Data points computed using information from the NACA 0015 airfoil .These facts are highlighted.
- Based on two sets of data in the YZ plane, two splines are built.
- At discrete intervals, normal vectors are produced based on the local curvature of the lower base spline.
- To print the wing segment, a spline is created to run through a collection of points.
- To create a set of concentric splines, two splines are separated by the typical 0.5 mm spacing up to the height of the chord line.

- A non-linear toolpath with a perpendicular orientation is created by nonplanar layers below and above the chord line.

**Formation of Support Structure:** Below the chord line, support material for concave-shaped nonplanar layers, both planar and nonplanar, is deposited. To eliminate the staircase effect at the lower surface of the wing, two nonplanar layers are utilised. After printing the wing segment, the support structure's planar and nonplanar layers are removed.

## Toolpath Generation

With a 50% infill density, the toolpath for planar layers is formed in a zigzag pattern. This toolpath is transformed into quick commands for the ABB robot via MATLAB software. The extruder orientation, which is predetermined to be perpendicular to the printing surface, is generated simultaneously with the toolpath for each nonplanar layer. This will guarantee even material deposition and remove the possibility of running into already-deposited materials. Following the non-linear toolpath causes the nozzle orientation to shift while still travelling at a constant speed. Two different types of non-linear toolpaths are created by MATLAB programmes to deposit material along non-planar layers. The orientation of the extruder is determined based on the curvature of each spline, and each data point on each spline is handled as a motion target. The programme is used to create each type of toolpath and connect the data points in two different ways. The two most crucial facts are that each spline's data points are linked together to form a non-linear toolpath and that there is a 0.5 mm gap between each one in the X direction. The MATLAB software employs this toolpath to print a nonplanar layer with a concave form.

In the second type of nonlinear toolpath, the data points of nearby splines are joined along the X axis to form the toolpath. While neighbouring toolpaths advance progressively in the direction of Y with a distance of 0.5 mm, they have distinct orientations from the extruder's continuous orientation as it moves along the toolpath. The MATLAB application is then used to connect the toolpath in a zigzag pattern to produce a concave-shaped nonplanar layer. To provide strong bonding between adjacent nonplanar layers, nonlinear toolpaths are produced for them. The top nonplanar layer is covered by the second kind, while the outer layers of the wing segment and bottom nonplanar support layer are covered by the first.

## Simulation

Using Robot Studio simulation software, the quick programme created for the support structure and wing segments is reproduced. A model of the extruder is created, built, and mounted to the flange of a simulated robot. A work object is defined at a specified height, and TCP is defined at the first nozzle's tip. There are two levels of validation for the TCP movement simulation. The robotic cell is synchronised with the fast programme, and the simulation environment displays the toolpath. The nonlinear toolpath went along the nonplanar layers with a continual change in orientation, whereas the nozzle tip of the extruder model travelled along the planar layers of the support structure with a fixed perpendicular orientation. The robotic cell is synchronised with the wing segment's fast programme, and the simulation environment displays the toolpath. When the "Play" option is chosen, the extruder model's nozzle tip moves continuously, changing orientations along each nonplanar layer's nonlinear toolpath. Both the support structure and the wing segment's rapid programme simulation results show that each target is within the robot's reach and that the extruder moved without experiencing singularity errors.

## 3D Printing of the Wing

According to simulation results, the quick software created to print the wing segment and support structure was reachable and singularity error-free. An industrial robot with a dual nozzle extruder attached to the flange performs the actual printing. PLA (Polylactic acid) is the substance used for deposition. The support structure is printed in one stage with yellow PLA material, while the wing segment is produced in two stages with black PLA material. By employing the suggested slicing technique, the NACA 0015 air foil-shaped wing segment with nonplanar layers is successfully printed using an industrial robot. The wing section is printed using black PLA with a 100% infill density after the support material is printed in the first stage with a 50% infill density. The upper and lower surfaces of the wing segment have a smooth surface finish, which eliminates the staircase effect and gives the wing segment high mechanical qualities. In order to prevent warpage, thermal plasters were employed to attach the support structure to the heat bed. A permanent support structure constructed of metal can be employed in place of a temporary one. When the wing segment is produced, this will guarantee the support structure's fixed location.

## Robotic FDM Process

In order to produce components faster, the suggested robotic FDM approach has addressed some of the issues with the traditional FDM process. Producing jigs, fixtures, moulds, and patterns can be done with the environmentally friendly, non-toxic, and safe PLA material that is being employed. Lightweight components, testable functional prototypes, and functionally rated spare parts can all be produced using an industrial robot and the dual nozzle FDM technique. The printed pieces can be utilised in the medical industry for custom prosthetics, surgical guidance, implant testing, and visualisation aids. Industries including aircraft, consumer products, architecture, and automotive can all benefit from this technology.

## SUMMARY

Two novel methods for printing objects simultaneously on various planes and curved surfaces were provided, and they were put to the test by printing test objects. The outcomes and a variety of potential future projects are presented. Using data from earlier research projects, a brief literature on 3D printing technology, its parameters, and applications has been developed. An explanation has been given for the case study on the aerofoil wing produced via industrial robotics programming. An illustration of a flow chart that can display both three-phase and multi-plane operations has been provided.

### Abbreviations

ABS- Acrylonitrile Butadiene Styrene
ANOVA- Analysis of Variance
CAD-Computer Aided Design
DOF-Δεγρεεσ Οφ Φρεεδομ
FDM- Fused Deposition Modelling
FEM – Finite Element Method
NACA - National Advisory Committee for Aeronautics
PLA-Polylactic Acid-
STL- Standard Triangle Language
TCP - Tool Center Point

## REFERENCES


Babu, B. S., Kamalakannan, J., Meenatchi, N., Karthik, S., & Boopathi, S. (2022). Economic impacts and reliability evaluation of battery by adopting Electric Vehicle. *IEEE Explore*, (pp. 1–6).

Badekila, A. K., Kini, S., & Jaiswal, A. K. (2021). Fabrication techniques of biomimetic scaffolds in three-dimensional cell culture: A review. *Journal of Cellular Physiology*, *236*(2), 741–762. doi:10.1002/jcp.29935 PMID:32657458

Boopathi, S. (2019). Experimental investigation and parameter analysis of LPG refrigeration system using Taguchi method. *SN Applied Sciences*, *1*(8), 892. doi:10.100742452-019-0925-2

Boopathi, S. (2021). Improving of Green Sand-Mould Quality using Taguchi Technique. *Journal of Engineering Research*, 1–12. doi:10.36909/jer.14079

Boopathi, S. (2022). (in press). Cryogenically treated and untreated stainless steel grade 317 in sustainable wire electrical discharge machining process: A comparative study. *Environmental Science and Pollution Research International*, 1–10. doi:10.100711356-022-22843-x PMID:36057706

Boopathi, S. (2023). An Investigation on Friction Stir Processing of Aluminum Alloy-Boron Carbide Surface Composite. In *Materials Horizons: From Nature to Nanomaterials* (pp. 249–257). Springer., doi:10.1007/978-981-19-7146-4_14

Boopathi, S., Arigela, S. H., Raman, R., Indhumathi, C., Kavitha, V., & Bhatt, B. C. (2022). Prominent Rule Control-based Internet of Things: Poultry Farm Management System. *IEEE Explore*, (pp. 1–6). IEEE.

Boopathi, S., Balasubramani, V., & Sanjeev Kumar, R. (2023). Influences of various natural fibers on the mechanical and drilling characteristics of coir-fiber-based hybrid epoxy composites. *Engineering Research Express*, *5*(1), 15002–15020. doi:10.1088/2631-8695/acb132

Boopathi, S., Haribalaji, V., Mageswari, M., & Asif, M. M. (2022). Influences of Boron Carbide Particles on the Wear Rate and Tensile Strength of Aa2014 Surface Composite Fabricated By Friction-Stir Processing. *Materiali in Tehnologije*, *56*(3), 263–270. doi:10.17222/mit.2022.409

Boopathi, S., & Khare, R. KG, J. C., Muni, T. V., & Khare, S. (2023). Additive Manufacturing Developments in the Medical Engineering Field. In Development, Properties, and Industrial Applications of 3D Printed Polymer Composites (pp. 86–106). IGI Global.

Boopathi, S., Siva Kumar, P. K., & Meena, R. S. J., S. I., P., S. K., & Sudhakar, M. (2023). Sustainable Developments of Modern Soil-Less Agro-Cultivation Systems. In Human Agro-Energy Optimization for Business and Industry (pp. 69–87). IGI Global. doi:10.4018/978-1-6684-4118-3.ch004

Boopathi, S., & Sivakumar, K. (2014). Study of water assisted dry wire-cut electrical discharge machining. *Indian Journal of Engineering and Materials Sciences*, *21*(1), 75–82.

Boopathi, S., Thillaivanan, A., Azeem, M. A., Shanmugam, P., & Pramod, V. R. (2022). Experimental investigation on abrasive water jet machining of neem wood plastic composite. *Functional Composites and Structures*, *4*(2), 025001. doi:10.1088/2631-6331/ac6152

Datta, P., Dey, M., Ataie, Z., Unutmaz, D., & Ozbolat, I. T. (2020). 3D bioprinting for reconstituting the cancer microenvironment. *NPJ Precision Oncology*, *4*(1), 1–13. doi:10.103841698-020-0121-2 PMID:32793806

Derby, B. (2012). Printing and prototyping of tissues and scaffolds. *Science*, *338*(6109), 921–926. doi:10.1126cience.1226340 PMID:23161993

Hua, L., Lei, T., Qian, H., Zhang, Y., Hu, Y., & Lei, P. (2021). 3D-printed porous tantalum: Recent application in various drug delivery systems to repair hard tissue defects. *Expert Opinion on Drug Delivery*, *18*(5), 625–634. doi:10.1080/17425247.2021.1860015 PMID:33270470

Jeevanantham, Y. A., Saravanan, A., Vanitha, V., Boopathi, S., & Kumar, D. P. (2022). Implementation of Internet-of Things (IoT) in Soil Irrigation System. *IEEE Explore*, (pp. 1–5).

Jiao, Z., Wang, C., Yang, Q., & Wang, X. (2018). Preparation and characterization of UV-curable urethane acrylate oligomers modified with cycloaliphatic epoxide resin. *Journal of Coatings Technology and Research*, *15*(2), 251–258. doi:10.100711998-017-9994-6

Kannan, E., Trabelsi, Y., Boopathi, S., & Alagesan, S. (2022). Influences of cryogenically treated work material on near-dry wire-cut electrical discharge machining process. *Surface Topography : Metrology and Properties*, *10*(1), 015027. doi:10.1088/2051-672X/ac53e1

Kaur, R., & Kautish, S. (2022). Multimodal sentiment analysis: A survey and comparison. *Research Anthology on Implementing Sentiment Analysis Across Multiple Disciplines*, 1846-1870.

Kautish, S., Peng, S. L., & Obaid, A. J. (Eds.). (2021). *Computational intelligence techniques for combating COVID-19*. Springer International Publishing. doi:10.1007/978-3-030-68936-0

Kautish, S., Reyana, A., & Vidyarthi, A. (2022). SDMTA: Attack detection and mitigation mechanism for DDoS vulnerabilities in hybrid cloud environment. *IEEE Transactions on Industrial Informatics*, *18*(9), 6455–6463. doi:10.1109/TII.2022.3146290

Klein, G. T., Lu, Y., & Wang, M. Y. (2013). 3D printing and neurosurgery—Ready for prime time? *World Neurosurgery, 80*(3–4), 233–235. doi:10.1016/j.wneu.2013.07.009 PMID:23871811

Kumara, V., Mohanaprakash, T. A., Fairooz, S., Jamal, K., Babu, T., & B., S. (2023). Experimental Study on a Reliable Smart Hydroponics System. In *Human Agro-Energy Optimization for Business and Industry* (pp. 27–45). IGI Global. doi:10.4018/978-1-6684-4118-3.ch002

Lawlor, K. T., Vanslambrouck, J. M., Higgins, J. W., Chambon, A., Bishard, K., Arndt, D., Er, P. X., Wilson, S. B., Howden, S. E., Tan, K. S., Li, F., Hale, L. J., Shepherd, B., Pentoney, S., Presnell, S. C., Chen, A. E., & Little, M. H. (2021). Cellular extrusion bioprinting improves kidney organoid reproducibility and conformation. *Nature Materials, 20*(2), 260–271. doi:10.103841563-020-00853-9 PMID:33230326

Li, X., Liu, B., Pei, B., Chen, J., Zhou, D., Peng, J., Zhang, X., Jia, W., & Xu, T. (2020). Inkjet Bioprinting of Biomaterials. *Chemical Reviews, 120*(19), 10793–10833. doi:10.1021/acs.chemrev.0c00008 PMID:32902959

Madhu, G., Govardhan, A., Ravi, V., Kautish, S., Srinivas, B. S., Chaudhary, T., & Kumar, M. (2022). DSCN-net: A deep Siamese capsule neural network model for automatic diagnosis of malaria parasites detection. *Multimedia Tools and Applications, 81*(23), 34105–34127. doi:10.100711042-022-13008-6

Magalhães, L. S. S. M., Santos, F. E. P., de Maria Vaz Elias, C., Afewerki, S., Sousa, G. F., Furtado, A. S. A., Marciano, F. R., & Lobo, A. O. (2020). Printing 3D hydrogel structures employing low-cost stereolithography technology. *Journal of Functional Biomaterials, 11*(1), 12. doi:10.3390/jfb11010012 PMID:32098431

Mahmood, M. A., Visan, A. I., Ristoscu, C., & Mihailescu, I. N. (2021). Artificial neural network algorithms for 3D printing. In Materials (Vol. 14, pp. 1–29). doi:10.3390/ma14010163

Moon, S., Hasan, S. K., Song, Y. S., Xu, F., Keles, H. O., Manzur, F., Mikkilineni, S., Hong, J. W., Nagatomi, J., Haeggstrom, E., Khademhosseini, A., & Demirci, U. (2010). Layer by layer three-dimensional tissue epitaxy by cell-laden hydrogel droplets. *Tissue Engineering. Part C, Methods, 16*(1), 157–166. doi:10.1089/ten.tec.2009.0179 PMID:19586367

Oladapo, B. I., Zahedi, S. A., & Adeoye, A. O. M. (2019). 3D printing of bone scaffolds with hybrid biomaterials. *Composites. Part B, Engineering, 158*, 428–436. doi:10.1016/j.compositesb.2018.09.065

Özcan, M., Hotza, D., Fredel, M. C., Cruz, A., & Volpato, C. A. M. (2021). Materials and manufacturing techniques for polymeric and ceramic scaffolds used in implant dentistry. *Journal of Composites Science*, *5*(3), 78. doi:10.3390/jcs5030078

Palaniappan, M., Tirlangi, S., Mohamed, M. J. S., Moorthy, R. M. S., Valeti, S. V., & Boopathi, S. (2023). Fused Deposition Modelling of Polylactic Acid (PLA)-Based Polymer Composites: A Case Study. In Development, Properties, and Industrial Applications of 3D Printed Polymer Composites (pp. 66–85). IGI Global.

S., P. K., Sampath, B., R., S. K., Babu, B. H., & N., A. (2022). Hydroponics, Aeroponics, and Aquaponics Technologies in Modern Agricultural Cultivation. In *Trends, Paradigms, and Advances in Mechatronics Engineering* (pp. 223–241). IGI Global. doi:10.4018/978-1-6684-5887-7.ch012

Sampath, B., & Myilsamy, S. (2021). Experimental investigation of a cryogenically cooled oxygen-mist near-dry wire-cut electrical discharge machining process. *Strojniski Vestnik. Jixie Gongcheng Xuebao*, *67*(6), 322–330. doi:10.5545v-jme.2021.7161

Sampath, B., Naveenkumar, N., Sampathkumar, P., Silambarasan, P., Venkadesh, A., & Sakthivel, M. (2021). Experimental comparative study of banana fiber composite with glass fiber composite material using Taguchi method. *Materials Today: Proceedings*, *49*(5), 1475–1480. doi:10.1016/j.matpr.2021.07.232

Sampath, B. C. S., & Myilsamy, S. (2022). Application of TOPSIS Optimization Technique in the Micro-Machining Process. In Trends, Paradigms, and Advances in Mechatronics Engineering (pp. 162–187). IGI Global. doi:10.4018/978-1-6684-5887-7.ch009

Sampathkumar, A., Rastogi, R., Arukonda, S., Shankar, A., Kautish, S., & Sivaram, M. (2020). An efficient hybrid methodology for detection of cancer-causing gene using CSC for micro array data. *Journal of Ambient Intelligence and Humanized Computing*, *11*(11), 4743–4751. doi:10.100712652-020-01731-7

Saravanan, M., Vasanth, M., Boopathi, S., Sureshkumar, M., & Haribalaji, V. (2022). Optimization of Quench Polish Quench (QPQ) Coating Process Using Taguchi Method. *Key Engineering Materials*, *935*, 83–91. doi:10.4028/p-z569vy

Senthil, T. S., Puviyarasan, M., Babu, S. R., Surakasi, R., Sampath, B., & Associates. (2023). Industrial Robot-Integrated Fused Deposition Modelling for the 3D Printing Process. In Development, Properties, and Industrial Applications of 3D Printed Polymer Composites (pp. 188–210). IGI Global.

Top, N., Şahin, İ., Gökçe, H., & Gökçe, H. (2021). Computer-aided design and additive manufacturing of bone scaffolds for tissue engineering: State of the art. *Journal of Materials Research*, *36*(19), 3725–3745. doi:10.155743578-021-00156-y

Trojovský, P., Dhasarathan, V., & Boopathi, S. (2023). Experimental investigations on cryogenic friction-stir welding of similar ZE42 magnesium alloys. *Alexandria Engineering Journal*, *66*(1), 1–14. doi:10.1016/j.aej.2022.12.007

Van Noort, R. (2012). The future of dental devices is digital. *Dental Materials*, *28*(1), 3–12. doi:10.1016/j.dental.2011.10.014 PMID:22119539

Xu, F., Celli, J., Rizvi, I., Moon, S., Hasan, T., & Demirci, U. (2011). A three-dimensional in vitro ovarian cancer coculture model using a high-throughput cell patterning platform. *Biotechnology Journal*, *6*(2), 204–212. doi:10.1002/biot.201000340 PMID:21298805

Yan, Q., Dong, H., Su, J., Han, J., Song, B., Wei, Q., & Shi, Y. (2018). A Review of 3D Printing Technology for Medical Applications. *Engineering*, *4*(5), 729–742. doi:10.1016/j.eng.2018.07.021

Yupapin, P., Trabelsi, Y., Nattappan, A., & Boopathi, S. (2022). Performance Improvement of Wire-Cut Electrical Discharge Machining Process Using Cryogenically Treated Super-Conductive State of Monel-K500 Alloy. *Iranian Journal of Science and Technology - Transactions of Mechanical Engineering*, 1–17. doi:10.1007/s40997-022-00513-0

# Compilation of References

Abcouwer, N., Daftry, S., del Sesto, T., Toupet, O., Ono, M., Venkatraman, S., Lanka, R., Song, J., & Yue, Y. (2021). Machine Learning Based Path Planning for Improved Rover Navigation. *2021 IEEE Aerospace Conference* (50100), (pp. 1–9). IEEE. 10.1109/AERO50100.2021.9438337

*About Joseph Engelberger - Father of Robotics.* (n.d.). Automate. https://www.automate.org/a3-content/joseph-engelberger-about

Aceto, G., Persico, V., & Pescapé, A. (2019). A survey on information and communication technologies for industry 4.0: State-of-the-art, taxonomies, perspectives, and challenges. *IEEE Communications Surveys and Tutorials, 21*(4), 3467–3501. doi:10.1109/COMST.2019.2938259

Agrawal, R. (2021). The future of space robots. *International Journal of Advance Research, Ideas and Innovations In Technology, 7*(3), 2429–2435. https://www.ijariit.com/manuscripts/v7i3/V7I3-2141.pdf

Agrawal, S. (2018). *The encyclopedia of medical robotics.* World Scientific. doi:10.1142/10770

Aguida, M. A., Ouchani, S., & Benmalek, M. (2021). An IoT-based Framework for an Optimal Monitoring and Control of Cyber-Physical Systems: Application on Biogas Production System. In *ACM International Conference Proceeding Series* (Vol. 1, Issue 1). Association for Computing Machinery. 10.1145/3494322.3494341

Ahmed, N., De, D., & Hussain, I. (2018). Internet of Things (IoT) for Smart Precision Agriculture and Farming in Rural Areas. *IEEE Internet of Things Journal, 5*(6), 4890–4899. doi:10.1109/JIOT.2018.2879579

Ahn, H. S., Yep, W., Lim, J., Ahn, B. K., Johanson, D. L., Hwang, E. J., Lee, M. H., Broadbent, E., & MacDonald, B. A. (2019). Hospital Receptionist Robot v2: Design for Enhancing Verbal Interaction with Social Skills. In *Proceedings of the 2019 28th IEEE International Conference on Robot and Human Interactive Communication (RO-MAN),* New Delhi, India. 10.1109/RO-MAN46459.2019.8956300

Akter, M., Zohra, F. T., & Das, A. K. (2017). Q-MAC: QoS and mobility aware optimal resource allocation for dynamic application offloading in mobile cloud computing. *ECCE 2017 - International Conference on Electrical, Computer and Communication Engineering,* (pp. 803–808). IEEE. 10.1109/ECACE.2017.7913013

Alavi, M., & Leidner, D. E. (2001). Review: Knowledge management and knowledge management systems: Conceptual foundations and research issues. *Management Information Systems Quarterly*, *25*(1), 107–136. doi:10.2307/3250961

Alemi, M., Meghdari, A., Ghanbarzadeh, A., Moghadam, L. J., & Ghanbarzadeh, A. (2014). Effect of Utilizing a Humanoid Robot as a Therapy-Assistant in Reducing Anger, Anxiety, and Depression. In *Proceedings of the 2014 Second RSI/ISM International Conference on Robotics and Mechatronics,* Tehran, Iran. 10.1109/ICRoM.2014.6990993

Alimo, R., Sam, D., Lakhmiri, D., Kahovec, B., & Divsalar, D. (2021). Automated data accountability for missions in Mars rover data. *2021 IEEE Aerospace Conference* (50100), (pp. 1–8). IEEE. 10.1109/AERO50100.2021.9438220

AlkaliY.RoutrayI.WhigP. (2022). Study of various methods for reliable, efficient and Secured IoT using Artificial Intelligence. SSRN 4020364. doi:10.2139/ssrn.4020364

Alzubi, J., Nayyar, A., & Kumar, A. (2018). Machine Learning from Theory to Algorithms: An Overview. *Journal of Physics: Conference Series*, *1142*(1), 12012. doi:10.1088/1742-6596/1142/1/012012

Anand, M., Velu, A., & Whig, P. (2022). Prediction of Loan Behaviour with Machine Learning Models for Secure Banking [JCSE]. *Journal of Computing Science and Engineering : JCSE*, *3*(1), 1–13.

*Apollo 11:Destination Sea of Tranquility - 1st Lunar Landing*. (n.d.). DLRARTICLE DLR. https://www.dlr.de/content/en/articles/moon-special/the-apollo-program/apollo-11.html

Arrouch, I., Ahmad, N. S., Goh, P., & Mohamad-Saleh, J. (2022). Close Proximity Time-to-collision Prediction for Autonomous Robot Navigation: An Exponential GPR Approach. *Alexandria Engineering Journal*, *61*(12), 11171–11183. doi:10.1016/j.aej.2022.04.041

Arumugam, R., Enti, V. R., Bingbing, L., Xiaojun, W., Baskaran, K., Kong, F. F., Kumar, A. S., Meng, K. D., & Kit, G. W. (2010). DAvinCi: A cloud computing framework for service robots. *Proceedings - IEEE International Conference on Robotics and Automation,* (pp. 3084–3089). IEEE. 10.1109/ROBOT.2010.5509469

AZoRobotics. (2013, June 6). *Robots in Space*. AZoRobotics.com. https://www.azorobotics.com/Article.aspx?ArticleID=132

Babaiasl, M., Hamed, S., Poorya, J., & Mojtaba, Y. (2016). A review of technological and clinical aspects of robotaided rehabilitation of upper-extremity after stroke. *Disabil Rehabil Assist Technol Taylor & Francis*, *11*(4), 263–280. doi:10.3109/17483107.2014.1002539 PMID:25600057

Babu, B. S., Kamalakannan, J., Meenatchi, N., Karthik, S., & Boopathi, S. (2022). Economic impacts and reliability evaluation of battery by adopting Electric Vehicle. *IEEE Explore*, (pp. 1–6).

Babu, B. S., Kamalakannan, J., Meenatchi, N., Karthik, S., & Boopathi, S. (2022). Economic impacts and reliability evaluation of battery by adopting Electric Vehicle. *IEEE Explore*, (pp. 1–6). IEEE.

Badekila, A. K., Kini, S., & Jaiswal, A. K. (2021). Fabrication techniques of biomimetic scaffolds in three-dimensional cell culture: A review. *Journal of Cellular Physiology*, *236*(2), 741–762. doi:10.1002/jcp.29935 PMID:32657458

Bagha, H., Yavari, A., & Georgakopoulos, D. (2022). Hybrid Sensing Platform for IoT-Based Precision Agriculture. *Future Internet*, *14*(8), 233. doi:10.3390/fi14080233

Banthia, V., Maddahi, Y., May, M., Blakley, D., Chang, Z., Gbur, A., Tu, C., & Sepehri, N. (2016). *Development of a Graphical User Interface for a Socially Interactive Robot: A Case Study Evaluation*. IEEE.

Barreiro, C. A., & Treglown, L. (2020). What makes an engaged employee? A facet-level approach to trait emotional intelligence as a predictor of employee engagement. *Personality and Individual Differences*, *159*, 109892. doi:10.1016/j.paid.2020.109892

Bartlett, M. E., Costescu, C., Baxter, P., & Thill, S. (2020). Requirements for Robotic Interpretation of Social Signals "in the Wild": Insights from Diagnostic Criteria of Autism Spectrum Disorder. *Information (Basel)*, *11*(2), 81. doi:10.3390/info11020081

Belda-Lois, J. M., Mena-del Horno, S., Bermejo-Bosch, I., Moreno, J. C., Pons, J. L., Farina, D., Iosa, M., Molinari, M., Tamburella, F., Ramos, A., Caria, A., Solis-Escalante, T., Brunner, C., & Rea, M. (2011). Rehabilitation of gait after stroke: A review towards a top-down approach. *J Neuroeng Rehabil BioMed Central Ltd*, *8*(1), 66. doi:10.1186/1743-0003-8-66 PMID:22165907

Bellegarda, G., & Nguyen, Q. (2021). Robust High-speed Running for Quadruped Robots via Deep Reinforcement Learning. arXiv:2103.06484

Ben-Ari, M., & Mondada, F. (2017). Robots and Their Applications. *Elements of Robotics*, 1–20. doi:10.1007/978-3-319-62533-1_1

Bengio, Y. (2012, June). *Deep learning of representations for unsupervised and transfer learning*. In *Proceedings of ICML workshop on unsupervised and transfer learning* (pp. 17-36). JMLR.

Benitez-Quiroz, C. F., Srinivasan, R., Feng, Q., Wang, Y., & Martinez, A. M. (2017). Emotionet challenge: Recognition of facial expressions of emotion in the wild. arXiv preprint arXiv:1703.01210

Beraldo, G., Menegatti, E., Tommasi, V. D., Mancin, R., & Benini, F. (2019). A Preliminary Investigation of Using Humanoid Social Robots as Non-Pharmacological Techniques with Children. In *Proceedings of the IEEE International Conference on Advanced Robotics and Its Social Impacts (ARSO)*, Beijing, China. 10.1109/ARSO46408.2019.8948760

Bhatia, A. B., & Mittal, K. (2022). Measuring Computational Psychometrics Analysis Motivational Level in Learner's using Different Parameters through Deep Learning Algorithm. [JCSE]. *Journal of Computing Science and Engineering : JCSE*, *3*(2), 120–136.

Bhat, M. A., Adil, A. W., Sikander, B. M., Lone, Y., & Malik, J. A. (2019). Waste Management Technology for Sustainable Agriculture. In *Waste Management Technology for Sustainable Agriculture* (pp. 156–176). Issue September. doi:10.4018/978-1-7998-0031-6.ch009

Bhatnagar, A., & Mittal, K. (2022). Computational Psychometrics Analysis of Learners' *Motivational Level Using Different Parameters. In International Conference on Innovative Computing and Communications* (pp. 493-507). Springer, Singapore.

Bhushan, M., & Goel, S. (2016). Improving software product line using an ontological approach. *Sadhana*, *41*(12), 1381–1391. doi:10.100712046-016-0571-y

Bhushan, M., Goel, S., & Kaur, K. (2018). Analyzing inconsistencies in software product lines using an ontological rule-based approach. *Journal of Systems and Software*, *137*, 605–617. doi:10.1016/j.jss.2017.06.002

Bhushan, M., Goel, S., & Kumar, A. (2018). Improving quality of software product line by analysing inconsistencies in feature models using an ontological rule-based approach. *Expert Systems: International Journal of Knowledge Engineering and Neural Networks*, *35*(3), e12256. doi:10.1111/exsy.12256

Bhushan, M., Goel, S., Kumar, A., & Negi, A. (2017). Managing software product line using an ontological rule-based framework. *2017 International Conference on Infocom Technologies and Unmanned Systems (Trends and Future Directions) (ICTUS)*, (pp. 376–382). IEEE. 10.1109/ICTUS.2017.8286036

Bledt, G., Powell, M. J., Katz, B., & Di Carlo, J. (2018). MIT cheetah 3: Design and control of a robust, dynamic quadruped robot. *International Conference on Intelligent Robots and Systems (IROS)*. IEEE.

Bliley Technologies. (2019). *How are Robots Used in Space Exploration?* Bliley.com. https://blog.bliley.com/robots-used-in-space-exploration

Boopathi, S. (2021). Improving of Green Sand-Mould Quality using Taguchi Technique. *Journal of Engineering Research*. doi:10.36909/jer.14079

Boopathi, S., & Khare, R. K. G., J. C., Muni, T. V., & Khare, S. (2023). Additive Manufacturing Developments in the Medical Engineering Field. In Development, Properties, and Industrial Applications of 3D Printed Polymer Composites (pp. 86–106). IGI Global.

Boopathi, S., & Khare, R. KG, J. C., Muni, T. V., & Khare, S. (2023). Additive Manufacturing Developments in the Medical Engineering Field. In Development, Properties, and Industrial Applications of 3D Printed Polymer Composites (pp. 86–106). IGI Global.

Boopathi, S., & Myilsamy, S. (2021). Material removal rate and surface roughness study on Near-dry wire electrical discharge Machining process. *Materials Today: Proceedings*, *45*, 8149–8156. doi:10.1016/j.matpr.2021.02.267

Boopathi, S., Arigela, S. H., Raman, R., Indhumathi, C., Kavitha, V., & Bhatt, B. C. (2022). Prominent Rule Control-based Internet of Things: Poultry Farm Management System. *IEEE Explore*, (pp. 1–6).

Boopathi, S., Arigela, S. H., Raman, R., Indhumathi, C., Kavitha, V., & Bhatt, B. C. (2022). Prominent Rule Control-based Internet of Things: Poultry Farm Management System. *IEEE Explore*, (pp. 1–6). IEEE.

Boopathi, S., Siva Kumar, P. K., & Meena, R. S. J., S. I., P., S. K., & Sudhakar, M. (2023). Sustainable Developments of Modern Soil-Less Agro-Cultivation Systems. In Human Agro-Energy Optimization for Business and Industry (pp. 69–87). IGI Global. doi:10.4018/978-1-6684-4118-3.ch004

Boopathi, S. (2019). Experimental investigation and parameter analysis of LPG refrigeration system using Taguchi method. *SN Applied Sciences*, *1*(8), 892. doi:10.100742452-019-0925-2

Boopathi, S. (2022a). An experimental investigation of Quench Polish Quench (QPQ) coating on AISI 4150 steel. *Engineering Research Express*, *4*(4), 045009. doi:10.1088/2631-8695/ac9ddd

Boopathi, S. (2022a). An Extensive Review on Sustainable Developments of Dry and Near-Dry Electrical Discharge Machining Processes. *Journal of Manufacturing Science and Engineering*, *144*(5), 1–37. doi:10.1115/1.4052527

Boopathi, S. (2022b). An investigation on gas emission concentration and relative emission rate of the near-dry wire-cut electrical discharge machining process. *Environmental Science and Pollution Research International*, *29*(57), 86237–86246. doi:10.100711356-021-17658-1 PMID:34837614

Boopathi, S. (2022c). Experimental investigation and multi-objective optimization of cryogenic Friction-stir-welding of AA2014 and AZ31B alloys using MOORA technique. *Materials Today. Communications*, *33*, 104937. doi:10.1016/j.mtcomm.2022.104937

Boopathi, S. (2022d). (in press). Cryogenically treated and untreated stainless steel grade 317 in sustainable wire electrical discharge machining process: A comparative study. *Environmental Science and Pollution Research International*, 1–10. doi:10.100711356-022-22843-x PMID:36057706

Boopathi, S. (2022e). Performance Improvement of Eco-Friendly Near-Dry Wire-Cut Electrical Discharge Machining Process Using Coconut Oil-Mist Dielectric Fluid. *Journal of Advanced Manufacturing Systems*, 1–20. doi:10.1142/S0219686723500178

Boopathi, S. (2023). An Investigation on Friction Stir Processing of Aluminum Alloy-Boron Carbide Surface Composite. In *Materials Horizons: From Nature to Nanomaterials* (pp. 249–257). Springer., doi:10.1007/978-981-19-7146-4_14

Boopathi, S., Balasubramani, V., Kumar, R. S., & Singh, G. R. (2021). The influence of human hair on kenaf and Grewia fiber-based hybrid natural composite material: An experimental study. *Functional Composites and Structures*, *3*(4), 045011–045021. doi:10.1088/2631-6331/ac3afc

Boopathi, S., Balasubramani, V., & Sanjeev Kumar, R. (2023). Influences of various natural fibers on the mechanical and drilling characteristics of coir-fiber-based hybrid epoxy composites. *Engineering Research Express*, *5*(1), 15002–15020. doi:10.1088/2631-8695/acb132

Boopathi, S., Haribalaji, V., Mageswari, M., & Asif, M. M. (2022). Influences of Boron Carbide Particles on the Wear Rate and Tensile Strength of Aa2014 Surface Composite Fabricated By Friction-Stir Processing. *Materiali in Tehnologije*, *56*(3), 263–270. doi:10.17222/mit.2022.409

Boopathi, S., Jeyakumar, M., Singh, G. R., King, F. L., Pandian, M., Subbiah, R., & Haribalaji, V. (2022). An experimental study on friction stir processing of aluminium alloy (AA-2024) and boron nitride (BNp) surface composite. *Materials Today: Proceedings*, *59*, 1094–1099. doi:10.1016/j.matpr.2022.02.435

Boopathi, S., Lewise, K. A. S., Subbiah, R., & Sivaraman, G. (2021). Near-dry wire-cut electrical discharge machining process using water-air-mist dielectric fluid: An experimental study. *Materials Today: Proceedings*, *49*(5), 1885–1890. doi:10.1016/j.matpr.2021.08.077

Boopathi, S., Myilsamy, S., & Sukkasamy, S. (2021). (Preprint). Experimental Investigation and Multi-Objective Optimization of Cryogenically Cooled Near-Dry Wire-Cut EDM Using TOPSIS Technique. *Ijamt*. Advance online publication. doi:10.21203/rs.3.rs-254117/v1

Boopathi, S., & Sivakumar, K. (2013). Experimental investigation and parameter optimization of near-dry wire-cut electrical discharge machining using multi-objective evolutionary algorithm. *International Journal of Advanced Manufacturing Technology*, *67*(9–12), 2639–2655. doi:10.100700170-012-4680-4

Boopathi, S., & Sivakumar, K. (2014). Study of water assisted dry wire-cut electrical discharge machining. *Indian Journal of Engineering and Materials Sciences*, *21*(1), 75–82.

Boopathi, S., Sivakumar, K., & Kalidas, R. (2012). Parametric Study of Dry WEDM Using Taguchi Method. *International Journal of Engineering Research and Development*, *2*(4), 1–6.

Boopathi, S., Thillaivanan, A., Azeem, M. A., Shanmugam, P., & Pramod, V. R. (2022). Experimental investigation on abrasive water jet machining of neem wood plastic composite. *Functional Composites and Structures*, *4*(2), 025001. doi:10.1088/2631-6331/ac6152

Boopathi, S., Thillaivanan, A., Pandian, M., Subbiah, R., & Shanmugam, P. (2021). Friction stir processing of boron carbide reinforced aluminium surface (Al-B4C) composite: Mechanical characteristics analysis. *Materials Today: Proceedings*, *50*(5), 2430–2435. doi:10.1016/j.matpr.2021.10.261

Boopathi, S., Venkatesan, G., & Anton Savio Lewise, K. (2023). Mechanical Properties Analysis of Kenaf–Grewia–Hair Fiber-Reinforced Composite. In *Lecture Notes in Mechanical Engineering* (pp. 101–110). Springer. doi:10.1007/978-981-16-9057-0_11

Bossone, E., Ranieri, B., Coscioni, E., & Baliga, R. R. (2019). Community health and prevention: It takes a village to reduce cardiovascular risk! Let us do it together! *European Journal of Preventive Cardiology*, *26*(17), 1840–1842. doi:10.1177/2047487319867505 PMID:31382807

Broekens, J., Heerink, M., & Rosendal, H. (2009). Assistive social robots in elderly care: A review. *Gerontechnology (Valkenswaard)*, *8*(2). doi:10.4017/gt.2009.08.02.002.00

Builtin. (2019). *What Is Robotics? Types Of Robots*. Built In. https://builtin.com/robotics

Burns, R., Jeon, M., & Park, C. H. (2018). Robotic Motion Learning Framework to Promote Social Engagement. *Applied Sciences (Basel, Switzerland), 8*(2), 241. doi:10.3390/app8020241 PMID:35582004

Bustamante Valles, K., Montes, S., Madrigal, M. J., Burciaga, A., Martínez, M. E., & Johnson, M. J. (2016). Technologyassisted stroke rehabilitation in Mexico: A pilot randomized trial comparing traditional therapy to circuit training in a robot/technology-assisted therapy gym. *Journal of Neuroengineering and Rehabilitation, 13*(1), 1–15. doi:10.118612984-016-0190-1 PMID:27634471

Caramihale, T., Popescu, D., & Ichim, L. (2018). Emotion classification using a tensorflow generative adversarial network implementation. *Symmetry, 10*(9), 414. doi:10.3390ym10090414

Carra, G., Argiolas, A., Bellissima, A., Niccolini, M., & Ragaglia, M. (2018). *Robotics in the construction industry: State of the art and future opportunities.* In ISARC. Proceedings of the International Symposium on Automation and Robotics in Construction (Vol. 35, pp. 1-8). IAARC Publications.

Carrillo, F. M. (2018) Physiotherapists' acceptance of a socially assistive robot in ongoing clinical deployment. In: *Proceedings of 27th IEEE international symposium on Robot and Human Interactive Communication, RO-MAN'18*, Nanjing China. 10.1109/ROMAN.2018.8525508

Chakravarthi, P. K., Yuvaraj, D., & Venkataramanan, V. (2022). IoT-based smart energy meter for smart grids. *ICDCS 2022 - 2022 6th International Conference on Devices, Circuits and Systems*, (pp. 360–363). 10.1109/ICDCS54290.2022.9780714

Chang, V., Baudier, P., Zhang, H., Xu, Q., Zhang, J., & Arami, M. (2020). How Blockchain can impact financial services–The overview, challenges and recommendations from expert interviewees. *Technological Forecasting and Social Change, 158*, 120166. doi:10.1016/j.techfore.2020.120166 PMID:32834134

Chaurasiya, P. K. (2020). A Detailed Study of Cattle Dung based Biogas Plant for Cooking Fuel and Organic Fertilizer. *September.* doi:10.13140/RG.2.2.18496.97282

Chen, C., Garrod, O. G. B., Zhan, J., Beskow, J., Schyns, P. G., & Jack, R. E. (2018). Reverse Engineering Psychologically Valid Facial Expressions of Emotion into Social Robots. In Proceedings of *the 2018 13th IEEE International Conference on Automatic Face & Gesture Recognition,* Xi'an, China. 10.1109/FG.2018.00072

Chen, G., & Lee, T. H. (2019). Predictive validity of self-efficacy and academic goal orientation in online learning. *Computers & Education, 128*, 296–309.

Chertok, B., & Siddiqi, A. (n.d.). *Rockets and People The Moon Race.* NASA. https://www.nasa.gov/pdf/621513main_RocketsPeopleVolume4-ebook.pdf

Chiacchio, F., Petropoulos, G., & Pichler, D. (2018). *The impact of industrial robots on EU employment and wages: A local labour market approach (No. 2018/02).* Bruegel working paper.

Chiodini, S., Torresin, L., Pertile, M., & Debei, S. (2020). Evaluation of 3D CNN semantic mapping for rover navigation. *2020 IEEE 7th International Workshop on Metrology for AeroSpace (MetroAeroSpace)*, (pp. 32–36). IEEE.

Chollet, F. (2017). Xception: Deep learning with depthwise separable convolutions. In Proceedings of the IEEE conference on computer vision and pattern recognition (pp. 1251-1258). IEEE. arXiv:1610.02357v3 10.1109/CVPR.2017.195

Chou, C. C., & Tsai, C. C. (2018). Effects of personalized game-based instruction on students' learning performance and motivation in mathematics. *Journal of Educational Technology & Society*, *21*(3), 42–54.

Chung, K., & Park, R. C. (2019). Chatbot-Based Heathcare Service with a Knowledge Base for Cloud Computing. *Clust. Comput. J. Netw. Softw. Tools Appl.*, *22*(S1), 1925–1937. doi:10.100710586-018-2334-5

Coloma, S., Martinez, C., Yalçın, B. C., & Olivares-Mendez, M. A. (2022). Enhancing Rover Teleoperation on the Moon With Proprioceptive Sensors and Machine Learning Techniques. *IEEE Robotics and Automation Letters*, *7*(4), 11434–11441. doi:10.1109/LRA.2022.3198794

Comparetti, A. (2011). Precision Agriculture: Past, Present and Future. *Agricultural Engineering and Environment*, *16*, 17.

Cooper, S., Fava, A. D., Vivas, C., Marchionni, L., & Ferro, F. (2020). ARI: The Social Assistive Robot and Companion. In *Proceedings of the 2020 29th IEEE International Conference on Robot and Human Interactive Communication (RO-MAN)*, Naples, Italy. 10.1109/RO-MAN47096.2020.9223470

Cresswell, K., Cunningham-Burley, S., & Sheikh, A. (2018). Health Care Robotics: Qualitative Exploration of Key Challenges and Future Directions. *Journal of Medical Internet Research*, *20*(7), e10410. doi:10.2196/10410 PMID:29973336

Daftry, S., Abcouwer, N., Sesto, T. D., Venkatraman, S., Song, J., Igel, L., Byon, A., Rosolia, U., Yue, Y., & Ono, M. (2022). MLNav: Learning to Safely Navigate on Martian Terrains. *IEEE Robotics and Automation Letters*, *7*(2), 5461–5468. doi:10.1109/LRA.2022.3156654

Darwin, C., & Prodger, P. (1998). *The expression of the emotions in man and animals*. Oxford University Press.

Datta, P., Dey, M., Ataie, Z., Unutmaz, D., & Ozbolat, I. T. (2020). 3D bioprinting for reconstituting the cancer microenvironment. *NPJ Precision Oncology*, *4*(1), 1–13. doi:10.103841698-020-0121-2 PMID:32793806

Davenport, T., Guha, A., Grewal, D., & Bressgott, T. (2019). How Artificial Intelligence Will Change the Future of Marketing. *Journal of the Academy of Marketing Science, 48*(1), 24–42. Springer. doi:10.1007/s11747-019-00696-0 de Kervenoael, R., Hasan, R., Schwob, A., & Goh, E. (2020). Leveraging human-robot interaction in hospitality services: Incorporating the role of perceived value, empathy, and information sharing into visitors' intentions to use social robots. *Tourism Management*, *78*, 104042. doi:10.1016/j.tourman.2019.104042

DeFazio, D., & Zhang, S. (2021). Leveraging Human Knowledge to Learn Quadruped Locomotion Policies. arXiv:2107.10969

Demir, K. A., Döven, G., & Sezen, B. (2019). Industry 5.0 and human-robot co-working. *Procedia Computer Science*, *158*, 688–695. doi:10.1016/j.procs.2019.09.104

Derby, B. (2012). Printing and prototyping of tissues and scaffolds. *Science*, *338*(6109), 921–926. doi:10.1126cience.1226340 PMID:23161993

Desai, R. (2022, August 29). *Robot.* Dr Rajiv Desai. https://drrajivdesaimd.com/2022/08/29/robot/

Dhall, R., & Agrawal, H. (2018). An improved energy efficient duty cycling algorithm for IoT based precision agriculture. *Procedia Computer Science*, *141*, 135–142. doi:10.1016/j.procs.2018.10.159

Diaz-Boladeras, M., Angulo, C., Domenech, M., Albo-Canals, J., Serrallonga, N., Raya, C., & Barco, A. (2016). Assessing Pediatrics Patients' Psychological States from Biomedical Signals in a Cloud of Social Robots. In *Proceedings of the XIV Mediterranean Conference on Medical and Biological Engineering and Computing,* (pp. 1179–1184). Springer. 10.1007/978-3-319-32703-7_230

Dinh Thai Bao. (2019). *ROBOT.* Academia.edu. https://www.academia.edu/19892932/ROBOT

Dodds, P., Martyn, K., & Brown, M. (2018). Infection Prevention and Control Challenges of Using a Therapeutic Robot. *Nursing Older People*, *30*(3), 34–40. doi:10.7748/nop.2018.e994 PMID:29569863

Domakonda, V. K., Farooq, S., Chinthamreddy, S., Puviarasi, R., Sudhakar, M., & Boopathi, S. (2022). Sustainable Developments of Hybrid Floating Solar Power Plants: Photovoltaic System. In Human Agro-Energy Optimization for Business and Industry (pp. 148–167). IGI Global.

Drachsler, H., & Goldhammer, F. (2020). Learning Analytics and e-assessment—towards computational psychometrics by combining psychometrics with learning analytics. In *Radical Solutions and Learning Analytics* (pp. 67–80). Springer. doi:10.1007/978-981-15-4526-9_5

Duan, Y., & Chen, X. (2016). *Rein Houthooft, John Schulman, and Pieter Abbeel. "Benchmarking deep reinforcement learning for continuous control.* Proceedings of the 33rd International Conference on International Conference on Machine Learning (pp. 1329–1338). *JMLR.org*, .

Dunjko, V., & Briegel, H. J. (2018). Machine learning & artificial intelligence in the quantum domain: A review of recent progress. *Reports on Progress in Physics*, *81*(7), 074001. doi:10.1088/1361-6633/aab406 PMID:29504942

Duo, N., Wang, Q., Lv, Q., Wei, H., & Zhang, P. (2019). A deep reinforcement learning based mapless navigation algorithm using continuous actions. *2019 International Conference on Robots & Intelligent System (ICRIS)*, (pp. 63–68). IEEE. 10.1109/ICRIS.2019.00025

Du, S., & Martinez, A. M. (2022). Compound facial expressions of emotion: From basic research to clinical applications. *Dialogues in Clinical Neuroscience*. PMID:26869845

Du, S., Tao, Y., & Martinez, A. M. (2014). Compound facial expressions of emotion. *Proceedings of the National Academy of Sciences of the United States of America*, *111*(15), E1454–E1462. doi:10.1073/pnas.1322355111 PMID:24706770

Ekman, P., & Friesen, W. V. (1971). Constants across cultures in the face and emotion. *Journal of Personality and Social Psychology*, *17*(2), 124–129. doi:10.1037/h0030377 PMID:5542557

Ellitan, L. (2020). Competing in the era of industrial revolution 4.0 and society 5.0. *Jurnal Maksipreneur: Manajemen, Koperasi, dan Entrepreneurship, 10*(1), 1-12.

Eriksson, Y. (2018). The Perception of Aging and Use of Robots. In *International Conference on Human Aspects of IT for the Aged Population,* (pp. 30–39). Springer: Cham, Switzerland. 10.1007/978-3-319-92034-4_3

Fabian Benitez-Quiroz, C., Srinivasan, R., & Martinez, A. M. (2016). Emotionet: An accurate, real-time algorithm for the automatic annotation of a million facial expressions in the wild. In *Proceedings of the IEEE conference on computer vision and pattern recognition* (pp. 5562-5570). IEEE. 10.1109/CVPR.2016.600

Ferreira, I., & Sequeira, J. When Children Interact with Robots: Ethics in the Monarch Project; IEEE: Piscatway, NJ, USA, 2014; ISBN 978- 981-4623-34-6.

Fong, C. J., & Lam, H. C. (2020). Applying artificial intelligence in education: A review of emerging trends and issues. *Educational Technology Research and Development*, *68*(6), 2709–2730.

Furfaro, R., Scorsoglio, A., Linares, R., & Massari, M. (2020). Adaptive generalized ZEM-ZEV feedback guidance for planetary landing via a deep reinforcement learning approach. *Acta Astronautica*, *171*, 156–171. doi:10.1016/j.actaastro.2020.02.051

Gaudet, B., Linares, R., & Furfaro, R. (2020). Adaptive guidance and integrated navigation with reinforcement meta-learning. *Acta Astronautica*, *169*, 180–190. doi:10.1016/j.actaastro.2020.01.007

Getson, C., & Nejat, G. (2021). Socially Assistive Robots Helping Older Adults through the Pandemic and Life after COVID-19. *Robotics*, *10*(3), 106. doi:10.3390/robotics10030106

Gnjatovic, M. (2018) Pilot corpus of childrobot interaction in therapeutic settings. In: *Proceedings of 8th IEEE international conference on Cognitive Infocommunications, CogInfoCom 2018*, Debrecen, Hungary. 10.1109/CogInfoCom.2017.8268252

Goncalves, D., & Arsenio, A. (2015). *Human-Driven Multi-Robot Design Process for Social Interactions with Children on Complex Environments*. IEEE. doi:10.1109/ICARA.2015.7081127

Gonzalez, R., Chandler, S., & Apostolopoulos, D. (2019). Characterization of machine learning algorithms for slippage estimation in planetary exploration rovers. *Journal of Terramechanics*, *82*, 23–34. doi:10.1016/j.jterra.2018.12.001

Goodfellow, I. (2016). Nips 2016 tutorial: Generative adversarial networks. arXiv preprint arXiv:1701.00160

Gore, U. C. (n.d.). *Univerzitet Crne Gore - O Univerzitetu*. UCG - Univerzitet Crne Gore. https://www.ucg.ac.me/skladiste/blog_13268/objava_56689/fajlovi/1_1%20Introduction%20to%20Robotics.ppt

Goris, K., Saldien, J., Vanderniepen, I., & Lefeber, D. (2009). The Huggable Robot Probo, a Multi-Disciplinary Research Platform. In *International Conference on Research and Education in Robotics*. Springer. 10.1007/978-3-642-03558-6_4

Gottlieb, J., & Anderson, D. L. (n.d.). *Kinds of Robots*. Mind.ilstu.edu. https://mind.ilstu.edu/curriculum/medical_robotics/kinds_of_robots.html

Guizzo, E. (2011). *Types of Robots - ROBOTS: Your Guide to the World of Robotics*. @Robotsapp; Robots. https://robots.ieee.org/learn/types-of-robots/

Gunes, H., & Schuller, B. (2013). Categorical and dimensional affect analysis in continuous input: Current trends and future directions. *Image and Vision Computing*, *31*(2), 120–136. doi:10.1016/j.imavis.2012.06.016

Guo, J., Lei, Z., Wan, J., Avots, E., Hajarolasvadi, N., Knyazev, B., Kuharenko, A., Jacques Junior, J. C. S., Baro, X., Demirel, H., Escalera, S., Allik, J., & Anbarjafari, G. (2018). Dominant and complementary emotion recognition from still images of faces. *IEEE Access : Practical Innovations, Open Solutions*, *6*, 26391–26403. doi:10.1109/ACCESS.2018.2831927

Guo, Y., Mi, Z., Yang, Y., & Obaidat, M. S. (2018). An energy sensitive computation offloading strategy in cloud robotic network based on ga. *IEEE Systems Journal*, *13*(3), 3513–3523. doi:10.1109/JSYST.2018.2830395

Haarnoja, T., Ha, S., Zhou, A., Tan, J., Tucker, G., & Levine, S. (2018). Learning to walk via deep reinforcement learning," arXiv:1812.11103

Haarnoja, T., Zhou, A., Abbeel, P., & Levine, S. (2018). Soft actor-critic: Off-policy maximum entropy deep reinforcement learning with a stochastic actor. arXiv preprint arXiv:1801.01290

Hafner, R., Hertweck, T., Kloppner, P., Bloesch, M., Neunert, N., Wulfmeier, W., Tunyasuyunakool, S., Hees, N., & Riedmiller, M. (2020). Towards general and autonomous learning of core skills: A case study in locomotion. arXiv preprint arXiv:2008.12228

Han, J. (2020, January 23). *4 Ways Robots Will Improve Our Standard of Living*. MarTech Series. https://martechseries.com/mts-insights/guest-authors/4-ways-robots-will-improve-standard-living/ https://jpl.nasa.gov.

Haribalaji, V., Boopathi, S., & Asif, M. M. (2021a). Optimization of friction stir welding process to join dissimilar AA2014 and AA7075 aluminum alloys. *Materials Today: Proceedings*, *50*, 2227–2234. doi:10.1016/j.matpr.2021.09.499

Haribalaji, V., Boopathi, S., Asif, M. M., Jeyakumar, M., Subbiah, R., & Lewise, K. A. S. (2021). Influences of Friction stir tool parameters for joining two similar AZ61A alloy plates. *Materials Today: Proceedings*, *50*(5), 2547–2553. doi:10.1016/j.matpr.2021.12.074

Heideker, A., Ottolini, D., Zyrianoff, I., Neto, A. T., Salmon Cinotti, T., & Kamienski, C. (2020). IoT-based Measurement for Smart Agriculture. *2020 IEEE International Workshop on Metrology for Agriculture and Forestry, MetroAgriFor 2020 - Proceedings*, (pp. 68–72.). IEEE doi:10.1109/MetroAgriFor50201.2020.927754610.1109/MetroAgriFor50201.2020.9277546

Higa, S., Iwashita, Y., Otsu, K., Ono, M., Lamarre, O., Didier, A., & Hoffmann, M. (2019). Vision-based estimation of driving energy for planetary rovers using deep learning and terramechanics. *IEEE Robotics and Automation Letters*, *4*(4), 3876–3883. doi:10.1109/LRA.2019.2928765

*History Of Robots Timeline* . (n.d.). Preceden. [REMOVED HYPERLINK FIELD]https://www.preceden.com/timelines/625227-history-of-robots

Holroyd, C. (2022). Technological innovation and building a 'super smart'society: Japan's vision of society 5.0. *Journal of Asian Public Policy*, *15*(1), 18–31. doi:10.1080/17516234.2020.1749340

Howard, A. G., Zhu, M., Chen, B., Kalenichenko, D., Wang, W., Weyand, T., Andreetto, M., & Adam, H. (2017). Mobilenets: Efficient convolutional neural networks for mobile vision applications. arXiv preprint arXiv:1704.04861

Hsu, Y. C., & Lin, J. F. (2018). Using educational data mining to improve student learning: A literature review. *Journal of Educational Technology & Society*, *21*(4), 74–86.

Hua, L., Lei, T., Qian, H., Zhang, Y., Hu, Y., & Lei, P. (2021). 3D-printed porous tantalum: Recent application in various drug delivery systems to repair hard tissue defects. *Expert Opinion on Drug Delivery*, *18*(5), 625–634. doi:10.1080/17425247.2021.1860015 PMID:33270470

Huang, R. H., Liu, D. J., & Tlili, A. (2019). Effects of integrating learning analytics into a flipped classroom on students' learning outcomes and satisfaction. *Internet and Higher Education*, *43*, 100705.

Hung, L., Gregorio, M., Mann, J., Wallsworth, C., Horne, N., Berndt, A., & Liu, C. (2019). Exploring the Perceptions of People with Dementia about the Social Robot PARO in a Hospital Setting. Dement. *Int. J. Soc. Res. Pract.*

Hutter, M., Gehring, C., Jud, D., Lauber, A., Bellicoso, C. D., Tsounis, V., Hwangbo, J., Bodie, K., Fankhauser, P., & Bloesch, M. (2016). *Anymal-a highly mobile and dynamic quadrupedal robot. International Conference on Intelligent Robots and Systems (IROS)*. IEEE. 10.1109/IROS.2016.7758092

Infineon. (n.d.). *Fundamentals of robotics*. Infineon Technologies. Www.infineon.com. https://www.infineon.com/cms/en/discoveries/fundamentals-robotics/

Jain, D., Iscen, A., & Caluwaerts, K. (2019 ). Hierarchical Reinforcement Learning for Quadruped Locomotion. *IEEE/RSJ International Conference on Intelligent Robots and Systems (IROS),* Macau.

Jannat, U. K., Kumar, M. M., & Islam, S. A. (2022). *A Case Study On The Cybersecurity Of Home Appliance Robots*. AGPE.

Jarraya, S. K., Masmoudi, M., & Hammami, M. (2020). Compound Emotion Recognition of Autistic Children During Meltdown Crisis Based on Deep Spatio-Temporal Analysis of Facial Geometric Features. *IEEE Access : Practical Innovations, Open Solutions*, *8*, 69311–69326. doi:10.1109/ACCESS.2020.2986654

Javaid, M., Haleem, A., Singh, R. P., & Suman, R. (2021). *Substantial capabilities of robotics in enhancing industry 4.0 implementation. Cognitive Robotics, 1, 58–75.* Sciencedirect. doi:10.1016/j.cogr.2021.06.001

Jeevanantham, Y. A., Saravanan, A., Vanitha, V., Boopathi, S., & Kumar, D. P. (2022). Implementation of Internet-of Things (IoT) in Soil Irrigation System. *IEEE Explore*, (pp. 1–5).

Jeevanantham, Y. A., Saravanan, A., Vanitha, V., Boopathi, S., & Kumar, D. P. (2022). Implementation of Internet-of Things (IoT) in Soil Irrigation System. *IEEE Explore*, (pp. 1–5). IEEE.

Jeon, H., Oh, H., & Lee, J. (2018). *Machine learning-based fast reading algorithm for future ICT-based education.* In: *2018 International Conference on Information and Communication Technology Convergence (ICTC)* (pp 771–775). . IEEE. 10.1109/ICTC.2018.8539447

Jeyakymar, M., Munusami, V., Shanmugam, P., Boopalan, N., Boopathi, S., & Sureshkumar, M. (2022). An investigation on wear loss and hardness of nitro-carburizing coated stainless-steel grade-316. *Materials Today: Proceedings*, *66*, 1398–1404. doi:10.1016/j.matpr.2022.05.223

Jiang, J., Zeng, X., Guzzetti, D., & You, Y. (2020). Path planning for asteroid hopping rovers with pre-trained deep reinforcement learning architectures. *Acta Astronautica*, *171*, 265–279. doi:10.1016/j.actaastro.2020.03.007

Jiao, Z., Wang, C., Yang, Q., & Wang, X. (2018). Preparation and characterization of UV-curable urethane acrylate oligomers modified with cycloaliphatic epoxide resin. *Journal of Coatings Technology and Research*, *15*(2), 251–258. doi:10.100711998-017-9994-6

Jin, B. E., & Shin, D. C. (2020). Changing the game to compete: Innovations in the fashion retail industry from the disruptive business model. *Business Horizons*, *63*(3), 301–311. doi:10.1016/j.bushor.2020.01.004

Jupalle, H., Kouser, S., Bhatia, A. B., Alam, N., Nadikattu, R. R., & Whig, P. (2022). Automation of human behaviors and its prediction using machine learning. *Microsystem Technologies*, *28*(8), 1–9. doi:10.100700542-022-05326-4

Kalidas, R., Boopathi, S., Sivakumar, K., & Mohankumar, P. (2012). Optimization of Machining Parameters of WEDM Process Based On the Taguchi Method. *IJEST, 6*(1\&2), 91.

Kang, H. S., Makimoto, K., Konno, R., & Koh, I. S. (2020). Review of outcome measures in PARO robot intervention studies for dementia care. *Geriatric Nursing*, *41*(3), 207–214. doi:10.1016/j.gerinurse.2019.09.003 PMID:31668459

Kang, S. (2020). A Study on Smart Homecare for Daily Living Ability and Safety Management of the Elderly. *Inf. Sci. Appl.*, *621*, 707–710. doi:10.1007/978-981-15-1465-4_72

Kannan, E., Trabelsi, Y., Boopathi, S., & Alagesan, S. (2022). Influences of cryogenically treated work material on near-dry wire-cut electrical discharge machining process. *Surface Topography : Metrology and Properties*, *10*(1), 015027. doi:10.1088/2051-672X/ac53e1

Kansal, N. J., & Chana, I. (2015). Artificial bee colony based energy-aware resource utilization technique for cloud computing. *Concurrency and Computation*, *27*(5), 1207–1225. doi:10.1002/cpe.3295

Kaur, R. & Kautish, S. (2022). Multimodal sentiment analysis: A survey and comparison. *Research Anthology on Implementing Sentiment Analysis Across Multiple Disciplines*, 1846-1870.

Kaur, R., & Kautish, S. (2022). Multimodal sentiment analysis: A survey and comparison. *Research Anthology on Implementing Sentiment Analysis Across Multiple Disciplines*, 1846-1870.

Kaur, R., & Kautish, S. (2022). Multimodal sentiment analysis: A survey and comparison. *Research Anthology on Implementing Sentiment Analysis Across Multiple Disciplines*, pp. 1846-1870.

Kautish, S., Peng, S. L., & Obaid, A. J. (Eds.). (2021). *Computational intelligence techniques for combating COVID-19*. Springer International Publishing. doi:10.1007/978-3-030-68936-0

Kautish, S., Reyana, A., & Vidyarthi, A. (2022). SDMTA: Attack detection and mitigation mechanism for DDoS vulnerabilities in hybrid cloud environment. *IEEE Transactions on Industrial Informatics*, *18*(9), 6455–6463. doi:10.1109/TII.2022.3146290

Kavitha, C., Malini, P. S. G., Charan, V., Manoj, N., Verma, A., & Boopathi, S. (2022). (in press). An experimental study on the hardness and wear rate of carbonitride coated stainless steel. *Materials Today: Proceedings*. doi:10.1016/j.matpr.2022.09.524

Khan, M. J., & Hong, K.-S. (2017). Hybrid EEG–fNIRSbased eight-command decoding for BCI: Application to quadcopter control. *Frontiers in Neurorobotics*, *11*, 6. doi:10.3389/fnbot.2017.00006 PMID:28261084

Kholiya, P. S., Kapoor, A., Rana, M., & Bhushan, M. (2021). Intelligent process automation: The future of digital transformation. *2021 10th International Conference on System Modeling & Advancement in Research Trends (SMART)*, (pp. 185–190). IEEE.

Khosravi, P., & Ghapanchi, A. H. (2016). Investigating the Effectiveness of Technologies Applied to Assist Seniors: A Systematic Literature Review. *International Journal of Medical Informatics*, *85*(1), 17–26. doi:10.1016/j.ijmedinf.2015.05.014 PMID:26216463

Kim, M. J., Kim, M., & Kim, H. (2019). An overview of recent applications of artificial intelligence in education: A systematic review. *Journal of Educational Technology & Society*, *22*(2), 1–16.

Klein, G. T., Lu, Y., & Wang, M. Y. (2013). 3D printing and neurosurgery—Ready for prime time? *World Neurosurgery*, *80*(3–4), 233–235. doi:10.1016/j.wneu.2013.07.009 PMID:23871811

Kober, J. & Peters, J. (2012). Reinforcement learning in robotics: A survey. In *Reinforcement Learning", pages* (pp. 579–610). Springer.

Kučak, D., Juričić, V., & Đambić, G. (2018). Machine Learning In Education-A Survey Of Current Research Trends. *Annals of DAAAM & Proceedings*, (pp. 29). Springer.

Kumar, A., Fu, Z., Pathak, D., & Malik, J. (2021). Rma: Rapid motor adaptation for legged robots. arXiv preprint arXiv:2107.04034. doi:10.15607/RSS.2021.XVII.011

Kumara, V., Mohanaprakash, T. A., Fairooz, S., Jamal, K., Babu, T., & B., S. (2023). Experimental Study on a Reliable Smart Hydroponics System. In *Human Agro-Energy Optimization for Business and Industry* (pp. 27–45). IGI Global. doi:10.4018/978-1-6684-4118-3.ch002

Kumar, V. Y., & Michae, L. N. (2012). Opportunities and challenges with autonomous micro aerial vehicles. *The International Journal of Robotics Research*, *31*(11), 1279–1291. doi:10.1177/0278364912455954

Larriba, F., Raya, C., Angulo, C., Albo-Canals, J., Diaz, M., & Boldu, R. (2016). Externalising Moods and Psychological States in a Cloud Based System to Enhance a Pet-Robot and Child's Interaction. *Biomedical Engineering Online*, *15*(S1), S72. doi:10.118612938-016-0180-3 PMID:27454772

Lawlor, K. T., Vanslambrouck, J. M., Higgins, J. W., Chambon, A., Bishard, K., Arndt, D., Er, P. X., Wilson, S. B., Howden, S. E., Tan, K. S., Li, F., Hale, L. J., Shepherd, B., Pentoney, S., Presnell, S. C., Chen, A. E., & Little, M. H. (2021). Cellular extrusion bioprinting improves kidney organoid reproducibility and conformation. *Nature Materials*, *20*(2), 260–271. doi:10.103841563-020-00853-9 PMID:33230326

Law, P. (2015). Social robotics in health-care service: the case of rehabilitation programmes in Hong Kong. In J. Vincent, (Ed.), *Social robots from a human perspective* (pp. 55–65). Springer International Publishing. doi:10.1007/978-3-319-15672-9_5

Lee, K. C., Chai, C. S., & Tsai, C. C. (2018). Computational thinking research in education. *Educational Technology Research and Development*, *66*(4), 765–768.

Lee, K., & Shin, I. (2015). User mobility model based computation offloading decision for mobile cloud. *Journal of Computing Science and Engineering : JCSE*, *9*(3), 155–162. doi:10.5626/JCSE.2015.9.3.155

Lekdioui, K., Messoussi, R., Ruichek, Y., Chaabi, Y., & Touahni, R. (2017). Facial decomposition for expression recognition using texture/shape descriptors and SVM classifier. *Signal Processing Image Communication*, *58*, 300–312. doi:10.1016/j.image.2017.08.001

Lewis, M., & Canamero, L. (2013). Are Discrete Emotions Useful in Human-Robot Interaction? Feedback from Motion Capture Analysis. In *Proceedings of the 2013 Humaine Association Conference on Affective Computing and Intelligent Interaction*, Geneva, Switzerland. 10.1109/ACII.2013.23

Li, J., Wu, T., Xu, Z., & Gu, X. (2014). A pilot study of post-total knee replacement gait rehabilitation using lower limbs robot-assisted training system. *European Journal of Orthopaedic Surgery & Traumatology*, *24*(2), 203–208. doi:10.100700590-012-1159-9 PMID:23412304

Liliana, D., Basaruddin, T., & Widyanto, M. (2018). Mixed facial emotion recognition using active appearance model and hidden conditional random fields. *International Journal of Pure and Applied Mathematics, 118*, 3159-3167.

Lim, Y. K., Liao, Z., Petridis, S., & Pantic, M. (2018). Transfer learning for action unit recognition.

Lindquist, K. A., Wager, T. D., Kober, H., Bliss-Moreau, E., & Barrett, L. F. (2012). The brain basis of emotion: A meta-analytic review. *Behavioral and Brain Sciences*, *35*(3), 121–143. doi:10.1017/S0140525X11000446 PMID:22617651

Li, S., & Deng, W. (2019). Blended emotion in-the-wild: Multi-label facial expression recognition using crowdsourced annotations and deep locality feature learning. *International Journal of Computer Vision*, *127*(6), 884–906. doi:10.100711263-018-1131-1

Li, S., & Deng, W. (2020). Deep facial expression recognition: A survey. *IEEE Transactions on Affective Computing*, *13*(3), 1195–1215. doi:10.1109/TAFFC.2020.2981446

Li, S., Deng, W., & Du, J. (2017). Reliable crowdsourcing and deep locality-preserving learning for expression recognition in the wild. In *Proceedings of the IEEE conference on computer vision and pattern recognition* (pp. 2852-2861). 10.1109/CVPR.2017.277

Liu, P., Han, S., Meng, Z., & Tong, Y. (2014). Facial expression recognition via a boosted deep belief network. In *Proceedings of the IEEE conference on computer vision and pattern recognition* (pp. 1805-1812). IEEE. 10.1109/CVPR.2014.233

Li, X., Liu, B., Pei, B., Chen, J., Zhou, D., Peng, J., Zhang, X., Jia, W., & Xu, T. (2020). Inkjet Bioprinting of Biomaterials. *Chemical Reviews*, *120*(19), 10793–10833. doi:10.1021/acs.chemrev.0c00008 PMID:32902959

Li, Y., Lu, Y., Li, J., & Lu, G. (2019, October). Separate loss for basic and compound facial expression recognition in the wild. In *Asian Conference on Machine Learning* (pp. 897-911). PMLR.

Logan, D. E., Breazeal, C., Goodwin, M. S., Jeong, S., O'Connell, B., Smith-Freedman, D., Heathers, J., & Weinstock, P. (2019). Social Robots for Hospitalized Children. *Pediatrics*, *144*(1), e20181511. doi:10.1542/peds.2018-1511 PMID:31243158

Lozano-Perez, T. (1983). Robot programming. *Proceedings of the IEEE*, *71*(7), 821–841. doi:10.1109/PROC.1983.12681

Lucey, P., Cohn, J. F., Kanade, T., Saragih, J., Ambadar, Z., & Matthews, I. (2010, June). The extended cohn-kanade dataset (ck+): A complete dataset for action unit and emotion-specified expression. In *2010 IEEE computer society conference on computer vision and pattern recognition-workshops* (pp. 94-101). IEEE.

*Lunokhod 1*. (2021, December 19). Wikipedia. https://en.wikipedia.org/wiki/Lunokhod_1

Lyons, M., Kamachi, M., & Gyoba, J. (1998). *The Japanese Female Facial Expression (JAFFE)*. Dataset.

Madhu, G., Govardhan, A., Ravi, V., Kautish, S., Srinivas, B. S., Chaudhary, T., & Kumar, M. (2022). DSCN-net: A deep Siamese capsule neural network model for automatic diagnosis of malaria parasites detection. *Multimedia Tools and Applications*, *81*(23), 34105–34127. doi:10.100711042-022-13008-6

Magalhães, L. S. S. M., Santos, F. E. P., de Maria Vaz Elias, C., Afewerki, S., Sousa, G. F., Furtado, A. S. A., Marciano, F. R., & Lobo, A. O. (2020). Printing 3D hydrogel structures employing low-cost stereolithography technology. *Journal of Functional Biomaterials*, *11*(1), 12. doi:10.3390/jfb11010012 PMID:32098431

Mahmood, M. A., Visan, A. I., Ristoscu, C., & Mihailescu, I. N. (2021). Artificial neural network algorithms for 3D printing. In Materials (Vol. 14, pp. 1–29). doi:10.3390/ma14010163

Mangusli, D. (2022, October 13). *The Emerging Role Of Robotics In Making Healthcare Accessible*. BW Defence. https://bwhealthcareworld.businessworld.in/article/The-Emerging-Role-Of-Robotics-In-Making-Healthcare-Accessible-/13-10-2022-450332/

Mania, H., Guy, A., & Recht, B. (2018). Simple random search provides a competitive approach to reinforcement learning. arXiv:1803.07055v1

Maris, G., & Bechger, T. (2019). Bayesian network modeling for psychometric applications. *Frontiers in Psychology*, *10*, 438.

Marquiss, R. (2018). *Five Types of Industrial Robots And How To Choose The Best Fit*. Valin. com. https://www.valin.com/resources/articles/five-types-of-industrial-robots-and-how-to-choose-the-best-fit

Matthies, L., Daftry, S., Tepsuporn, S., Cheng, Y., Atha, D., Swan, R. M., Ravichandar, S., & Ono, M. (2022). Lunar rover localization using craters as landmarks. *2022 IEEE Aerospace Conference (AERO)*, (pp. 1–17). 10.1109/AERO53065.2022.9843714

Meghdari, A., Alemi, M., Khamooshi, M., Amoozandeh, A., Shariati, A., & Mozafari, B. (2016). Conceptual Design of a Social Robot for Pediatric Hospitals. In *Proceedings of the 2016 4th International Conference on Robotics and Mechatronics (ICROM)*, Tehran, Iran. 10.1109/ICRoM.2016.7886804

Mekonnen, Y., Namuduri, S., Burton, L., Sarwat, A., & Bhansali, S. (2020). Review—Machine Learning Techniques in Wireless Sensor Network Based Precision Agriculture. *Journal of the Electrochemical Society*, *167*(3), 037522. doi:10.1149/2.0222003JES

Mell, P., & Grance, T. (2011). The NIST-National Institute of Standars and Technology- Definition of Cloud Computing. *NIST Special Publication 800-145*, 1–3.

Miao, Q., Zhang, M., Cao, J., & Xie, S. Q. (2018). Reviewing high-level control techniques on robot-assisted upper-limb rehabilitation. *Advanced Robotics*, *32*(24), 1253–1268. doi:10.1080/01691864.2018.1546617

Mitchell, R. (n.d.). *Advantages and disadvantages of robots*. FutureLearn. https://www.futurelearn.com/info/courses/begin-robotics/0/steps/2845

*Mobile Robots*. (2020, December 24). GeeksforGeeks. https://www.geeksforgeeks.org/mobile-robots/

Moerman, C. J., & Jansens, R. M. L. (2020). Using Social Robot PLEO to Enhance the Wellbeing of Hospitalised Children. *Journal of Child Health Care*. PMID:32840383

Moerman, C. J., van der Heide, L., & Heerink, M. (2019). Social Robots to Support Children's Wellbeing under Medical Treatment: A Systematic State-of-the-Art Review. *Journal of Child Health Care*, *23*(4), 596–612. doi:10.1177/1367493518803031 PMID:30394806

Moon, S., Hasan, S. K., Song, Y. S., Xu, F., Keles, H. O., Manzur, F., Mikkilineni, S., Hong, J. W., Nagatomi, J., Haeggstrom, E., Khademhosseini, A., & Demirci, U. (2010). Layer by layer three-dimensional tissue epitaxy by cell-laden hydrogel droplets. *Tissue Engineering. Part C, Methods*, *16*(1), 157–166. doi:10.1089/ten.tec.2009.0179 PMID:19586367

Mortlock, T., & Kassas, Z. M. (2021). Assessing Machine Learning for LEO Satellite Orbit Determination in Simultaneous Tracking and Navigation. 2021 IEEE Aerospace Conference (50100), (pp. 1–8). IEEE. 10.1109/AERO50100.2021.9438144

Murphy, R. R., Gandudi, V. B. M., Amin, T., Clendenin, A., & Moats, J. (2022). An analysis of international use of robots for COVID-19. *Robotics and Autonomous Systems*, *148*, 103922. doi:10.1016/j.robot.2021.103922 PMID:34803220

Myilsamy, S., Boopathi, S., & Yuvaraj, D. (2021). A study on cryogenically treated molybdenum wire electrode. *Materials Today: Proceedings*, *45*(9), 8130–8135. doi:10.1016/j.matpr.2021.02.049

Myilsamy, S., & Sampath, B. (2021). Experimental comparison of near-dry and cryogenically cooled near-dry machining in wire-cut electrical discharge machining processes. *Surface Topography : Metrology and Properties*, *9*(3), 035015. doi:10.1088/2051-672X/ac15e0

Nate. Kohl and Peter. Stone, "Policy gradient reinforcement learning for fast quadrupedal locomotion," *IEEE International Conference on Robotics and Automation, 2004. Proceedings. ICRA'04.*

Neerincx, M. A., van Vught, W., Blanson Henkemans, O., Oleari, E., Broekens, J., Peters, R., Kaptein, F., Demiris, Y., Kiefer, B., Fumagalli, D., & Bierman, B. (2019). Socio-Cognitive Engineering of a Robotic Partner for Child's Diabetes Self-Management. *Frontiers in Robotics and AI*, *6*, 118. doi:10.3389/frobt.2019.00118 PMID:33501133

Negi, A., & Kaur, K. (2017). Method to resolve software product line errors. *Information, Communication and Computing Technology: Second International Conference, ICICCT 2017*, New Delhi, India.

Nergui, M., Komekine, K., Nagai, H., & Otake, M. (2014). *Development of Laughter Motion on the Cognitive Robot "Bono-02" Assisting Group Conversation*. IEEE. doi:10.1109/U-MEDIA.2014.30

Ngo, H. Q. T., Le, V. N., Thien, V. D. N., Nguyen, T. P., & Nguyen, H. (2020). Develop the Socially Human-Aware Navigation System Using Dynamic Window Approach and Optimize Cost Function for Autonomous Medical Robot. *Advances in Mechanical Engineering*, *12*(12), 12. doi:10.1177/1687814020979430

Nguyen, C., & Nguyen, Q. (2022). Contact-timing and Trajectory Optimization for 3D Jumping on Quadruped Robots. arXiv:2110.06764. doi:10.1109/IROS47612.2022.9981284

Niyetkaliyev, A. S., Hussain, S., Ghayesh, M. H., & Alici, G. (2017). Review on design and control aspects of robotic shoulder rehabilitation orthoses. *IEEE Transactions on Human-Machine Systems*, *47*(6), 1134–1145. doi:10.1109/THMS.2017.2700634

Norman, J. (n.d.). *Geroge Devol Invents Unimate, the First Industrial Robot*. History of Information. Www.historyofinformation.com. https://www.historyofinformation.com/detail.php?entryid=4071

Nouaille, L., Laribi, M. A., Nelson, C. A., Zeghloul, S., & Poisson, G. (2017). Review of Kinematics for Minimally Invasive Surgery and Tele-Echography Robots. *Journal of Medical Devices*, *11*(4), 040802. doi:10.1115/1.4037053

Obi, F., Ugwuishiwu, B., & Nwakaire, J. (2016). Agricultural Waste Concept, Generation, Utilization and Management. *Nigerian Journal of Technology*, *35*(4), 957. doi:10.4314/njt.v35i4.34

Oladapo, B. I., Zahedi, S. A., & Adeoye, A. O. M. (2019). 3D printing of bone scaffolds with hybrid biomaterials. *Composites. Part B, Engineering*, *158*, 428–436. doi:10.1016/j.compositesb.2018.09.065

Ono, S., Obo, T., Kiong, L. C., & Kubota, N. (2015). Robot Communication Based on Relational Trust Model. In *IECON 2015-41st Annual Conference of the IEEE Industrial Electronics Society*. IEEE. 10.1109/IECON.2015.7392941

Ono, M., Rothrock, B., Otsu, K., Higa, S., Iwashita, Y., Didier, A., Islam, T., Laporte, C., Sun, V., & Stack, K. (2020). Maars: Machine learning-based analytics for automated rover systems. *2020 IEEE Aerospace Conference*, (pp. 1–17). 10.1109/AERO47225.2020.9172271

Osunmakinde, I., & Ramharuk, V. (2014). Development of a survivable cloud multi-robot framework for heterogeneous environments. *International Journal of Advanced Robotic Systems*, *11*(10), 1–22. doi:10.5772/58891

Özcan, M., Hotza, D., Fredel, M. C., Cruz, A., & Volpato, C. A. M. (2021). Materials and manufacturing techniques for polymeric and ceramic scaffolds used in implant dentistry. *Journal of Composites Science*, *5*(3), 78. doi:10.3390/jcs5030078

Palaniappan, M., Tirlangi, S., Mohamed, M. J. S., Moorthy, R. M. S., Valeti, S. V., & Boopathi, S. (2023). Fused Deposition Modelling of Polylactic Acid (PLA)-Based Polymer Composites: A Case Study. In Development, Properties, and Industrial Applications of 3D Printed Polymer Composites (pp. 66–85). IGI Global.

Pal, S., Mishra, N., Bhushan, M., Kholiya, P. S., Rana, M., & Negi, A. (2022). Deep Learning Techniques for Prediction and Diagnosis of Diabetes Mellitus. *2022 International Mobile and Embedded Technology Conference (MECON)*, (pp. 588–593). IEEE. 10.1109/MECON53876.2022.9752176

Pandey, R. K., Karmakar, S., Ramakrishnan, A. G., & Saha, N. (2019). Improving facial emotion recognition systems using gradient and laplacian images. arXiv preprint arXiv:1902.05411

Pandey, P., Pompili, D., & Yi, J. (2015). Dynamic Collaboration between Networked Robots and Clouds in Resource-Constrained Environments. *IEEE Transactions on Automation Science and Engineering*, *12*(2), 471–480. doi:10.1109/TASE.2015.2406115

Panel, E. (2022, November 16). *Council Post: 16 Ways Robots And RPA May Soon Change How We Live And Work*. Forbes. https://www.forbes.com/sites/forbestechcouncil/2022/11/16/16-ways-robots-and-rpa-may-soon-change-how-we-live-and-work/?sh=10821e1b3a0d

Panico, F., Cordasco, G., Vogel, C., Trojano, L., & Esposito, A. (2020). Ethical Issues in Assistive Ambient Living Technologies for Ageing Well. *Multimedia Tools and Applications*, *79*(47-48), 36077–36089. doi:10.100711042-020-09313-7

Pantic, M., Valstar, M., Rademaker, R., & Maat, L. (2005, July). Web-based database for facial expression analysis. In *2005 IEEE international conference on multimedia and Expo* (pp. 5-pp). IEEE.

Papadopoulos, I., Koulouglioti, C., & Ali, S. (2018). Views of nurses and other health and social care workers on the use of assistive humanoid and animal-like robots in health and social care: A scoping review. *Contemporary Nurse*, *54*(4-5), 425–442. doi:10.1080/10376178.2018.1519374 PMID:30200824

Partners, T. I. (2022, September 22). *Medical Robots Market 2022 Recovering from Covid-19 Outbreak*. Open PR. Www.openpr.com. https://www.openpr.com/news/2744760/medical-robots-market-2022-recovering-from-covid-19-outbreak

Patil, R. V., Thote, A. M., & Chavan, S. T. (2022). Pre-Detection and Classification of Coronavirus Disease by Artificial Intelligence and Computer Vision. In Pandemic Detection and Analysis Through Smart Computing Technologies (pp. 169-194). Apple Academic Press. doi:10.1201/9781003281610-8

Pelánek, R. (2016). Applications of the Elo rating system in adaptive educational systems. *Computers & Education*, *98*, 169–179. doi:10.1016/j.compedu.2016.03.017

Perez, J. A., Deligianni, F., Ravi, D., & Yang, G.-Z. (2017). *Artificial Intelligence and Robotics*. UKRAS White Papers., doi:10.31256/WP2017.1

Pflueger, M., Agha, A., & Sukhatme, G. S. (2019). Rover-IRL: Inverse Reinforcement Learning With Soft Value Iteration Networks for Planetary Rover Path Planning. *IEEE Robotics and Automation Letters*, *4*(2), 1387–1394. doi:10.1109/LRA.2019.2895892

Piau, A. (2019) Use of a robotic walking aid in rehabilitation to reduce fear of falling is feasible and acceptable from the end user's perspective: a randomised comparative study. Maturitas Elsevier 120:40–46. doi:.maturitas.2018.11.008 doi:10.1016/j

*Pierre Jaquet-Droz - Complete Biography, History and Inventions.* (2021, January 4). History Computer. https://history-computer.com/pierre-jaquet-droz-complete-biography/

Piñero-Fuentes, E., Canas-Moreno, S., Rios-Navarro, A., Delbruck, T., & Linares-Barranco, A. (2021). *Autonomous driving of a rover-like robot using neuromorphic computing.* Advances in Computational Intelligence: 16th International Work-Conference on Artificial Neural Networks, IWANN 2021, Virtual Event, Proceedings, Part II, (pp. 57–68). IEEE.

Pollock, B. (2019, August 5). *How Robots Will Save the Human Race*. Automation World. https://www.automationworld.com/factory/robotics/blog/13320046/how-robots-will-save-the-human-race

Poverly, E. (2022, April 26). *The Future of Robotics: How Will Robots Change The World?* Revolutionized. https://revolutionized.com/the-future-of-robotics/

Prinz, R., Bulbul, R., Scholz, J., Eder, M., & Steinbauer-Wagner, G. (2022). Off-Road Navigation Maps for Robotic Platforms using Convolutional Neural Networks. *AGILE: GIScience Series*, *3*, 55. doi:10.5194/agile-giss-3-55-2022

Quaglia, S. (2022, June 22). *Scientists unveil bionic robo-fish to remove microplastics from seas*. The Guardian. https://www.theguardian.com/environment/2022/jun/22/scientists-unveil-bionic-robo-fish-to-remove-microplastics-from-seas

Radford, A., Metz, L., & Chintala, S. (2015). Unsupervised representation learning with deep convolutional generative adversarial networks. arXiv preprint arXiv:1511.06434.

Rajeswaran, A., Lowrey, K., Todorov, E., & Kakade, S. (2017). Towards generalization and simplicity in continuous control. arXiv preprint arXiv:1703.02660

Rana, M., & Bhushan, M. (2022). Advancements in Healthcare Services using Deep Learning Techniques. *2022 International Mobile and Embedded Technology Conference (MECON)*, (pp. 157–161). IEEE. 10.1109/MECON53876.2022.9752020

Ren, Z., Lai, J., Wu, Z., & Xie, S. (2021). Deep neural networks-based real-time optimal navigation for an automatic guided vehicle with static and dynamic obstacles. *Neurocomputing*, *443*, 329–344. doi:10.1016/j.neucom.2021.02.034

Revina, I. M., & Emmanuel, W. S. (2021). A survey on human face expression recognition techniques. *Journal of King Saud University-Computer and Information Sciences*, *33*(6), 619–628. doi:10.1016/j.jksuci.2018.09.002

Reynoso, R. (2021, May 25). *A Complete History of Artificial Intelligence*. AI & Machine Learning Operationalization Category. https://www.g2.com/articles/history-of-artificial-intelligence

Rezapour, E. (2015, March). *Model-based Locomotion Control of Underactuated Snake Robots*. Research Gate. https://www.researchgate.net/publication/273770473_Model-based_Locomotion_Control_of_Underactuated_Snake_Robots

Riaz, M. N., Shen, Y., Sohail, M., & Guo, M. (2020). Exnet: An efficient approach for emotion recognition in the wild. *Sensors (Basel)*, *20*(4), 1087. doi:10.339020041087 PMID:32079319

Robinson, H., MacDonald, B., Kerse, N., & Broadbent, E. (2013). The Psychosocial Effects of a Companion Robot: A Randomized Controlled Trial. *Journal of the American Medical Directors Association*, *14*(9), 661–667. doi:10.1016/j.jamda.2013.02.007 PMID:23545466

*Robot Technologies from A to Z at a Glance*. (2021, September 2). Mobilefest. https://www.mobilefest.net/robot-technologies-from-a-to-z-at-a-glance/

*Robotics Career*. (2021, June 7). Robotics Career. Www.roboticscareer.org. https://www.roboticscareer.org/news-and-events/news/23037

Robotics, P. (2020, December 25). *The Origin, History and Story of Robotics*. Medium. https://medium.com/@plusrobotics07/the-origin-history-and-story-of-robotics-ab2f060a4259#:~:text=One%20of%20the%20first%20instances

*Robots can be Classified by their Application or their Kinematics*. (n.d.). Design Tech Products. Www.designtechproducts.com. https://www.designtechproducts.com/articles/classification-of-robots

*Robots in Space - Universe Today*. (2009, November 2). Universe Today. https://www.universetoday.com/43750/robots-in-space/

*RobotWorx*. (2019). RobotWorx. https://www.robots.com/faq/what-are-the-main-types-of-robots

Rouaix, N., Retru-Chavastel, L., Rigaud, A.-S., Monnet, C., Lenoir, H., & Pino, M. (2017). Affective and Engagement Issues in the Conception and Assessment of a Robot-Assisted Psychomotor Therapy for Persons with Dementia. *Frontiers in Psychology*, *8*, 950. doi:10.3389/fpsyg.2017.00950 PMID:28713296

S., P. K., Sampath, B., R., S. K., Babu, B. H., & N., A. (2022). Hydroponics, Aeroponics, and Aquaponics Technologies in Modern Agricultural Cultivation. In *Trends, Paradigms, and Advances in Mechatronics Engineering* (pp. 223–241). IGI Global. doi:10.4018/978-1-6684-5887-7.ch012

Samikannu, R., Koshariya, A. K., Poornima, E., Ramesh, S., Kumar, A., & Boopathi, S. (2022). Sustainable Development in Modern Aquaponics Cultivation Systems Using IoT Technologies. In *Human Agro-Energy Optimization for Business and Industry* (pp. 105–127). IGI Global.

Sampath, B. C. S., & Myilsamy, S. (2022). Application of TOPSIS Optimization Technique in the Micro-Machining Process. In Trends, Paradigms, and Advances in Mechatronics Engineering (pp. 162–187). IGI Global. doi:10.4018/978-1-6684-5887-7.ch009

Sampath, B., & Myilsamy, S. (2021). Experimental investigation of a cryogenically cooled oxygen-mist near-dry wire-cut electrical discharge machining process. *Strojniski Vestnik. Jixie Gongcheng Xuebao*, *67*(6), 322–330. doi:10.5545v-jme.2021.7161

Sampath, B., Naveenkumar, N., Sampathkumar, P., Silambarasan, P., Venkadesh, A., & Sakthivel, M. (2021). Experimental comparative study of banana fiber composite with glass fiber composite material using Taguchi method. *Materials Today: Proceedings*, *49*(5), 1475–1480. doi:10.1016/j.matpr.2021.07.232

Sampath, B., Pandian, M., Deepa, D., & Subbiah, R. (2022). Operating parameters prediction of liquefied petroleum gas refrigerator using simulated annealing algorithm. *AIP Conference Proceedings*, *2460*(1), 070003. doi:10.1063/5.0095601

Sampathkumar, A., Rastogi, R., Arukonda, S., Shankar, A., Kautish, S., & Sivaram, M. (2020). An efficient hybrid methodology for detection of cancer-causing gene using CSC for micro array data. *Journal of Ambient Intelligence and Humanized Computing*, *11*(11), 4743–4751. doi:10.100712652-020-01731-7

Sarabia, M., Young, N., Canavan, K., Edginton, T., Demiris, Y., & Vizcaychipi, M. P. (2018). Assistive Robotic Technology to Combat Social Isolation in Acute Hospital Settings. *International Journal of Social Robotics*, *10*(5), 607–620. doi:10.100712369-017-0421-z

Saravanan, M., Vasanth, M., Boopathi, S., Sureshkumar, M., & Haribalaji, V. (2022). Optimization of Quench Polish Quench (QPQ) Coating Process Using Taguchi Method. *Key Engineering Materials*, *935*, 83–91. doi:10.4028/p-z569vy

Scassellati, B., Admoni, H., & Mataric, M. (2012). *Robots for Use in Autism Research*. *14*, 275–294.

Schröder, J., Truijen, S., Van Criekinge, T. A., & Saeys, W. (2019). Feasibility and effectiveness of repetitive gait training early after stroke: A systematic review and meta-analysis. *Journal of Rehabilitation Medicine*, *51*(2), 78–88. doi:10.2340/16501977-2505 PMID:30516821

Schulman, J., Moritz, P., Levine, S., Jordan, M., & Abbeel, P. (2015). High-dimensional continuous control using generalized advantage estimation. arXiv preprint arXiv:1506.02438.

Schulman, J., Wolski, F., Dhariwal, P., Radford, A., & Klimov, O. (2017). Proximal policy optimization. *Algorithms. CoRR, abs/1707.06347.*

Senthil, T. S., Puviyarasan, M., Babu, S. R., Surakasi, R., Sampath, B., & Associates. (2023). Industrial Robot-Integrated Fused Deposition Modelling for the 3D Printing Process. In Development, Properties, and Industrial Applications of 3D Printed Polymer Composites (pp. 188–210). IGI Global.

Sequeira, J. S., & Ferreira, I. A. (2016). *Lessons from the MOnarCH Project*. IEEE. doi:10.5220/0005998102410248

Shafi, U., Mumtaz, R., García-Nieto, J., Hassan, S. A., Zaidi, S. A. R., & Iqbal, N. (2019). Precision agriculture techniques and practices: From considerations to applications. *Sensors (Switzerland)*, *19*(17), 1–25. doi:10.339019173796 PMID:31480709

Shah, R., & Chircu, A. (2018). IOT AND AI IN HEALTHCARE: A SYSTEMATIC LITERATURE REVIEW. *Issues in Information Systems*, *19*(3), 33–41. https://iacis.org/iis/2018/3_iis_2018_33-41.pdf

Shamsuddin, S. (2017) Preliminary study on the use of therapeutic seal robot for patients with depression. In: *Proceedings of IEEE 4th International Symposium on Robotics and Intelligent Sensors: empowering robots with smart sensors*, Tokyo, Japan. 10.1109/IRIS.2016.8066065

Shao, Z., Liu, Z., Cai, J., Wu, Y., & Ma, L. (2019). Facial action unit detection using attention and relation learning. *IEEE Transactions on Affective Computing*, *13*(3), 1274–1289. doi:10.1109/TAFFC.2019.2948635

Sharif, M. S., & Alsibai, M. H. (2017). *Medical Data Analysis Based on Nao Robot: An Automated Approach Towards Robotic Real-Time Interaction with Human Body*. IEEE.

Shayan, A. M., Sarmadi, A., Pirastehzad, A., Moradi, H., & Soleiman, P. (2016). RoboParrot 2.0: A Multi-Purpose Social Robot. In *Proceedings of the 2016 4th International Conference on Robotics and Mechatronics*, Tehran, Iran. 10.1109/ICRoM.2016.7886775

Shen, Y., Guo, D., Long, F., Mateos, L. A., Ding, H., Xiu, Z., Hellman, R. B., King, A., Chen, S., Zhang, C., & Tan, H. (2021). Robots Under COVID-19 Pandemic: A Comprehensive Survey. *IEEE Access : Practical Innovations, Open Solutions*, *9*, 1590–1615. doi:10.1109/ACCESS.2020.3045792 PMID:34976569

Sheridan, T. B. (2020). A Review of Recent Research in Social Robotics. *Current Opinion in Psychology*, *36*, 7–12. doi:10.1016/j.copsyc.2020.01.003 PMID:32294577

Shi, H., Shi, L., Xu, M., & Hwang, K.-S. (2020). End-to-End Navigation Strategy With Deep Reinforcement Learning for Mobile Robots. *IEEE Transactions on Industrial Informatics*, *16*(4), 2393–2402. doi:10.1109/TII.2019.2936167

Simonyan, K., & Zisserman, A. (2014). Very deep convolutional networks for large-scale image recognition. arXiv preprint arXiv:1409.1556

Simshaw, D., Terry, N. P., Hauser, K., & Cummings, M. L. (2016). Regulating Healthcare Robots: Maximizing Opportunities While Minimizing Risks. *Publisher*. https://hdl.handle.net/1805/11587

Singh, R. K., Berkvens, R., & Weyn, M. (2021). AgriFusion: An Architecture for IoT and Emerging Technologies Based on a Precision Agriculture Survey. *IEEE Access : Practical Innovations, Open Solutions*, *9*, 136253–136283. doi:10.1109/ACCESS.2021.3116814

Smith, L., Kew, J. C., Peng, X. B., Ha, S., Tan, J., & Levine, S. (2021). Legged robots that keep on learning: Fine-tuning locomotion policies in the real world. arXiv preprint arXiv:2110.05457.

Song, I. Y., Song, M., Timakum, T., Ryu, S. R., & Lee, H. (2018). The landscape of smart aging: Topics, applications, and agenda. *Data & Knowledge Engineering*, *115*, 68–79. doi:10.1016/j.datak.2018.02.003

Stanford University. (2019). *Robotics: A Brief History*. Stanford.edu. https://cs.stanford.edu/people/eroberts/courses/soco/projects/1998-99/robotics/history.html

Staroverov, A., Yudin, D. A., Belkin, I., Adeshkin, V., Solomentsev, Y. K., & Panov, A. I. (2020). Real-Time Object Navigation With Deep Neural Networks and Hierarchical Reinforcement Learning. *IEEE Access : Practical Innovations, Open Solutions*, *8*, 195608–195621. doi:10.1109/ACCESS.2020.3034524

Stenmark, M., Malec, J., Nilsson, K., & Robertsson, A. (2015). On distributed knowledge bases for robotized small-batch assembly. *IEEE Transactions on Automation Science and Engineering*, *12*(2), 519–528. doi:10.1109/TASE.2015.2408264

Studer, S., Bui, T. B., Drescher, C., Hanuschkin, A., Winkler, L., Peters, S., & Müller, K. R. (2021). Towards CRISP-ML (Q): a machine learning process model with quality assurance methodology. *Machine Learning and Knowledge Extraction, 3*(2), 392-413. https://blog.devgenius.io/predicting-tesla-stocks-tsla-using-python-pycaret-45af9ed47de9

Sun, Y., Guan, L., Wu, M., Gao, Y., & Chang, Z. (2020). Vehicular navigation based on the fusion of 3D-RISS and machine learning enhanced visual data in challenging environments. *Electronics (Basel)*, *9*(1), 193. doi:10.3390/electronics9010193

Suwa, M. (1978). A preliminary note on pattern recognition of human emotional expression. *Proc. IJPR*, (pp. 408-410). IEEE.

Szegedy, C., Ioffe, S., Vanhoucke, V., & Alemi, A. A. (2017, February). Inception-v4, inception-resnet and the impact of residual connections on learning. In *Thirty-first AAAI conference on artificial intelligence*. 10.1609/aaai.v31i1.11231

Szegedy, C., Vanhoucke, V., Ioffe, S., Shlens, J., & Wojna, Z. (2016). Rethinking the inception architecture for computer vision. In *Proceedings of the IEEE conference on computer vision and pattern recognition* (pp. 2818-2826). IEEE. 10.1109/CVPR.2016.308

Takebayashi, T., Miyagusuku, R., & Ozaki, K. (2021). Development of Magnetic-Based Navigation by Constructing Maps Using Machine Learning for Autonomous Mobile Robots in Real Environments. In Sensors (Vol. 21, Issue 12). doi:10.339021123972

Tan, J., Zhang, T., Coumans, E., Iscen, A., Bai, Y., Hafner, D., Bohez, S., & Vanhoucke, V. (2018). Sim-to-Real: Learning Agile Locomotion For Quadruped Robots. arXiv:1804.10332v2

Tang, J., Zhou, Z., & Yu, Y. (2017) A hybrid computer interface for robot arm control. In: *Proceedings of 8th International Conference on Information Technology in Medicine and Education, ITME 2016*, Fuzhou, China. 10.1109/ITME.2016.0088

Taylor, C. (2019, June 26). *Robots could take over 20 million jobs by 2030, study claims*. CNBC; CNBC. https://www.cnbc.com/2019/06/26/robots-could-take-over-20-million-jobs-by-2030-study-claims.html

Terra, J. (2022, April 18). *The Future of Robotics: Everything You Need to Know in 2022*. Simplilearn. https://www.simplilearn.com/future-of-robotics-article

*The Invention of the Industrial Robot*. (n.d.). Invent.org. Www.invent.org. https://www.invent.org/blog/inventors/George-Devol-Industrial-Robot

Timothy, P., Lillicrap, J., Hunt, J., Pritzel, A., Heess, N., Erez, T., Tassa, Y., Silver, D., & Wierstra, D. (2015). Continuous control with deep reinforcement learning. arXiv preprint arXiv:1509.02971, 2015

Tomar, U. Chakroborty, N. Sharma, H., & Whig, P. (2021). AI based Smart Agricuture System. *Transactions on Latest Trends in Artificial Intelligence*, 2(2).

*Top 10 Applications of Robotics in 2020.* (2020, November 2). GeeksforGeeks. https://www.geeksforgeeks.org/top-10-applications-of-robotics-in-2020/

Top, N., Şahin, İ., Gökçe, H., & Gökçe, H. (2021). Computer-aided design and additive manufacturing of bone scaffolds for tissue engineering: State of the art. *Journal of Materials Research, 36*(19), 3725–3745. doi:10.155743578-021-00156-y

Trojovský, P., Dhasarathan, V., & Boopathi, S. (2023). Experimental investigations on cryogenic friction-stir welding of similar ZE42 magnesium alloys. *Alexandria Engineering Journal, 66*(1), 1–14. doi:10.1016/j.aej.2022.12.007

Tucker, M. R., Olivier, J., Pagel, A., Bleuler, H., Bouri, M., Lambercy, O., Millán, J. del R., Riener, R., Vallery, H., & Gassert, R. (2015). Control strategies for active lower extremity prosthetics and orthotics: A review. *Journal of Neuroengineering and Rehabilitation, 12*(1), 1. doi:10.1186/1743-0003-12-1 PMID:25557982

Tuisku, O., Pekkarinen, S., Hennala, L., & Melkas, H. (2018). Robots do not replace a nurse with a beating heart. *Information Technology & People, 32*(1), 47–67. doi:10.1108/ITP-06-2018-0277

Turja, T., & Parviainen, J.(2020). The Use of Affective Care Robots Calls Forth Value-Based Consideration. In *Proceedings of the 2020 29th IEEE International Conference on Robot and Human Interactive Communication (RO-MAN),* Naples, Italy.

*Types of Robot.* (n.d.). Javapoint. Www.javatpoint.com. https://www.javatpoint.com/types-of-robot

*Types of Robots and Industry Applications of Robotics Technology*. (n.d.). Intel. https://www.intel.com/content/www/us/en/robotics/types-and-applications.html

*Types of Robots and Their Applications - Guardforce.* (2019). Guardforce.com.hk. https://www.guardforce.com.hk/en/news/blog_115/Types-of-Robots-and-Their-Applications_1729

Uma, S. (2018). Latest Research Trends and Challenges of Computational Intelligence Using Artificial Intelligence and Augmented Reality. *Computational Intelligence and Sustainable Systems*, 43–59. doi:10.1007/978-3-030-02674-5_3

Uma, S., & Suganthi, J. (2015). *A Smart and Dynamic Decision Support System for Nonlinear Environments.* Recent Advances in Intelligent Technologies and Information Systems. https://www.igi-global.com/chapter/a-smart-and-dynamic-decision-support-system-for-nonlinear-environments/125509

Uma, S., & T, S. (2019). *The Potential of Cognitive Social Mining Applications in Data Analytics for Protecting Social Media Users for National and International Security.* Cognitive Social Mining Applications in Data Analytics and Forensics. https://www.igi-global.com/chapter/the-potential-of-cognitive-social-mining-applications-in-data-analytics-for-protecting-social-media-users-for-national-and-international-security/218389

Uma, S., & Chitra, A. (2012). Pattern recognition using enhanced non-linear time-series models for predicting dynamic real-time decision making environments. *International Journal of Business Information Systems, 11*(1), 69. doi:10.1504/IJBIS.2012.048342

Uma, S., Chitra, A., & Suganthi, J. (2013). Design of a non-linear time series prediction model for daily electricity demand forecasting. *International Journal of Business Innovation and Research, 7*(3), 298. doi:10.1504/IJBIR.2013.053633

Uma, S., & Suganthi, J. (2012). An Intelligent and Dynamic Decision Support System for Nonlinear Environments. *International Journal of Intelligent Information Technologies, 8*(4), 43–61. doi:10.4018/jiit.2012100104

Valles-Peris, N., Angulo, C., & Domenech, M. (2018). Children's Imaginaries of Human-Robot Interaction in Healthcare. *International Journal of Environmental Research and Public Health, 15*(5), 970. doi:10.3390/ijerph15050970 PMID:29757221

Van Noort, R. (2012). The future of dental devices is digital. *Dental Materials, 28*(1), 3–12. doi:10.1016/j.dental.2011.10.014 PMID:22119539

Vandemeulebroucke, T., Dierckx de Casterlé, B., & Gastmans, C. (2018). The use of care robots in aged care: A systematic review of argument-based ethics literature. *Archives of Gerontology and Geriatrics, 74*, 15–25. doi:10.1016/j.archger.2017.08.014 PMID:28926749

Vanitha, S. K. R., & Boopathi, S. (2023). Artificial Intelligence Techniques in Water Purification and Utilization. In *Human Agro-Energy Optimization for Business and Industry* (pp. 202–218). IGI Global. doi:10.4018/978-1-6684-4118-3.ch010

Veerabahu, M. (2022, November 7). *The role of embedded vision in powering autonomous robotics.* Times of India Blog. https://timesofindia.indiatimes.com/blogs/voices/the-role-of-embedded-vision-in-powering-autonomous-robotics/

Vennila, T., Karuna, M. S., Srivastava, B. K., Venugopal, J., Surakasi, R., & Sampath, B. (2022). New Strategies in Treatment and Enzymatic Processes: Ethanol Production From Sugarcane Bagasse. In Human Agro-Energy Optimization for Business and Industry (pp. 219–240). IGI Global.

Verma, U., Garg, C., Bhushan, M., Samant, P., Kumar, A., & Negi, A. (2022). Prediction of students' academic performance using Machine Learning Techniques. *2022 International Mobile and Embedded Technology Conference (MECON)*, (pp. 151–156). IEEE. 10.1109/MECON53876.2022.9751956

*Viking 1 and 2, NASA's first Mars landers.* (n.d.). The Planetary Society. https://www.planetary.org/space-missions/vikingReik doi:10.1145/3127874Vincent Wang, X., & Wang, L. (2021). A literature survey of the robotic technologies during the COVID-19 pandemic. *Journal of Manufacturing Systems, 60*, 823–836. Advance online publication. doi:10.1016/j.jmsy.2021.02.005 PMID:33612914

Visca, M., Powell, R., Gao, Y., & Fallah, S. (2022). Deep meta-learning energy-aware path planner for unmanned ground vehicles in unknown terrains. *IEEE Access : Practical Innovations, Open Solutions, 10*, 30055–30068. doi:10.1109/ACCESS.2022.3155161

Vitanza, A., D'Onofrio, G., Ricciardi, F., Sancarlo, D., Greco, A., & Giuliani, F. (2019). Assistive Robots for the Elderly: Innovative Tools to Gather Health Relevant Data. *Data Science for Healthcare*, 195–215. doi:10.1007/978-3-030-05249-2_7

Von Davier, A. A., Deonovic, B., Yudelson, M., Polyak, S. T., & Woo, A. (2019). Computational psychometrics approach to holistic learning and assessment systems. In *Frontiers in Education* (p. 69). Frontiers. doi:10.3389/feduc.2019.00069

Vulpi, F., Milella, A., Marani, R., & Reina, G. (2021). Recurrent and convolutional neural networks for deep terrain classification by autonomous robots. *Journal of Terramechanics, 96*, 119–131. doi:10.1016/j.jterra.2020.12.002

Wang, W., Lin, L., Fan, Z., & Liu, J. (2022). Semi-Supervised Learning for Mars Imagery Classification and Segmentation. *ACM Transactions on Multimedia Computing Communications and Applications.*

Wang, Z., Zhao, Z., Min, G., Huang, X., Ni, Q., & Wang, R. (2018). User mobility aware task assignment for Mobile Edge Computing. *Future Generation Computer Systems, 85*, 1–8. doi:10.1016/j.future.2018.02.014

Whig, P., & Ahmad, S. N. (2014d). Simulation of linear dynamic macro model of photo catalytic sensor in SPICE. *COMPEL: The International Journal for Computation and Mathematics in Electrical and Electronic Engineering.*

Whig, P., Kouser, S., Velu, A., & Nadikattu, R. R. (2022). Fog-IoT-Assisted-Based Smart Agriculture Application. In Demystifying Federated Learning for Blockchain and Industrial Internet of Things. IGI Global.

Whig, P., Nadikattu, R. R., & Velu, A. (2022). COVID-19 pandemic analysis using application of AI. Healthcare Monitoring and Data Analysis Using IoT: Technologies and Applications, (p. 1). IET. doi:10.1049/PBHE038E_ch1

Whig, P., Nadikattu, R. R., & Velu, A. (2022). COVID-19 pandemic analysis using application of AI. *Healthcare Monitoring and Data Analysis Using IoT: Technologies and Applications*, 1.

Whig, P., Velu, A., & Bhatia, A. B. (2022). Protect Nature and Reduce the Carbon Footprint with an Application of Blockchain for IIoT. In Demystifying Federated Learning for Blockchain and Industrial Internet of Things. IGI Global.

Whig, P., Velu, A., & Naddikatu, R. R. (2022). The Economic Impact of AI-Enabled Blockchain in 6G-Based Industry. In AI and Blockchain Technology in 6G Wireless Network. Springer.

Whig, P., Velu, A., & Nadikattu, R. R. (2022). Blockchain Platform to Resolve Security Issues in IoT and Smart Networks. In AI-Enabled Agile Internet of Things for Sustainable FinTech Ecosystems. IGI Global.

Whig, P., Velu, A., & Ready, R. (2022). Demystifying Federated Learning in Artificial Intelligence with Human-Computer Interaction. In Demystifying Federated Learning for Blockchain and Industrial Internet of Things. IGI Global.

Whig, P., Velu, A., & Sharma, P. (2022). Demystifying Federated Learning for Blockchain: A Case Study. In Demystifying Federated Learning for Blockchain and Industrial Internet of Things. IGI Global. doi:10.4018/978-1-6684-3733-9.ch008

Whig, P., Kouser, S., Velu, A., & Nadikattu, R. R. (2022). Fog-IoT-Assisted-Based Smart Agriculture Application. In *Demystifying Federated Learning for Blockchain and Industrial Internet of Things* (pp. 74–93). IGI Global. doi:10.4018/978-1-6684-3733-9.ch005

Whig, P., Velu, A., & Bhatia, A. B. (2022). Protect Nature and Reduce the Carbon Footprint With an Application of Blockchain for IIoT. In *Demystifying Federated Learning for Blockchain and Industrial Internet of Things* (pp. 123–142). IGI Global. doi:10.4018/978-1-6684-3733-9.ch007

Whig, P., Velu, A., & Naddikatu, R. R. (2022). The Economic Impact of AI-Enabled Blockchain in 6G-Based Industry. In *AI and Blockchain Technology in 6G Wireless Network* (pp. 205–224). Springer. doi:10.1007/978-981-19-2868-0_10

Whig, P., Velu, A., & Nadikattu, R. R. (2022). Blockchain Platform to Resolve Security Issues in IoT and Smart Networks. In *AI-Enabled Agile Internet of Things for Sustainable FinTech Ecosystems* (pp. 46–65). IGI Global. doi:10.4018/978-1-6684-4176-3.ch003

Whig, P., Velu, A., & Ready, R. (2022). Demystifying Federated Learning in Artificial Intelligence With Human-Computer Interaction. In *Demystifying Federated Learning for Blockchain and Industrial Internet of Things* (pp. 94–122). IGI Global. doi:10.4018/978-1-6684-3733-9.ch006

WHO. (2021, April 23). *COVID-19 continues to disrupt essential health services in 90% of countries*. Www.who.int. https://www.who.int/news/item/23-04-2021-covid-19-continues-to-disrupt-essential-health-services-in-90-of-countries

Wolfram Research. (2018). *FER-2013*. Wolfram Data Repository.

World Commission on the Ethics of Scientific Knowledge and Technology (COMEST). (2017). *Report of COMEST on Robotics Ethics*. SHS/YES/COMEST-10/17/2 REV. www.facebook.com/elprocus.

Wu, B., WK, P. R., Ludivig, P., Chung, A. S., & Seabrook, T. (2019). Absolute localization through orbital maps and surface perspective imagery: A synthetic lunar dataset and neural network approach. *2019 IEEE/RSJ International Conference on Intelligent Robots and Systems (IROS)*, (pp. 3262–3267). IEEE. 10.1109/IROS40897.2019.8968124

Wu, C. H., Chen, C. H., & Huang, Y. M. (2018). Using artificial intelligence to analyze and evaluate students' online learning behaviors. *Journal of Educational Technology & Society, 21*(4), 58–69. https://www.learndatasci.com/tutorials/introduction-pycaret-machine-learning/

Wu, J. H., Tennyson, R. D., & Hsia, T. L. (2010). A study of student satisfaction in a blended e-learning system environment. *Computers & Education, 55*(1), 155–164. doi:10.1016/j.compedu.2009.12.012

Wynsberghe, A. (2013). van A Method for Integrating Ethics into the Design of Robots. *Ind. Robot. Int. J. Robot. Res. Appl., 40*(5), 433–440. doi:10.1108/IR-12-2012-451

Xia, Q., Liang, W., Xu, Z., & Zhou, B. (2014). Online algorithms for location-aware task offloading in two-tiered mobile cloud environments. *Proceedings - 2014 IEEE/ACM 7th International Conference on Utility and Cloud Computing, UCC 2014*, (pp. 109–116). IEEE. 10.1109/UCC.2014.19

Xu, F., Celli, J., Rizvi, I., Moon, S., Hasan, T., & Demirci, U. (2011). A three-dimensional in vitro ovarian cancer coculture model using a high-throughput cell patterning platform. *Biotechnology Journal, 6*(2), 204–212. doi:10.1002/biot.201000340 PMID:21298805

Xu, F., Liu, F., Jin, H., & Vasilakos, A. V. (2013). Managing performance overhead of virtual machines in cloud computing: A survey, state of the art, and future directions. *Proceedings of the IEEE, 102*(1), 11–31. doi:10.1109/JPROC.2013.2287711

Y. Alkali, I. Routray, & Whig, P. (2022). Strategy for Reliable, Efficient and Secure IoT Using Artificial Intelligence. *IUP Journal of Computer Sciences, 16*(2).

Yan, W. J., Li, S., Que, C., Pei, J., & Deng, W. (2020). RAF-AU Database: In-the-Wild Facial Expressions with Subjective Emotion Judgement and Objective AU Annotations. In *Proceedings of the Asian Conference on Computer Vision*. Springer.

Yan, Q., Dong, H., Su, J., Han, J., Song, B., Wei, Q., & Shi, Y. (2018). A Review of 3D Printing Technology for Medical Applications. *Engineering, 4*(5), 729–742. doi:10.1016/j.eng.2018.07.021

Yupapin, P., Trabelsi, Y., Nattappan, A., & Boopathi, S. (2022). Performance Improvement of Wire-Cut Electrical Discharge Machining Process Using Cryogenically Treated Super-Conductive State of Monel-K500 Alloy. *Iranian Journal of Science and Technology - Transactions of Mechanical Engineering*, 1–17. doi:10.1007/s40997-022-00513-0

Yu, R., Hui, E., Lee, J., Poon, D., Ng, A., Sit, K., Ip, K., Yeung, F., Wong, M., Shibata, T., & Woo, J. (2015). Use of a Therapeutic, Socially Assistive Pet Robot (PARO) in Improving Mood and Stimulating Social Interaction and Communication for People With Dementia: Study Protocol for a Randomized Controlled Trial. *JMIR Research Protocols, 4*(2), e45. doi:10.2196/resprot.4189 PMID:25934173

Zhang, X., Liu, Y., Hu, D., & Liu, L. (2021). A Maze Robot Autonomous Navigation Method Based on Curiosity and Reinforcement Learning. *The 7th Int. Workshop on Advanced Computational Intelligence and Intelligent Informatics (IWACIII 2021), Article, M1-6*. IEEE.

Zhang, L., Fu, D., Liu, J., Ngai, E. C. H., & Zhu, W. (2017). On Energy-Efficient Offloading in Mobile Cloud for Real-Time Video Applications. *IEEE Transactions on Circuits and Systems for Video Technology*, *27*(1), 170–181. doi:10.1109/TCSVT.2016.2539690

Zhang, M., Davies, T. C., & Xie, S. (2013). Effectiveness of robot-assisted therapy on ankle rehabilitation – a systematic review. *Journal of Neuroengineering and Rehabilitation*, *10*(1), 30. Advance online publication. doi:10.1186/1743-0003-10-30 PMID:23517734

Zhao, G., & Pietikainen, M. (2007). Dynamic texture recognition using local binary patterns with an application to facial expressions. *IEEE Transactions on Pattern Analysis and Machine Intelligence*, *29*(6), 915–928. doi:10.1109/TPAMI.2007.1110 PMID:17431293

Zhu, D., Liu, S., Jiang, W., Gao, C., Wu, T., Wang, Q., & Guo, G. (2019). *Ugan: Untraceable gan for multi-domain face translation.*

Ziya, T. A. N., & KARAKOSE, M. (2020). Comparative study for deep reinforcement learning with CNN, RNN, and LSTM in autonomous navigation. *2020 International Conference on Data Analytics for Business and Industry: Way Towards a Sustainable Economy (ICDABI)*, (pp. 1–5). IEEE.

# About the Contributors

**Sandeep Kautish** is an academician by choice and comes along with 18+ Years of work experience in academics including about 09 years in academic administration in various institutions in India and abroad. He earned his bachelor's, master's, and doctorate degrees in Computer Science on Intelligent Systems in Social Networks. He holds a Diploma in Management also. His areas of research interest are Decision Support Systems, Business Analytics, Machine Learning, Data Mining, and Information Systems. He has 46 publications in his Scopus account and his research works have been published in reputed journals i.e. Springer, Elsevier, Taylor & Francis, Hindawi, and IGI Global with high impact factor and SCI/SCIE/Scopus/WoS index. His research papers can be found at IEEE Transaction on Industrial Informatics (Impact Factor: 10.215), IEEE Access (SCI, Impact Factor: 3.715), Computer Standard & Interface (SCI, Elsevier, Impact Factor: 3.721), Journal of Ambient Intelligence and Humanized Computing (SCIE, Springer, Impact Factor: 3.662)), Wireless Personal Communications (SCIE, Springer, Impact Factor: 2.017), Multimedia Tools and Applications (Springer, SCIE, Impact Factor: 2.577).

**Nirbhay Kumar Chaubey** (SMIEEE, SMACM, LMCSI) received his Ph.D in Computer Science from Gujarat University, Ahmedabad, India. He is currently an Associate Professor of Computer Science at S.S.Agrawal Institute of Computer Science, Navsari (Code 554), Gujarat Technological University (GTU), Ahmedabad, Gujarat, India. Before that, he was working as an Assistant Professor of Computer Science, at Institute of Science & Technology for Advanced Studies & Research (ISTAR), Vallabh Vidyanagar, Gujarat Technological University, Ahmedabad, Gujarat, India. His research interests lie in the areas of wireless networks, Quantum Computing, Cyber Security, Cloud Computing, Mobile Computing and Sensor Network etc. He has published several research papers in peered reviewed International Journals and Conferences and has been a member of editorial board, technical program committee, reviewer for various Transactions and Journal of Computers and Communications. Chaubey received IEEE Outstanding Volunteer Award- Year 2015(Asia Pacific), Gujarat Technological University Pedagogical Innovation Awards (PIA) -2015, IEEE Outstanding Branch Counselor Award - Year 2010 (Asia Pacific).

**S B Goyal** had completed PhD in the Computer Science & Engineering in 2012 from India and served many institutions in many different academic and administrative positions. He is holding 19+ years experience at national and international level. He has peerless inquisitiveness and enthusiasm to get abreast with the latest development in the IT field. He has good command over Industry Revolution (IR) 4.0 technologies like Big-Data, Data Science, Artificial Intelligence & Blockchain, computer networking, deep learning etc. He is the first one to introduce IR 4.0 including Blockchain technology in the academic curriculum in Malaysian Universities. He had participated in many panel discussions on IR 4.0 technologies at academia as well as industry platforms. He is holding 19+ years' experience in academia at National & International level. He is serving as a reviewer or guest editors in many Journals published by Inderscience, IGI Global, Springer. He is contributing as a Co-Editor in many Scopus books. He had contributed in many Scopus/ SCI Journal/ conferences. Currently, Dr Goyal is associated as a Director, Faculty of Information Technology, City University, MALAYSIA.

* * *

**Ashima Bhatia** is a Research Scholar (JIMS NCR) & Assistant Professor, VIPS - TC, Delhi, India

**Megha Bhushan** is an Assistant Professor in the School of Computing, DIT University, Dehradun, India. She has received her ME and Ph.D. degrees from Thapar University, Punjab, India. She was awarded with a fellowship by UGC, Government of India, in 2014. In 2017, she was a recipient of Grace Hopper Celebration India (GHCI) fellowship. She has published 5 national patents and 1 international patent has been granted. She has published many research articles in international journals and conferences of repute. Further, she is the editor of many edited books with different publishers such as CRC Press, Taylor & Francis Group, Wiley-Scrivener and Bentham Science. Her research interests include Artificial Intelligence, Knowledge representation, Expert systems, and Software quality. She is also the reviewer and editorial board member of many international journals.

**S.Boopathi** () completed his undergraduate in Mechanical Engineering and postgraduate in the field of Computer-Aided Design. He completed his Ph.D. from Anna University and his field of research includes Manufacturing and optimization. He published 60 more research articles in Internationally Peer-reviewed journals, one Patent grant, and three published patents.He has 16 more years of academic and research experiences in the various Engineering Colleges in Tamilnadu, India..

**Rohan Appasaheb Borgalli**, currently pursuing Ph.D. under Fr. Conceicao Rodrigues College of Engineering, Mumbai University. He received his M.Tech degree in Digital Systems from Motilal Nehru National Institute of Technology (MNNIT), Allahabad, in 2013. Working as an Asst. Prof. in Shah & Anchor Kutchhi Engineering College, Mumbai. His research interests include artificial intelligence, machine learning, human computer interaction and deep learning.

**Sye Arif Islam** currently works at the computer science, computer application, cyber security and IT, Karpagam Academy of Higher Education as a Doctoral Scholar. Syed Arif Islam does research in Software Engineering.

**Umma Khatuna Jannat** is a Doctoral Scholar at the Karpagam Academy of Higher Education. I am focusing on artificial intelligence, machine learning, deep learning, natural language processing, and cyber security in my current work. My work involves the investigation, development, and implementation of novel computational and machine learning algorithms and applications for big data integration and data mining.

**Mohan Kumar** currently works at the Computer science, Computer Application & Information technology, Karpagam Academy of Higher Education. Mohan kumar does research in Software Engineering.

**Shirshendu Layek** holds masters and doctoral degrees from IIT (ISM) Dhanbad. Currently, he is serving as an Assistant Professor in the School of Computing, DIT University, Dehradun. His research interests are artificial intelligence, machine learning, deep learning, computer vision and applications of artificial intelligence in various engineering fields.

**Radhika Mahajan** is an Assistant Professor, Vivekananda Institute of Professional Studies – Technical Campus, Delhi, India

**Kavita Mittal** is an Associate Professor, Department of Computer Science, JIMS NCR, Haryana, India

**Akash Mohanty** is working as Professor, in the School of Mechanical Engineering, Vellore Institute of Technology, Vellore, India. He has about 13+ years of teaching and 6+ years of industrial experience. He received his bachelor's degree in science with distinction from Utkal University, Odisha, mechanical engineering from The Institute of Engineers (India), and M.Tech. degree in plastics engineering from CIPET-Bhubaneswar, Biju Patnaik University of Technology, Rourkela, Odisha. He obtained his PhD degree from the department of Mechanical Engineering, IIT-BHU, Varanasi. He has published 75 research papers in refereed international journals and at various national/international conferences. He has also had research exposure at IIT Kanpur while studying for his master's degree. He has received several best paper awards for his research papers at various international conferences. His areas of research include solid mechanics, fracture and fatigue, damage and failure analysis, structural composites, metal/polymer matrix hybrid composites, life time estimation and stress analysis of thermal barrier coatings, and FEA. He is an fellow member of IE (I).

**Arun Negi** is currently a Manager at Deloitte USI, Hyderabad, India. He has completed a course in Business Management from IIM, Ahmedabad and has obtained B. Tech degree from Jawaharlal Nehru University, New Delhi, India. He has 13+ years of diverse experience in cyber risk services. He has worked on various network security technologies and platforms for Fortune 500 clients. His experience includes cyber security audits, gap assessments, network security audits, cloud migrations and project management. He is currently oriented towards developing multi cloud skills and has achieved certifications in Oracle Cloud and AWS. He has published one national patent and one international patent has been granted. He has published many research articles in international journals and conferences of repute. His research areas include Artificial Intelligence, Software Product Line, Cloud Computing, and Cyber Security.

**Rahamathunnisa U** graduated B.E Computer Science in Anna University and M.E in Multimedia Technology in College of Engineering, Anna University, Chennai. Received Phd in the area of Multimedia Networks. Currently works as an Associate Professor(Sr) in School of Information Technology & Engineering, Vellore Institute of Technology, Vellore. Her area of specialization includes Image Processing, Video Processing and Multimedia Networks

**M Rajkumar** is an Assistant professor senior grade 2,Department of Smart Computing, School of Information Technology and Engineering, Vellore Institute of Technology, VIT University, Vellore Campus, Tamil Nadu 632014, India

**Maanas Singal** is currently pursuing M.Tech from DIT University, Dehradun after completing his Bachelors' in Computer Science and Engineering from Kalinga University, Raipur in 2021. His areas of interest include Machine Learning, DevOps, and Cloud Computing.

**Sunil Surve**, completed his PhD from VJTI, Mumbai in robotics and artificial intelligence. Working as a Professor in Fr. Conceicao Rodrigues College of Engineering, Mumbai. He published more than 30 research papers in various journals and conferences. His research interests include machine learning, robotics and computer architecture.

**S. Uma** is Professor in the Department of Computer Science and Engineering at Hindusthan College of Engineering and Technology, Coimbatore, Tamil Nadu, India. She received her B.E., degree in Computer Science and Engineering (CSE) in First Class with Distinction from P.S.G. College of Technology, M.S.,(By Res.,)., degree from Anna University, Chennai, Tamil Nadu. She received her Ph.D., in the faculty of Information and Communication Engineering from Anna University, Chennai with High Commendation. She has 31 years of academic experience and organized many National and International seminars, workshops and conferences. She has published more than 100 research papers in National and International Conferences, Journals, Book Chapters, and Books. She has published more than 20 patents. She is a recognized Supervisor of Anna University, Chennai and has supervised several research projects. She is a potential reviewer of International Journals by leading publishers such as Springer, IGI Global, IET, AICV etc., and Member of Professional Bodies like ISTE, CSI, IEEE, IAENG, etc., She is a recipient of "Bharath Jyoti", "Certificate of Excellence" and "Best Citizen of India" Awards. Her research interests are Pattern Recognition and Analysis of Non Linear Time Series Data, AI and Digital Analytics

**V. Saranya** is Associate Professor in the Department of Computer Science and Engineering at Hindusthan College of Engineering and Technology, Coimbatore. She received her B.Tech., IT degree from Hindusthan college of Engineering and Technology, Coimbatore, M.E under Software Engineering at Anna University, Coimbatore and PhD under Information and Communication Engineering at Anna University, Chennai. She has more than 10 years of teaching experience and has published several research papers. She is a reviewer for journals like Springer, IGI Global, Bentham Science Publishers, Bulletin of Electrical Engineering and Informatics, and so on. She is a recognized supervisor of Anna University, Chennai. Her research interests are Mobile Computing, Data Analytics, Software Engineering and Mobile Networks.

# Index

## H



## I



## K



## L



## M



## O



## P



## Q



## R



## S

## T



## V



## W

www.ingramcontent.com/pod-product-compliance
Lightning Source LLC
LaVergne TN
LVHW081339300926
849161LV00008B/2021